Practical Guide to Geo-Engineering

GEOTECHNICAL, GEOLOGICAL AND EARTHQUAKE ENGINEERING

Volume 29

For further volumes:
http://www.springer.com/series/6011

Milutin Srbulov

Practical Guide to Geo-Engineering

With Equations, Tables, Graphs and Check Lists

Springer

Milutin Srbulov
Isleworth, United Kingdom

ISSN 1573-6059 ISSN 1872-4671 (electronic)
ISBN 978-94-017-7897-8 ISBN 978-94-017-8638-6 (eBook)
DOI 10.1007/978-94-017-8638-6
Springer Dordrecht Heidelberg New York London

Printed on acid-free paper

Springer is part of Springer Science+Business Media (www.springer.com)

Preface

The idea to write this volume came from publications that are very useful in practice containing equations, tables and graphs for various engineering disciplines (e.g. Gieck and Gieck 1997, Young and Budynas 2002, Simons and Menzies 1977, Carter and Bentley 1991). Only bulky and lengthy manuals and reference books exist for geotechnics/ground engineering. This volume can be used both by established professionals as a handy reminder and by non-geotechnical engineers or beginners in geotechnics as a starting point. Those readers who find that the lists, equations, tables and graphs provided in this volume are not sufficient for their needs are referred to the existing manuals and reference books in geo-engineering (e.g. Burland et al. 2012).

This volume is intended to cover all the major topics in geo-engineering although rather briefly and focuses on engineering practice rather than research. Guidance given in relevant codes and standards are favoured when codes and standards provide recommendations concerning different topics. When such recommendations are not given or they are very brief, then the simple methods that are suggested in this volume can be used for rough estimates in initial stages and comparisons with the results of more sophisticated methods in detailed stages of engineering considerations. A modern and concise style of presentation is favoured using bullet points. The equations that are presented are suitable for hand calculations or at most a spread sheet as are the formulas given in codes and standards.

It should be acknowledged that geo-engineering is both a science (when analysing natural phenomena) and an art (when approximating actual reality) as initially pointed by Terzaghi, and that it is of multi-disciplinary nature involving engineering geology, engineering seismology, structures, physics and chemistry but also economical, legal, political, environmental and social issues.

Isleworth, UK Milutin Srbulov

References

Burland J, Chapman T, Skinner H, Brown M (eds) (2012) ICE manual of geotechnical engineering, 2 vol. set. The Institution of Civil Engineers, UK
Carter M, Bentley SP (1991) Correlations for soil properties. Pentech Press, London
Gieck K, Gieck R (1997) Engineering formulas, 7th edn. McGraw-Hill, New York
Simons NE, Menzies BK (1977) A short course in foundation engineering. Butterworth's, Bristol
Young WC, Budynas RG (2002) Roark's formulas for stress and strain, 7th edn. McGraw-Hill, New York

Acknowledgements

I was honoured and privileged to work with Prof. Nicholas Ambraseys as his assistant (i.e. associate after the completion of my Ph.D. in 1994) on a number of research projects supported by the Engineering and Physical Science Research Council of the United Kingdom and by the EPOCH program of the Community of European Countries at Imperial College in London (UK) during the period 1991–1997. The simplified approach used in our research is directly applicable to routine engineering practice. My initial involvement with earthquake engineering was a starting point for further work concerning soil dynamics and ground vibration related issues while working on a number of projects worldwide for Mott MacDonald Ltd.

Dr. E.T.R. Dean of Caribbean Geotechnical Design Limited (Trinidad) and Soil Models Limited (UK) reviewed several of my papers and books and was of great help with his detailed and precise comments for the improvement of initial versions of the publications.

Several of my former colleagues influenced my professional carrier and hence this volume. Professor Milan Maksimovic persuaded me to switch profession from concrete structures to geotechnics right after my graduation. He pioneered studies of soil mechanics paid by Energoprojekt Co. at Imperial College in the UK. The M.Sc. soil mechanics study in 1984/85 enabled me to obtain the position of a research assistant later. Professor Petar Anagnosti (former principal geotechnical engineer of Energoprojekt Co. and former vice president of the International Society for Soil Mechanics and Foundation Engineering) and Mr. Aleksandar Bozovic (former technical director of Energoprojekt Co. and former chairman of the committee on seismic aspects of dam design of the International Commission of Large Dams) encouraged me to apply for the position of a research assistant at Imperial College.

The past editor of *European Earthquake Engineering* and *Ingegneria Sismica* journals Prof. Duilio Benedetti accepted kindly for publication 21 of my papers related to geotechnical earthquake engineering written in the period 1995 to end of 2011. Numerous publications of my papers encourage me to work more in the field of geotechnical earthquake engineering and related topics such as ground vibration. Dr. Massimo Tognetti from Patron Editore provided publisher's consent to use material from my papers published in the journals *European Earthquake Engineering* and *Ingegneria Sismica*.

Petra D. van Steenbergen – senior publishing editor for geosciences with the publisher Springer – showed interest in my proposals for writing books and arranged for the publishing agreements.

The initiative for writing books came from my wife Radmila, who had no time to do it herself but, instead, provided encouragement and brain stimulation. Our son Ivan helped with proof reading the initial text.

Contents

Chapter 1
Geo-investigations

Abstract This chapter contains basic information for the site investigations performed for construction purposes. Site investigation needs to provide information on:

- **Geological origin** (igneous, sedimentary, metamorphic for rock, transported by wind, streams, glaciers for soil) and global formations (tectonic folds and faults, sediment basins, landslides)
- **Past use** concerning presence of contamination and existing buried structures (of archaeological or industrial significance – foundations, underground chambers) including unexploded ordnance (UXO)
- **Current condition** (baseline state) in terms of topography, layering, state of strata (density of coarse soil, consolidation state, plasticity and consistency of fine soil, rock type and rock weathering extent), ground water and gas regime
- **Engineering properties** for prediction of ground response to intended development over its useful life.

1.1 Methodology and Stages of Site Investigations

The following stages should be followed in sequence:

- **Desk study** based on existing documents (technical and non-technical such as newspapers and memoirs), geological and topographic maps including historical survey maps
- **Site reconnaissance** (walkover) onshore for checking of the latest condition, details and access
- **Program for site investigations** in stages if possible (initial and detail)
- **Geophysical non-intrusive** from ground surface or remote sensing from air on regional scale
- **Boring and digging of pits and trenches** (shafts and galleries for dams), coring in rock and soil and sampling. Minimum recommended rock core diameter for

M. Srbulov, *Practical Guide to Geo-Engineering: With Equations, Tables, Graphs and Check Lists*, Geotechnical, Geological and Earthquake Engineering 29,
DOI 10.1007/978-94-017-8638-6_1, © Springer Science+Business Media Dordrecht 2014

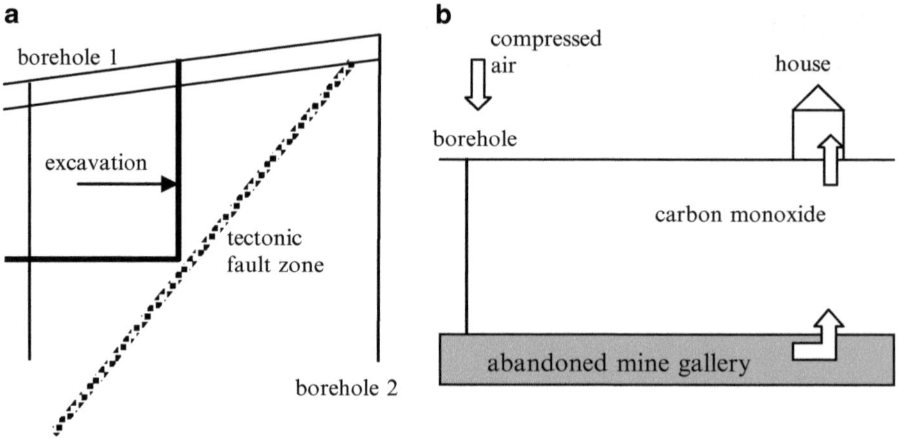

Fig. 1.1 Examples of serious consequences of non-integral site investigations: (**a**) use of boreholes without geophysical profiling resulted in non-detection of a tectonic fault zone that contributed to instability of an excavation later, (**b**) with no prior desk study, the use of compressed air for drilling through an abundant mine gallery caused displacement of carbon monoxide into an adjacent house and death of its occupants

testing is 50 mm and the length to width ratio of at least 3. Least disturbed soil samples are of cubical shape with about 0.3 m long side and obtained from pits, trenches, shafts and galleries, slightly disturbed soil cylindrical samples of 100 mm diameter preferably and with the length to width ratio of more than 2 are obtained by pushing into soil steel cylinders with the wall thickness up to 1.5 mm, more or less disturbed cylindrical samples are obtained by hammering and rotating of thick wall cylindrical samplers into soil and completely disturbed samples are obtained from excavation hips and tips of coarse grained non-cohesive soil and rock.

- **In situ and laboratory testing** are described in Chap. 2
- **Reporting** (field offshore only), factual and interpretative

Illustrative examples when non-integral site investigation caused serious consequences are shown in Fig. 1.1.

1.2 Non-intrusive Geophysical and Remote Sensing Methods

For long (roads, railways, tunnels, canals, dikes, pipelines) and large (dams, power plants) structures in order to detect caverns (in soluble rock: halite –salt, gypsum, chalk, limestone), abandoned tunnels, natural and mining cavities, tectonic and landslide discontinuities, natural gas and oil concentrations, metal objects within

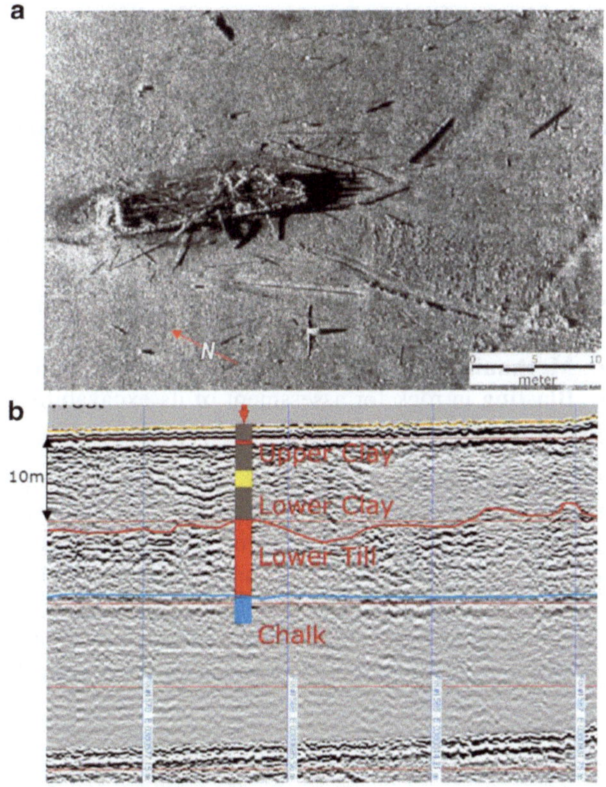

Fig. 1.2 Examples of results of geophysical survey: (**a**) side scan sonar image of a wooden ship wreck at sea bottom in total darkness, (**b**) ground profile obtained by an acoustic sparker offshore compared to a borehole log

ground (including UXO) and on water bottom (ship wrecks, pipelines, anchors, fishing gear), bathymetry over water and areal sensing for regional investigations:

- **Wave refraction** to a depth of about 20 m (e.g. ASTM D5777-00) for ground profiling and tectonic structure, wave velocity measurements and shallow gas detection
- **Wave reflection** for depths greater than about 20 m (e.g. ASTM D7128-05) and purposes as above
- **Electrical conductivity/resistivity** for assessment of corrosivity to metals, earthhening purposes and electro osmotic drainage (e.g. ASTM G57)
- **Electro-magnetic** for detection of metal objects (e.g. ASTM D6639-01)
- **Interferometric synthetic aperture radar** (InSAR) from satellites for detection of topographic changes on regional scale
- **Light detection and ranging** (LIDAR) from aeroplanes for ground topography mapping
- **Thermal infra red** (TIR) from aeroplanes for detection of heat sources
- **Multi beam echo sounding** (Swathe) from water surfaces for water bed profiling
- **Side scan sonar** from water surfaces for water bed topography

Illustrative examples of results of implementation of a side scan sonar and of a geophysical ground profiling offshore are shown in Fig. 1.2.

1.3 Intrusive by Boring and Digging

For calibration of non-intrusive methods, taking ground samples for laboratory testing, in-situ testing described in Chap. 2 and logging of ground profile and ground water and gas conditions.

- **Rotary boring** with coring most revealing and expensive
- **Augering** in soil when heavy drill rig access is limited for standard penetration testing and soil sampling
- **Rotary percussive drilling** in rock for testing of water permeability and mass stiffness
- **Mechanised digging** in soil for soil classification, testing and sampling
- **Blasting** in rock for assessment of its excavability, vibration and noise during exploitation

Illustrative sketches of different types of boring and drilling bits are shown in Fig. 1.3. Sizes of rock core sizes are given in Table 1.1.

Examples of unsafe practice are: cross contamination by drilling at contaminated sites containing asbestos, heavy metals and volatile hydrocarbons, drilling offshore through strata reach in hydrogen sulfide gas that is highly poisonous, digging into or drilling through buried high voltage cables, buried gas mains, entering unsupported excavations, handling polluted ground without protective equipment etc.

1.4 Extent of Site Investigations

Depends on available information initially and type, extent and development stage of structures for which the investigation is required. EN 1997-2 (2007) recommends the spacing and depth of site investigations. In general, an investigation depth

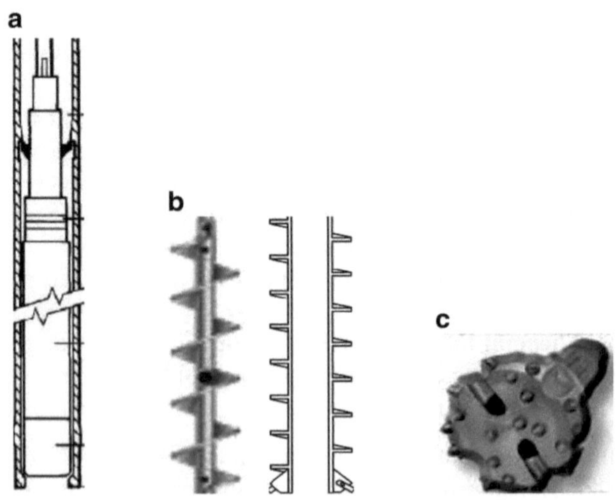

Fig. 1.3 Details of intrusive equipment: (**a**) cross section through a rotary boring tube with a core barrel inside, (**b**) side view of a solid stem auger and cross section through a hollow stem auger for taking soil samples and performing filed tests, (**c**) toe view of a drill bit for rotary percussion drilling in rock

Table 1.1 Rock core sizes

Core barrel type	Borehole diameter (mm)	Standard core size (mm)	Core size for rigid plastic liner (mm)	Casing size or type	Casing outside diameter (mm)	Casing inside diameter (mm)
According to BS 4019 (1996) standard sizes						
HWM	75.7	54.7	51	NX	88.9	76.2
HWF	98.8	76.2	72	HX	114.3	100.0
HWAF	99.5	70.9	–	HX	114.3	100.0
PWF	120.0	92.1	87	PX	139.7	122.3
SWF	145.4	112.8	107	SX	168.3	14.7
UWF	173.7	139.8	132	UX	193.7	176.2
Wireline sizes						
BQ	59.9	36.4	35			
NQ	75.7	47.6	45			
HQ	96.1	63.5	61			
PQ	122.7	85.0	82			
Geobore S	146.0	102.0	102	SX	168.3	147.7
Thinwall sizes						
TNX	75.7	60.8	–	NX	88.9	76.2
T2 66	66.1	51.9	–	74	74.3	67.3
T2 76	76.1	51.9	–	84	84.3	77.3
T2 86	86.1	71.9	68	98	98.0	89.0
T2 101	101.1	83.9	80	113	113.0	104.0
T6 116	116.1	92.9	89	128	128.0	118.0
T6 131	131.1	107.9	104	143	143.0	133.3

should extend into rock, stiff clay or dense sand if they are not located at great depths and to the longest distance and greatest depth where a structure can influence or be influenced by adjacent ground, water and gas conditions and by existing structures. In particular, the spacing of intrusive works is:

- in the range from about 15 to 50 m and extending to the top of a soil slope adjacent to investigated site and for a distance of 2 depths of an excavation in a level ground
- for high rise structures and civil engineering projects greater of 6 m and 3 times the smaller side length of the foundation measuring from the foundation lowest level
- for raft foundations and structures with several foundation elements greater than 1.5 times the smaller side of the structures measuring from the foundation lowest level
- for embankments and natural soil slopes greater of 6 m and 1.2 times the maximum embankment/ slope height measuring from the embankment/ slope lowest toe level
- for cuttings greater of 2 m and 0.4 times the cutting depth measuring from the cutting lowest level
- for roads and airfields greater than 2 m measuring from the sub-base lowest level

- for trenches and pipelines greater of 2 m and 1.5 times the width of excavation measuring from the excavation lowest level
- for tunnels and caverns greater than 2 times the width of cavern measuring from the toe lowest level
- for cut-off walls greater than 2 m measuring from the surface of the stratum impermeable to ground water
- for piles greater of pile group width, 5 m and 3 times the pile base diameter measuring from the pile toe levels

1.5 Reporting of Results

EN 1997-2:2007 specifies contents for factual reports. A factual report should contain information on:

- **The purpose and scope** of the investigations with a description of investigated site and its topography including water streams and vegetation, the planned structure and the stage of design
- **Limitations** to the original program of site investigations, if any
- **Past use** of the site particularly mining and industrial activities, if any
- **Site reconnaissance** findings
- **Adjacent structures**, if any, behaviour
- **Local relevant experience** in the area (e.g. depth of shallow foundations)
- **Geology, tectonic setting and ground water aquifers**
- **Land movements** (landslides, subsidence, heave)
- **Non-destructive and remote sensing** investigation results, if any
- **Ground water and gas** (particularly radon) data
- **Borehole/ test pit/ trenches/ galleries logs** with in situ test results and sampling locations including sample type and size as well as sampling dates
- **Laboratory test results** as individual sheets with dates of testing and as summary tables containing the results of all tests performed for every sample tested
- **Description of the standards/ methods used** for field works and laboratory testing
- **Survey data** of locations of site investigations on a topographic map with plan of the local structures (present and future)
- **Aerial photographs** if any
- **Seismic hazard severity**
- **Short descriptions of the equipment, methods and standards** used
- **Valid calibration certificates** of sensors and transducers used

An illustrative example of seismic hazard severity rating is shown in Fig. 1.4.

EN 1997-1:2004 specifies contents of interpretative reports. An interpretative report should contain:

- **Summary of finding** from the factual report
- **Ground profiles** with ground and ground water levels

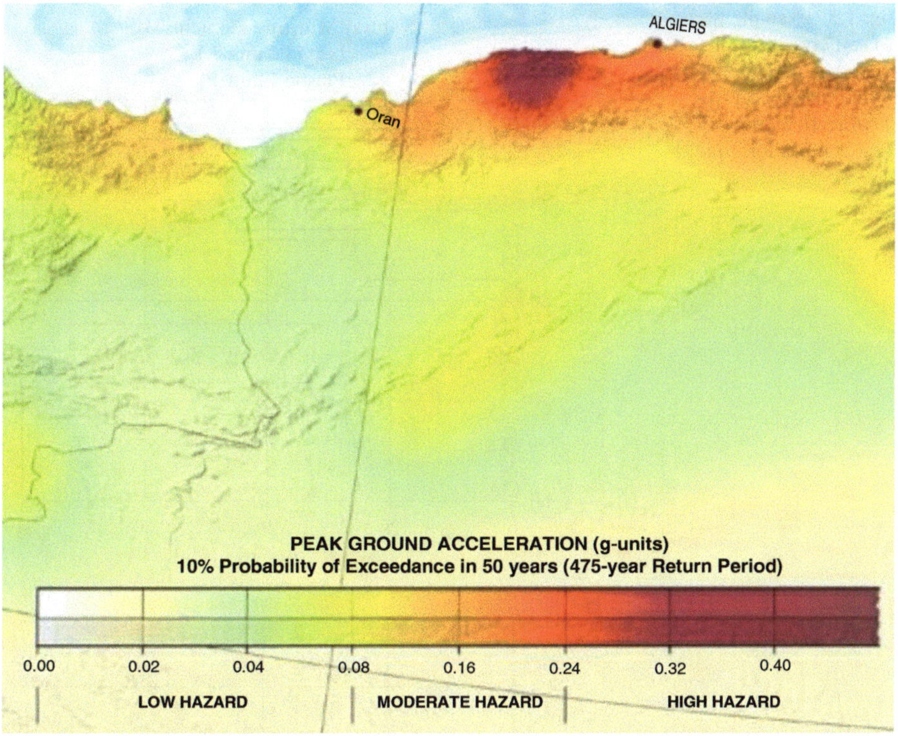

Fig. 1.4 An example of classification of severity of seismic hazard (Extracted from the European-Mediterranean Seismic Hazard Map, 2003, http://wija.ija.csic.es/gt/earthquakes/)

- **Summary graphs of test results** versus levels in nearly horizontally layered soil or for each individual layer for slopping layers
- **Graphs of correlations** between classification and engineering properties with added available published data for checking and comparison. The classification properties are: density for coarse soil, over consolidation ratio, plasticity and consistency indices for fine soil, rock mass rating (RMR) and/or quality (Q)
- **Tables of recommended engineering ground properties** (as a minimum: soil/rock type, unit density, shear strength parameters – undrained and drained for fine grained soil, modulus of compressibility: maximum value and its decrease with strain increase, coefficient of water permeability, rock unconfined compressive strength)
- **Ultimate and allowable bearing capacity and settlement**/horizontal displacement of shallow and deep foundations and retaining structures, safe slope inclination, etc. when required
- **Recommendations for additional investigation** and testing if necessary

Illustrative example plots of soil classification properties versus elevation and of engineering properties versus classification properties are shown in Fig. 1.5.

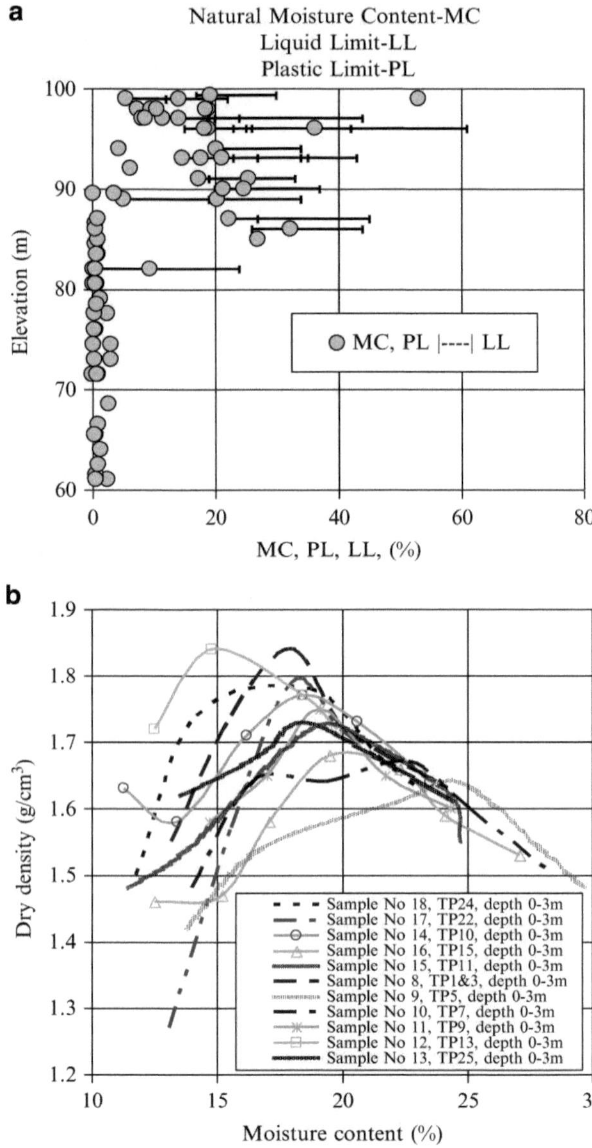

Fig. 1.5 Example plots of (**a**) soil classification properties versus elevation, (**b**) engineering versus classification property for a compaction trial

1.6 Summary

- **Methodology and stages of site investigations** listed in Sect. 1.1 should always be followed.
- **Types and extent of site investigations** are highly dependent on a particular case. An example of the bill of quantities for a small size site investigation for piled foundation is given in Table 1.2 for field works and Table 1.3 for

Table 1.2 An example of the bill of quantities for a small size field investigation

No	Description	Unit	Max. quantity	Unit price	Max. cost
1	Rotary drilling to 10 m depth, with core minimum diameter of 100 mm in soil and 50 mm in rock The unit price to include: Determinations of the coordinates and levels for ___ boreholes Reporting on the ground water levels in the borehole Digital colour photographs of the cores in boxes with clear labelling Borehole logs with visual in situ classification of soil and rock Keeping and securing of the cores for duration of 3 months Infill of the borehole by grout Equipment transport (mobilisation/demobilisation) and stand by Health and safety measures for the personnel as well as security of the working place and equipment	m′			
2	Rotary drilling to depth of 20 m, with core minimum diameter of 100 mm in soil and 50 mm in rock. Number of boreholes ____. The unit price to include all costs as mentioned for depths to 10 m	m′			
3	Rotary drilling to depth of 30 m, with core minimum diameter of 100 mm in soil and 50 mm in rock. Number of boreholes ____. The unit price to include all costs as mentioned for depths to 10 m	m′			
4	Taking, packing in plastic transparent bags, labelling, securing and transporting of disturbed samples from non-cohesive soil to the local laboratory of the contractor. The minimum sample mass 5 kg each	No			
5	Taking, packing in steel air tight containers or clinging film, aluminium foil and cardboard cylinders with vexed ends, labelling, securing and transporting of thin wall tube non-disturbed samples with the minimum diameter of 100 mm from cohesive soil to the local laboratory of the contractor. The minimum sample height to diameter ratio of 2.5	No			
6	Taking, packing, labelling, securing and transporting of rock core samples to the local laboratory of the contractor. The minimum sample length to diameter ratio of 3	No			
7	Taking, packing in 1 l bottle, labelling, securing and transporting of ground water samples to the local laboratory of the contractor	No			
8	Standard penetration test in non-cohesive soil at depth differences not exceeding 1.0 m with reporting of the results	No			

Notes: The national, ISO, BS, EN or ASTM relevant standards are applicable to all site investigations and testing. The actual standards used to be stated in the factual report. The calibration certificates of the SPT equipment used to be included with the factual report. Short description/type of the equipment and procedures used to be included in the factual report. All the personnel involved to be appropriately qualified and sufficiently experienced. Non-disturbed samples of cohesive soil and rock cores to be packed immediately after the extraction to preserve their natural moisture content. Geophysical and contamination tests (Sect. 4.9.2) not included here

Table 1.3 An example of the bill of quantities for a small size laboratory testing

No	Description	Unit	Max. quantity	Unit price	Max. cost
1	Grain size distribution (sieve only) of samples from each soil layer encountered	No			
2	Unit density and moisture content of all samples taken from cohesive soil	No			
3	Liquid and plastic limit of samples taken from cohesive soil				
4	pH, sulphate and chloride content on water samples	No			
5	pH, sulphate and chloride content on soil samples	No			
6	Unconsolidated undrained triaxial tests on the samples taken from cohesive soil	No			
	Cell pressure used ~0.5 overburden total stress at the sample depth				
7	Consolidated undrained triaxial test with pore water pressure measurement on the samples taken from cohesive soil. Cell pressures 50, 100, 200 kPa	No			
8	Unconfined compressive strength of rock core samples	No			
9	Factual report (paper and one electronic version) with the final borehole logs, summary tables and graphs for the laboratory test results	No			

Notes: The national, ISO, BS, EN or ASTM relevant standards are applicable to all laboratory testing. The actual standards used to be stated in the factual report. The calibration certificates of laboratory stress/strain/pressure/displacement gages used to be included with the factual report. Short description/type of the equipment and procedures used to be included in the factual report. All the personnel involved to be appropriately qualified and sufficiently experienced. Contamination tests not included

laboratory testing. An example of ground investigation health and safety risk assessment is given in Table 1.4.

- **Contents of factual and interpretative reports** are listed in Sect. 1.5

Table 1.4 An example of ground investigation health and safety risk assessment

Hazard/aspect	Consequence/ impact	Persons at risk	Initial risk level	If high risk, can the hazard be avoided?	Control measures	Residual risk level	Responsibility for implementing
Encountering services during excavation (by Contractor)	Physical injury or death	Site operatives	High	No	Consult service plans, Cable Avoidance Tool scan, use of hand dug pits, permit to dig system	Medium	Contractor
Contaminated ground – contact with hazardous materials in ground or groundwater, Leptospirosis	Sickness, feeling nauseous, skin irritations etc	Site operatives and visitors	Medium		Wear appropriate Personal Protection Equipment, including gloves, safety spectacles and protective clothing. Adopt good standards of hygiene. Carry Leoptspirosis card.	Low	Site operatives and contractors
Entry into shored excavations-Collapse of shored excavations and encountering explosive or asphyxiating ground gas.	Physical injury or death	Site operatives	High	No	Contractors to check quality of shoring and check for ground gas. No alone working.	Medium	Contractor and site operatives
Being struck by site vehicles, rotating drilling equipment etc	Physical injury or death	Site operatives	High	No	Attend contractor's site induction. Wear high-visibility jacket. Follow banksman's/machine operator's instructions. Do not stand too close to drilling rigs.	Medium	Contractor and site operatives
Moving traffic	Vehicle impact – physical injury or death.	Site operatives and visitors	High	No	Ensure correct Traffic Management measures in place. Park vehicles off the main road. Wear High visibility jackets if close to a road.	Medium	Site operatives

(continued)

Table 1.4 (continued)

Hazard/aspect	Consequence/ impact	Persons at risk	Initial risk level	If high risk, can the hazard be avoided?	Control measures	Residual risk level	Responsibility for implementing
Walking on uneven ground	Slips / Trips and falls	Site operatives and visitors	Medium		Wear appropriate footwear. Avoid carrying equipment over significant distance.	Low	Site operatives
Aggrieved landowners	Potential for physical violence	Site operatives	Medium		Client to confirm access is agreed. Do not enter property / leave immediately if requested to do so.	Low	Client/ Site operatives
Livestock	Physical injury	Site operatives	Medium		Only enter field with livestock if owner is present or livestock is benign. Livestock to be removed from area by owner if possible.	Low	Site operatives
Working close to water (canal and river)	Drowning	Site operatives	Medium		Appropriate foot wear to avoid slipping and stay away from the waters edge, avoid lone working.	Medium	Site operatives

References

ASTM D5777-00 Standard guide for using the seismic refraction method for subsurface investigation. American Society for Testing and Materials, Philadelphia, PA
ASTM D6639-01 Standard guide for using the frequency domain electromagnetic method for subsurface investigations. American Society for Testing and Materials, Philadelphia, PA
ASTM D7128-05 Standard guide for using the seismic-reflection method for shallow subsurface investigation. American Society for Testing and Materials, Philadelphia, PA
ASTM G57-06 Standard test method for field measurement of soil resistivity using the Wenner four electrode method. American Society for Testing and Materials, Philadelphia, PA
EN 1997-1:2004 Eurocode 7: geotechnical design – Part 1: General rules. European Committee for Standardization, Brussels
EN 1997-2:2007 Eurocode 7: geotechnical design – Part 2: Ground investigation and testing. European Committee for Standardization, Brussels

Chapter 2
Geo-testing

Abstract This chapter contains results of interpretations of most common field and laboratory tests performed in soil and rock for determination of engineering properties.

Field tests considered are:

- Standard penetration test (SPT) for estimation of mainly coarse grained soil density, shear strength and compressibility
- Cone penetration test (CPT) for determination of mainly coarse grained soil density, shear strength and compressibility
- Permeability to water of coarse grained soil using open pipe piezometers
- Permeability to water of rock mass using water pressure test (Lugeon test)
- Plate loading test for compressibility of soil sub-base
- California bearing ratio (CBR) for compressibility of soil sub-base
- Pressuremeter test in boreholes for soil and rock compressibility
- Geophysical survey (refraction, reflection, electrical resistivity, ground penetrating radar, electro-magnetic)

Laboratory tests considered are:

- Soil compressibility, swellability and collapsibility in oedometer
- Swellability of rock
- Soil static and cyclic shear strength and stiffness by simple shear apparatus
- Soil static and cyclic shear strength and stiffness by triaxial apparatus (unconsolidated undrained and consolidated undrained with measurement of excess pore water pressures and local strain on specimen side)
- Axial strength of rock cylinders in unconfined condition
- Abrasivity of rock fill (Los Angeles test)
- Transversal wave velocity by bender element
- Soil stiffness and damping at small to large strain in cyclic condition by resonant column apparatus
- Content of organics, sulphates, chlorides and carbonates, pH value

M. Srbulov, *Practical Guide to Geo-Engineering: With Equations, Tables, Graphs and Check Lists*, Geotechnical, Geological and Earthquake Engineering 29,
DOI 10.1007/978-94-017-8638-6_2,

2.1 Field Tests

2.1.1 Standard Penetration Test (SPT)

2.1.1.1 Definition

The test involves driving of a 51 mm outside diameter thick wall tubular sampler, Fig. 2.1, (or solid cone in gravel and weak rock) using 63.5 kg hammer falling from 0.76 m height within a borehole. The N_{SPT} denotes the number of blows required to achieve a penetration of 0.3 mm, after an initial seating drive of 0.15 m (ISO 22476-3; ASTM D1586; BS 1377-9). Disturbed ground samples for classification purposes

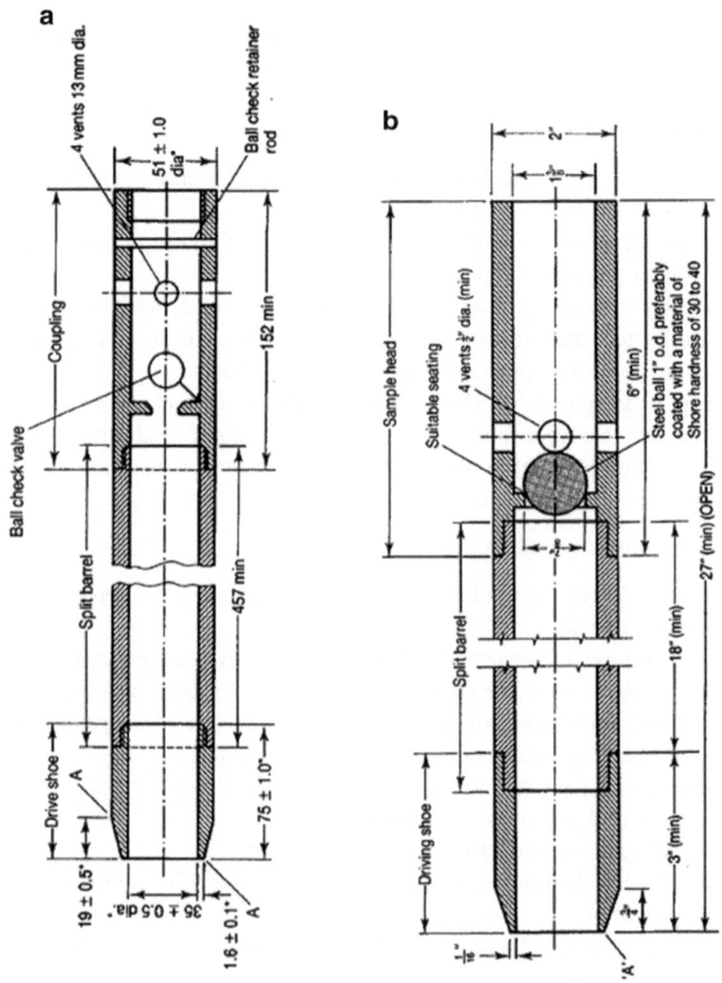

Fig. 2.1 SPT split-barrel samplers adopted from (**a**) BS 1377-9, (**b**) ASTM D1586

Table 2.1 Relative density D_r of coarse grained soil based on N_{SPT} and $(N_1)_{60}$

Description	Very loose	Loose	Medium dense	Dense	Very dense	References
D_r (%)	0–15	15–35	35–65	65–85	85–100	–
$(N_1)_{60}$	0–3	3–8	8–25	25–42	42–58	(1), (2), (3)
N_{SPT}	0–4	4–10	10–30	30–50	>50	(4)

Notes: (1) Skempton (1986), (2) Clayton (1995), (3) EN 1997-2:2007, (4) Terzaghi and Peck (1948). D_r defined in Chap. 3

are obtained from the test tube after its splitting. The (corrected) blow count N_{SPT} is used for estimation of ground engineering properties, e.g. Clayton (1995), Skempton (1986), Penetration Testing in the UK (1989).

2.1.1.2 Corrections (Multipliers) of Recorded N_{SPT}

- For depths of less than 3 m, 0.75 from EN 1998-5:2004
- For effective overburden pressure (total less water pressure) σ'_v in kPa, $0.5 \leq (100/\sigma'_v)^{1/2} \leq 2$ to obtain N_1 from EN 1998-5:2004
- For energy ratio ER specific to the resting equipment to 60 % of theoretical free-fall hammer energy, ER/0.6 to obtain N_{60} from EN 1998-5:2004
- For borehole diameter, rod length, sampling method by Skempton (1986)
- For the use of a solid cone instead of tube, 3/4 on average, 1/3 lower bound, 3/2 upper bound for $N_{SPT} < 40$ from Cearns and McKenzie (1989)

2.1.1.3 Correlations for Coarse Grained Soil

- **Relative density D_r** according to Table 2.1
- **Angle of friction ϕ** (degrees) in static and normally consolidated condition from Peck et al. (1953) graph

$$\begin{aligned} \phi &\approx 28^\circ + \frac{(N_1)_{60} - 3}{36} \cdot 12.5^\circ, 3 \leq (N_1)_{60} \leq 39 \\ \phi &\approx 40.5^\circ + \frac{(N_1)_{60} - 39}{21} \cdot 4.5^\circ, 39 \leq (N_1)_{60} \leq 60 \end{aligned} \tag{2.1}$$

Kulhaway and Mayne (1990) approximated the correlation provided by Schmertmann (1975) for ϕ (degrees) as

$$\phi = \arctan\left[\frac{N_{60}}{12.2 + 20.3 \cdot \left(\frac{\sigma'_{vo}}{p_a}\right)}\right]^{0.34} \tag{2.2}$$

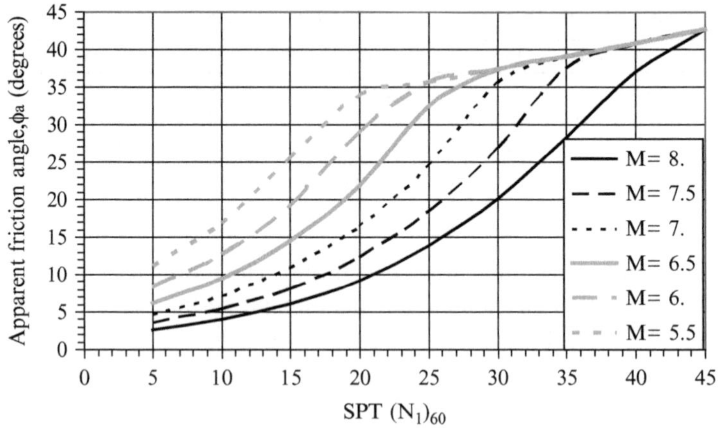

Fig. 2.2 Apparent friction angles ϕ_a during earthquakes for 5 % fines content and different earthquake magnitudes M when excess pore water pressure build-up has not been considered explicitly

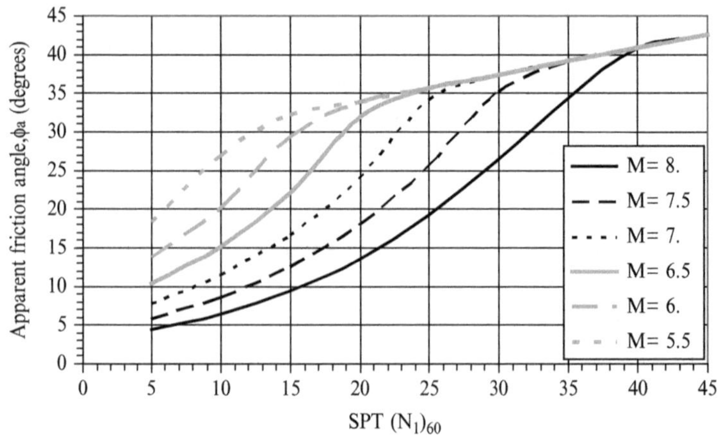

Fig. 2.3 Apparent friction angles ϕ_a during earthquakes for 15 % fines content and different earthquake magnitudes M when excess pore water pressure build-up has not been considered explicitly

σ'_{vo} is effective overburden pressure (total less water pressure)
p_a is atmospheric pressure ~100 kPa

Hatanaka and Uchida (1996) provided a simple correlation for ϕ (degrees)

$$\phi = \sqrt{20 \cdot (N_1)_{60}} + 20 \tag{2.3}$$

- **Apparent angle of friction ϕ_a** (degrees) due to the effect of excess pore water pressure development during earthquakes are shown in Figs. 2.2, 2.3, and 2.4

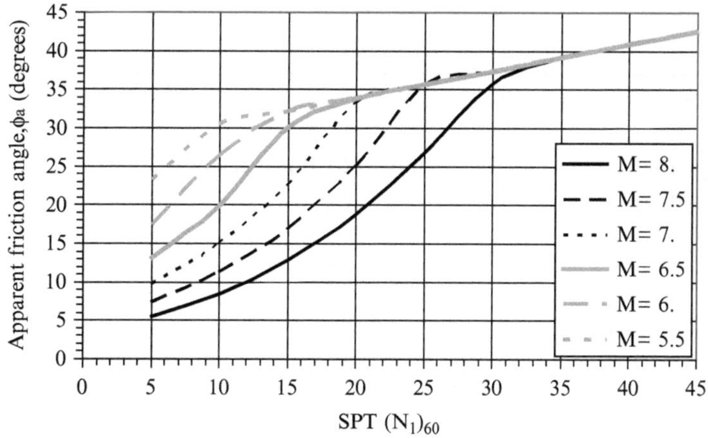

Fig. 2.4 Apparent friction angles ϕ_a during earthquakes for 35 % fines content and different earthquake magnitudes M when excess pore water pressure build-up has not been considered explicitly

based on expression $\arctan(\tau_e/\sigma'_{vo})$ where τ_e/σ' is given in EN 1998-5:2004 based on Seed et al. (1985) as boundaries between liquefied and non-liquefied sand of different fines content for earthquake surface wave magnitude $M_s = 7.5$. EN 1998-5:2004 uses magnitude correction factor CM according to Ambraseys (1988). Simplest fitting formulas for the values of τ_e/σ'_{vo} are used.

$$\begin{aligned}
&\phi_a = \arctan\left\{[71 \cdot \exp(-0.57 \cdot M_s)] \cdot \left[0.0413 \cdot e^{0.084 \cdot (N_1)_{60}}\right]\right\} \le \phi, \\
&\textit{for fines content of } 5\,\% \\
&\phi_a = \arctan\left\{[71 \cdot \exp(-0.57 \cdot M_s)] \cdot \left[0.0708 \cdot e^{0.0771 \cdot (N_1)_{60}}\right]\right\} \le \phi, \\
&\textit{for fines content of } 15\,\% \\
&\phi_a = \arctan\left\{[71 \cdot \exp(-0.57 \cdot M_s)] \cdot \left[0.086 \cdot e^{0.0855 \cdot (N_1)_{60}}\right]\right\} \le \phi, \\
&\textit{for fines content of } 35\,\% \\
&5.5 \le M_s \le 8.0
\end{aligned} \tag{2.4}$$

M_s is the surface wave magnitude of an earthquake

- **Maximum axial stiffness modulus E_{max}** (MPa) at small strain ($<10^{-6}$) from Stroud (1989) using data from a wide range of spread footings, raft foundations and large scale plate tests.

$$\begin{aligned}
E_{\max} &= 2 \cdot N_{60} \ \textit{normally consolidated sand,} \\
E_{\max} &= 16 \cdot N_{60} \ \textit{over-consolidated sand\&gravel}
\end{aligned} \tag{2.5}$$

- **Coefficient of volume compressibility m_v** (m^2MN^{-1}) at σ'_v at 100 kPa based on fitted data by Burland and Burbidge (1985)

$$m_v = 25.4 \cdot N_{SPT}{}^{-2.35}, 5 \le N_{SPT} \le 75 \tag{2.6}$$

Table 2.2 Chalk strength types based on N_{60}, Clayton (1995)

	Very weak	Weak	Moderately weak	Moderately strong to very strong
N_{60}	0–25	25–100	100–250	>250

- **Transversal wave velocity V_t** (Japan Road Association 2003)
  Coarse grained soil

$$V_t = 80 \cdot {N_{SPT}}^{1/3}, 1 \leq N_{SPT} \leq 50 \tag{2.7}$$

  Fine grained soil

$$V_t = 100 \cdot {N_{SPT}}^{1/3}, 1 \leq N_{SPT} \leq 25 \tag{2.8}$$

2.1.1.4 Correlations for Chalk

- Strength type according to Table 2.2
- **Mass compressive strength σ** (MPa) from Stroud (1989) based on loading plates or piles to failure

$$\sigma = 0.05 \cdot N_{60}, 5 \leq N_{60} \leq 80 \tag{2.9}$$

- **Ultimate end bearing stress q_u** (kPa) of piles from Lord et al. (2002)

$$\begin{aligned} q_u &= 200 \cdot N_{SPT} \quad \textit{bored \& CFA piles} \\ q_u &= 250 \cdot N_{SPT} \quad \textit{driven cast-in-place piles} \\ q_u &= 300 \cdot N_{SPT} \quad \textit{driven pre-formed piles} \end{aligned} \tag{2.10}$$

- **Maximum axial stiffness modulus E_{max}** (MPa) at small strain ($<10^{-6}$) from Stroud (1989) using data from in situ geophysical tests and by extrapolation of results beneath a tank.

$$E_{\max} = (65 \ \textit{to} \ 150) \cdot N_{60} \tag{2.11}$$

2.1.1.5 Correlations for Weak Rock Except Chalk

- Strength type according to Table 2.3
- **Unconfined compressive strength UCS** (MPa) for marl, mudstone and sandstone, based on data by Stroud (1989) gathered from back analysis of pile and pressuremeter tests using the simplest fitting formula

$$UCS = 0.649 \cdot e^{0.0067 \cdot N_{60}}, 60 \leq N_{60} \leq 550 \tag{2.12}$$

Table 2.3 Weak rock except chalk strength types based on N_{60}, Clayton (1995)

	Very weak	Weak	Moderately strong to very strong
N_{60}	0–80	80–200	>200

Table 2.4 Consistency and undrained shear strength c_u (kPa) of fine grained soil based on N_{SPT} according to Terzaghi and Peck (1974)

Description	Very soft	Soft	Medium	Stiff	Very stiff	Hard
N_{SPT}	<2	2–4	4–8	8–15	15–30	>30
c_u	<12.5	12.5–25	25–50	50–100	100–200	>200

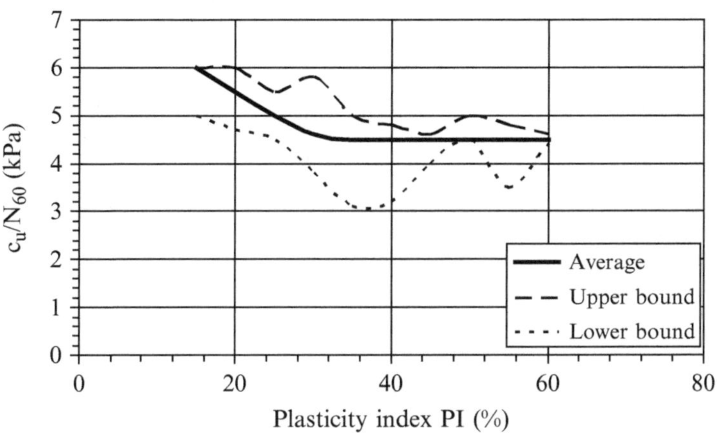

Fig. 2.5 Dependence of the ratio c_u/N_{60} on plasticity index for over consolidated clay

- **Maximum axial stiffness modulus E_{max}** (MPa) at small strain ($<10^{-6}$) from Stroud (1989) using extrapolation of data from a number of case histories of spread and piled foundations.

$$E_{\max} = 8 \cdot N_{60} \tag{2.13}$$

2.1.1.6 Correlations for Fine Grained Soil When Laboratory Test Results Are Not Available

- **Consistency** according to Table 2.4
- **Undrained shear strength c_u** according to Table 2.4. Stroud and Butler's (1975) range of the ratio c_u/N_{60} for over consolidated clay is shown in Fig. 2.5.
- **Maximum axial stiffness modulus E_{max}** (MPa) in drained condition at small strain ($<10^{-6}$) from Stroud (1989) based on the results for a number of raft and spread foundations on over consolidated clay.

$$\begin{aligned} E_{\max} &= 6.6 \cdot N_{60} \; \textit{for plasticity index } PI = 15\% \\ E_{\max} &= 4.4 \cdot N_{60} \; \textit{for plasticity index } PI = 50\% \end{aligned} \tag{2.14}$$

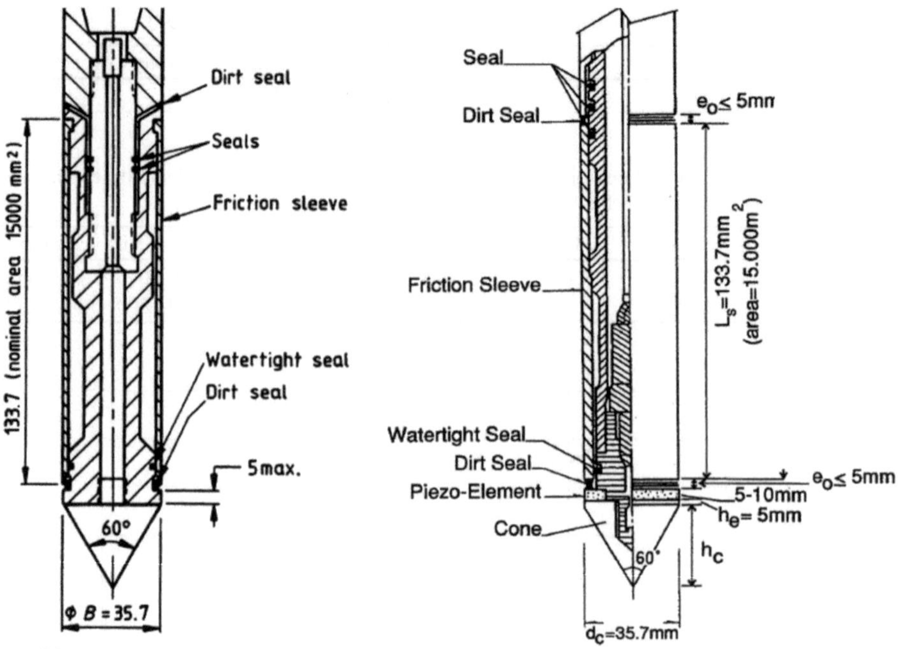

Fig. 2.6 CPT probes adopted from (**a**) BS 1377-9, (**b**) ASTM D5778

2.1.2 *Cone Penetration Test (CPT)*

2.1.2.1 Definition

The test consists of pushing a cone (Fig. 2.6) vertically into soil using a series of push rods in order to measure cone resistance q_c to penetration. Applied load at the top is 10, 20 kN or even 30 kN. Diameter of the base of the cone is usually 35.7 ± 0.3 mm, the area of the base of the cone is 10 cm^2, apex angle of cone is 60 ± 5°, rate of penetration 20 ± 5 mm/s. A cylindrical shaft or friction sleeve (with peripheral area usually of 150 cm^2) is frequently mounted above the cone for measurement of side friction. Modern cones with electrical transducers have also installed a piezometer for measurement of pore water pressure at the level of the base of the cone and in some cases a miniature velocity meter above the friction sleeve for measurement of velocities of longitudinal and transversal soil body waves. The applicable codes are ISO 22476-1, 12; BS 1377-9 and ASTM D5778. Specific references are Lunne et al. (1997), Meight (1987).

2.1.2.2 Corrections and Definitions

- **Total cone resistance q_t**

$$q_t = q_c + u \cdot (1 - a) \tag{2.15}$$

Table 2.5 Maximum values of friction ratio R_f and pore pressure ratio B_q (Robertson et al. 1986)

Soil type	Gravely sand to sand	Silty sand to sandy silt	Sandy silt to clayey silt	Clayey silt to silty clay
R_f	<2 %	<3 %	<4 %	<5 %
B_q	<0.1	<0.2	<0.3	<0.8

q_c is cone resistance
u is pore pressure at the cone base
a is the cone area ratio.

- **Friction ratio R_f.** Typical limits of R_f are shown in Table 2.5

$$R_f = \frac{f_s}{q_c} \tag{2.16}$$

f_s is side friction

- **Pore pressure ratio B_q.** Typical limits of B_q are shown in Table 2.5

$$B_q = \frac{\Delta u}{q_t - \sigma_{vo}} \tag{2.17}$$

Δu is excess pore pressure above in situ equilibrium pore water pressure
σ_{vo} is in situ total vertical stress

- **Normalized cone resistance Q_t**

$$Q_t = \frac{q_t - \sigma_{vo}}{\sigma'_{vo}} \tag{2.18}$$

σ'_{vo} is effective vertical stress $= \sigma_{vo} - u_o$
u_o is water pressure at the level where q_t is measured

- **Normalized friction ratio F_r**

$$F_r = \frac{f_s}{q_t - \sigma_{vo}} \tag{2.19}$$

An example of the use of Q_t and F_r for soil classification is shown in Fig. 2.7.

- **Ratio between normalized cone resistance q_c and N_{60}** (Robertson 1990)

$$\frac{q_c/p_a}{N_{60}} = 8.5 \cdot \left(1 - \frac{I_c}{4.6}\right)$$
$$I_c = \sqrt{(3.47 - \log Q_t)^2 + (\log F_r + 1.22)^2} \tag{2.20}$$

p_a is atmospheric pressure in the same unit as q_c

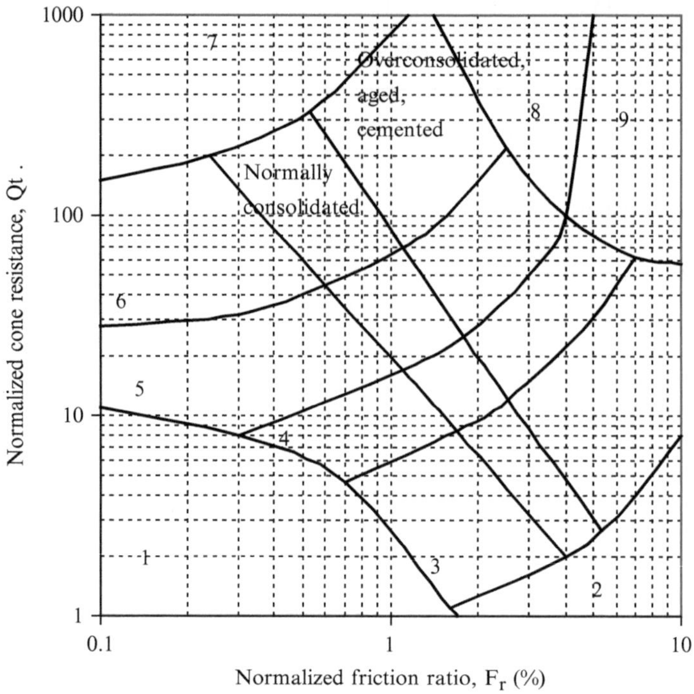

1. Sensitive, fine grained
2. Organic soil - peat
3. Clays - silty clay to clay
4. Silt mixtures - clayey silt to silty clay
5. Sand mixtures - silty sand to sandy silt
6. Sands- clean sand to silty sand
7. Gravelly sand to dense sand
8. Very stiff sand to clayey sand*
9. Very stiff, fine grained*

* Heavily overconsolidated or cemented

Fig. 2.7 Soil types from normalized cone penetration test data (Adopted from Robertson 1990)

Kulhaway and Mayne (1990) suggested the following correlations with respect to the average particle size diameter D_{50} (mm) and the content FC (%) of fines with diameter <0.06 mm.

$$\frac{q_c/p_a}{N_{SPT}} = 5.44 \cdot {D_{50}}^{0.26}, 0.001 \leq D_{50} \leq 1(10)\,\text{mm}$$
$$\frac{q_c/p_a}{N_{SPT}} = 4.25 - \frac{FC}{41.3} \pm 0.89 \cdot S.D. \qquad (2.21)$$

p_a is the atmospheric pressure
$S.D.$ is the number of standard deviations

- **Cone diameter** has a small influence on the CPT results as showed by Almeida et al. (1992) for the cone diameters of 35.7 and 25.7 mm.

Table 2.6 Relative density D_r and friction angles ϕ of quartz and feldspar coarse grained soil based on CPT from EN 1997-2:2007 based on Bergdahl et al. (1993)

Description	Very loose	Loose	Medium dense	Dense	Very dense
Cone q_c (MPa)	0–2.5	2.5–5.0	5.0–10.0	10.0–20.0	>20.0
Sand friction angle ϕ (degrees)	29–32	32–35	35–37	37–40	40–42

Notes: For silty soil ϕ in the above table should be reduced by 3° and for gravels 2° should be added. D_r defined in Chap. 3

2.1.2.3 Correlations for Coarse Grained Soil

- **Relative density D_r** according to Table 2.6. Baldi et al. (1986) suggested formulas for Ticino sand. For moderately compressible, normally consolidated unaged and uncemented silica sands

$$D_r = \frac{1}{2.41} \cdot \ln_e \frac{q_c}{157 \cdot \sigma_v'^{0.55}} \tag{2.22}$$

σ'_v is effective vertical stress (total less water pressure) in kPa
q_c is in kPa.

- **Friction angle ϕ** (degrees) according to Table 2.6. For poorly graded sand above ground water from DIN 4094-1 (2002) based on Stenzel and Melzer (1978)

$$\phi = 13.5 \cdot \log(q_c) + 23,5\,\text{MPa} \le q_c \le 28\,\text{MPa} \tag{2.23}$$

Kulhaway and Mayne (1990) provided the correlation of a graphs by Robertson and Campanella (1983) among vertical effective stress σ'_{vo} (kPa), q_c and ϕ (degrees) as

$$\phi = \arctan\left[0.1 + 0.38 \cdot \log_{10} \cdot \left(\frac{q_c}{\sigma'_{vo}}\right)\right] \tag{2.24}$$

- **Young's modulus of elasticity E'** from EN 1997-2:2007 based on Schmertmann (1970)

$$\begin{aligned} E' &= 2.5 \cdot q_c \; for \;\; circular \; \& \; square \;\; foundations \\ E' &= 3.5 \cdot q_c \; for \;\; strip \;\; foundations \end{aligned} \tag{2.25}$$

- **Drained constrained modulus M_o** from calibration chamber tests equivalent to the tangent modulus from oedometer tests for normally consolidated unaged and uncemented predominantly silica sand at the effective vertical stress σ'_{vo} recommended by Lunne and Christophersen (1983)

$$\begin{aligned} M_o &= 4 \cdot q_c \ \ for \ \ q_c < 10\,\text{MPa} \\ M_o &= 2 \cdot q_c + 20\,\text{MPa} \ \ for \ \ 10\,\text{MPa} < q_c < 50\,\text{MPa} \\ M_o &= 120\,\text{MPa} \ \ for \ \ q_c > 50\,\text{MPa} \end{aligned} \tag{2.26}$$

For over consolidated sand Lunne and Christophersen (1983) recommended

$$\begin{aligned} M_o &= 5 \cdot q_c \ \ for \ \ q_c < 50\,\text{MPa} \\ M_o &= 250\,\text{MPa} \ \ for \ \ q_c > 50\,\text{MPa} \end{aligned} \tag{2.27}$$

For an additional stress $\Delta\sigma'_v$, Lunne and Christophersen (1983) recommended Janbu's (1963) formulation to compute M for the stress range σ'_{vo} to $\sigma'_{vo} + \Delta\sigma'_v$

$$M = M_o \sqrt{\frac{\sigma'_{vo} + \Delta\sigma'_v/2}{\sigma'_{vo}}} \tag{2.28}$$

For silty soil, Senneset et al. (1988) suggested the following expressions

$$\begin{aligned} M_o &= 2 \cdot q_t \ \ (\text{MPa}) \ \ for \ \ q_t < 2.5\,\text{MPa} \\ M_o &= 4 \cdot q_t - 5 \ \ (\text{MPa}) \ \ 2.5\,\text{MPa} < q_t < 5\,\text{MPa} \end{aligned} \tag{2.29}$$

- **Maximum shear stiffness modulus G_{max}**

  Rix and Stokoe (1992) suggested the relationship for uncemented quartz sand

$$\left(\frac{G_{\max}}{q_c}\right)_{average} = 1{,}634 \cdot \left(\frac{\sqrt{\sigma'_v}}{q_c}\right)^{0.75}, 0.0003 \le \frac{\sqrt{\sigma'_v}}{q_c} \le 0.005 \tag{2.30}$$

G_{max} (kPa) is the maximum shear modulus (at small strain $<10^{-6}$)
σ'_v (kPa) is the vertical effective stress
q_c (kPa)

The scatter range is the average ±0.5 times the average

Baldi et al. (1989) plotted a relationship between normalized maximum shear modulus and q_c, which fitted formula is

$$\frac{G_{\max}}{q_c} = \frac{273.7}{\left(q_c/\sqrt{\sigma'_v \cdot p_a}\right)^{0.762}}, 20 \le \frac{q_c}{\sqrt{\sigma'_v \cdot p_a}} \le 300 \tag{2.31}$$

G_{max} the maximum shear modulus at small strain $<10^{-6}$
σ'_v is vertical effective stress
p_a is the atmospheric pressure

Table 2.7 The values of coefficients a_m according to Sanglerat (1972)

Soil type	CPT q_c (MPa)	α_m
Low plasticity clay (CL)	$q_c < 0.7$ MPa	$3 < \alpha_m < 8$
	$0.7 < q_c < 2.0$ MPa	$2 < \alpha_m < 5$
	$q_c > 2.0$ MPa	$1 < \alpha_m < 2.5$
Low plasticity silt (ML)	$q_c > 2$ MPa	$3 < \alpha_m < 6$
	$q_c < 2$ MPa	$1 < \alpha_m < 3$
Highly plastic silt and clay (MH, CH)	$q_c < 2$ MPa	$2 < \alpha_m < 6$
Organic soil (OL)	$q_c < 1.2$ MPa	$2 < \alpha_m < 8$
Peat and organic clay (P_t, OH)	$q_c < 0.7$ MPa	
	50 % < w < 100 %	$1.5 < \alpha_m < 4$
	100 % < w < 200 %	$1 < \alpha_m < 1.5$
	w > 200 %	$0.4 < \alpha_m < 1$

Note: w is water content

2.1.2.4 Correlations for Fine Grained Soil When Laboratory Test Results Are Not Available

- **Undrained shear strength c_u** based on comparisons of the results from 865 mm diameter plate tests and CPT tests reported by Marsland and Powell (1988) and Powell and Quarterman (1988); smaller N_{kt} are for smaller plasticity indices and larger N_{kt} for greater plasticity indices

$$\begin{aligned} c_u &= \frac{q_t - \sigma_{vo}}{N_{kt}}, \\ 11 \le N_{kt} &\le 20 \ \ discontinuities\ \ space < cone\ \ diameter \\ 20 \le N_{kt} &\le 30 \ \ discontinuities\ \ space > cone\ \ diameter \end{aligned} \tag{2.32}$$

- **Over consolidation ratio** (*OCR*) from Lunne et al. (1997)

$$OCR = k \cdot \frac{q_t - \sigma_{vo}}{\sigma'_{vo}}, k\sim 0.3, 0.2 < k < 0.5 \tag{2.33}$$

σ_{vo} and σ'_{vo} are total and effective (total less water pressure) vertical stresses

- **Constrained modulus *M*** according to Kulhaway and Mayne (1990)

$$M = 8.25 \cdot (q_t - \sigma_{vo}) \tag{2.34}$$

Sanglerat (1972) suggested the following expression

$$M = \alpha_m \cdot q_c \tag{2.35}$$

The values of α_m are given in Table 2.7.

2.1.2.5 Correlations for Chalk

- **Ratios between CPT and SPT** results q_c/N_{SPT} of 0.5 on average, 0.3 for lower bund, 0.7 for upper bound according to Power (1982), where q_c is in MPa and N_{SPT} is blows per 300 mm penetration.

2.1.3 Permeability to Water of Coarse Grained Soil Using Open Tube Piezometers

The coefficient k of water permeability of soil is calculated from measured decrease of water level in time in an open tube piezometer after addition of water to the tube. Product of the coefficient k and hydraulic gradient equals velocity of water flow through soil. The test is standardised e.g. BS 5930 (1999). According to Hvorslev (1951)

$$k = \frac{A}{F \cdot (t_2 - t_1)} \cdot \ln_e \frac{H_1}{H_2}$$

$$F = \frac{2.32 \cdot \pi \cdot L}{\ln_e \left[1.1 \cdot L/D + \sqrt{1 + 1.1 \cdot (L/D)^2} \right]} \quad (2.36)$$

$H_{1,2}$ are the variable water heights above a common datum measured at times $t_{1,2}$

A is the internal cross-sectional area of piezometer tube or of a granular filter material surrounding the piezometer tube.

L is the length of perforated part of a piezometer tube.

D is the diameter of perforated part of a piezometer tube. For large L/D ratio or large horizontal water permeability in comparison with the vertical permeability, the coefficient of horizontal water permeability k_h is obtained.

2.1.4 Permeability to Water of Rock Mass Using Water Pressure Test (Lugeon Test)

Lugeon (1933) test comprises the measurement of the volume of water that can flow from an uncased section of borehole between two packers in a unit time under unit pressure, e.g. BS 5930 (1999). A rock mass has a permeability of 1 Lugeon unit if, under water height of 100 m above ground water level, accepts 1 l/min of water flow over a 1 m length of borehole. The borehole diameter is usually 76 mm and the test results are not very sensitive to change in borehole diameter unless the length of borehole tested is small. It is generally accepted that 1 Lugeon ~10^{-7} m/s. An approximate formula for calculation of the coefficient of water permeability of rock k (m/s) is

$$k = \frac{Q}{2 \cdot \pi \cdot H \cdot L} \cdot \ln_e \frac{L}{r} \tag{2.37}$$

Q is the rate (m^3/s) of water injection
H is the pressure head (m) of water in the test Section
L is the length (m) of the test section
r is the radius (m) of the test Section

2.1.5 Plate Loading Test for Compressibility of Soil Sub-base

The test is performed on subgrade soil and compacted pavement components, in either the compact condition or the natural state, and provides data for use in the evaluation and design of rigid and flexible-type airport and highway pavements, raft foundations and alike conditions. The test is standardised by ASTM D1195 and future ISO 22476-13. According to EN 1997-2 (2007), the coefficient of sub-grade reaction k_s is

$$k_s = \frac{\Delta p}{\Delta s} \tag{2.38}$$

Δp is pressure range applied by test plate on soil
Δs is the settlement range of test plate due to Δp

The soil stiffness modulus E_{PLT} corresponding to the stress level $p + \Delta p$ to depth of maximum of two plate diameters

$$E_{PLT} = \frac{\Delta p}{\Delta s} \cdot \frac{\pi \cdot b}{4} \cdot \left(1 - \nu^2\right) \tag{2.39}$$

b is the diameter of the test plate
ν is Poisson's ratio (~0.3 for coarse grained soil, =0.5 for fine grained soil in undrained condition)

Equivalent soil springs are still used in practice because of their simplicity. Elastic soil spring is defined as the ratio between applied force and achieved displacement in the direction of the force. From Eqs. (2.38) and (2.39),

$$SS_e = \frac{b \cdot E_{(PLT)}}{(1 - \nu^2)} \tag{2.40}$$

$E_{(PLT)}$ is the soil stiffness modulus
SS_e is the elastic equivalent soil spring
ν is Poisson's ratio (~0.3 for coarse grained soil, =0.5 for fine grained soil in undrained condition)

2.1.6 *California Bearing Ratio (CBR) for Compressibility of Soil Sub-base*

The test is used for evaluation of potential strength of subgrade, sub-base and base coarse material for use in road and airfield pavements. The test is standardized for field condition (ASTM D4429-09a) and for laboratory (ASTM D1883-07e2). Powell et al. (1984) provided the following relationship between elastic deformation modulus E (MPa) and CBR (%)

$$E = 17.6 \cdot CBR^{0.64}, 2\,\% \leq CBR \leq 12\,\% \tag{2.41}$$

2.1.7 *Pressuremeter Test in Boreholes for Soil and Rock Compressibility*

The test for soil is standardized in ASTM D4719 or maximum expansion pressure of 5.0 MPa. The modulus of ground stiffness E_p (kPa) in the horizontal direction for vertical boreholes is

$$E_p = 2 \cdot (1 + \nu) \cdot (V_o + V_m) \cdot \frac{\Delta P}{\Delta V} \tag{2.42}$$

ν is Poisson's ratio ~0.33
V_o (cm^3) is volume of the measuring portion of the uninflated probe at zero volume reading
V_m (cm^3) is corrected volume reading in the centre portion where the volume increase ΔV is linearly dependent on the pressure increase ΔP
ΔP (kPa) is corrected pressure increase in the centre part of the linear portion of the pressure-volume function
ΔV (cm^3) is corrected volume increase in the centre part of the linear portion of the pressure-volume function, corresponding to ΔP pressure increase

2.1.8 *Geophysical Survey*

Geophysical survey is performed mainly for ground profiling, measurements of velocities of ground waves propagation and electrical resistivity and for locating of obstructions and cavities underground. Usefulness of considered geophysical methods is given in Table 2.8. The methods are standardized, for e.g.

- Geophysical crosshole testing ASTM D4428
- Geophysical refraction ASTM D5777
- Geophysical reflection-land ASTM D7128
- Resistivity to current flow in the field ASTM G57
- Ground penetrating radar ASTM D6432
- Electro-magnetic ASTM D6639

The maximum shear stiffness modulus G_{max} and maximum axial stiffness modulus E_{max} ($<10^{-6}$) are calculated from measured transversal V_t and longitudinal V_l wave velocities

$$\begin{aligned} G_{\max} &= \rho \cdot V_t^2 \\ E_{\max} &= \rho \cdot V_l^2 \end{aligned} \tag{2.43}$$

ρ is ground unit density

Poisson's ratio ν

$$\nu = \frac{\left(\frac{V_l}{V_t}\right)^2 - 2}{2 \cdot \left(\frac{V_l}{V_t}\right)^2 - 2} \tag{2.44}$$

Correlation between transversal V_t and longitudinal V_l wave velocities depending on Poisson's ratio ν

$$V_l = V_t \cdot \sqrt{\frac{2 - 2 \cdot \nu}{1 - 2 \cdot \nu}} \tag{2.45}$$

Typical ranges of electrical resistivities of common ground are given in Table 2.9. Electrical resistivity may be used to assess the corrodibility of soil to ferrous materials if laboratory testing of content of pH, chlorides and sulphates in ground and ground water has not been performed. The likelihood of severe corrosion usually decreases as the resistivity rises (appendix H of BS 5493 1977). Soil resistivity <10 ohm-m indicates severe corrosivity and >100 ohm-m slight corrosivity. Typical ranges of transversal wave velocities V_t, standard penetration test blow counts N_{SPT} and undrained shear strength c_u for various ground types are shown in Fig. 2.8.

Table 2.8 Usefulness of considered engineering geophysical methods (Adopted from BS 5930 1999)

Type	Depth to bedrock	Strati graphy	Litho logy	Fractured zones	Fault displace ments	Maximum elastic modulus	Density	Rippa bility	Cavity detection	Buried artefacts	Ground water explora tion	Water quality	Porosity	Buried channels	Clay pockets in lime stone	Sand and gravel
Refraction	A	A	b	b	A	b	c	A	d	d	c	e	e	A	d	c
Reflection-onshore	c	c	c	d	c	e	e	e	c	d	c	e	e	d	e	e
Reflection-offshore	A	A	c	c	A	e	e	d	e	c	e	e	e	A	e	e
Electrical resistivity	A	b	b	c	c	e	e	d	c	d	A	b	b	b	e	b
Ground probing radar	c	b	d	c	b	e	e	e	b	A	c	d	d	c	c	d
Electromagnetic and resistivity profiling	b	c	c	A	d	e	e	e	b	b	A	A	d	b	A	b

Notes:

e – not considered applicable

d – limited use

c – used but not best suited

b – excellent but requires further development

A – excellent

Table 2.9 Typical ranges of electrical resistivities of common ground (Adopted from McDowell et al. 2002)

Type	Dense limestone	Porous limestone	Sandstone	Sand	Hard shale	Soft shale	Clay	Metamorphic rock	Igneous rock
Resistivity range (ohm-m)	10^3 to 5×10^8	50 to 5×10^3	20 to 2×10^3	20 to 10^3	8 to 5×10^2	0.4 to 9	0.4 to 10	12 to 9×10^5	90 to 10^6

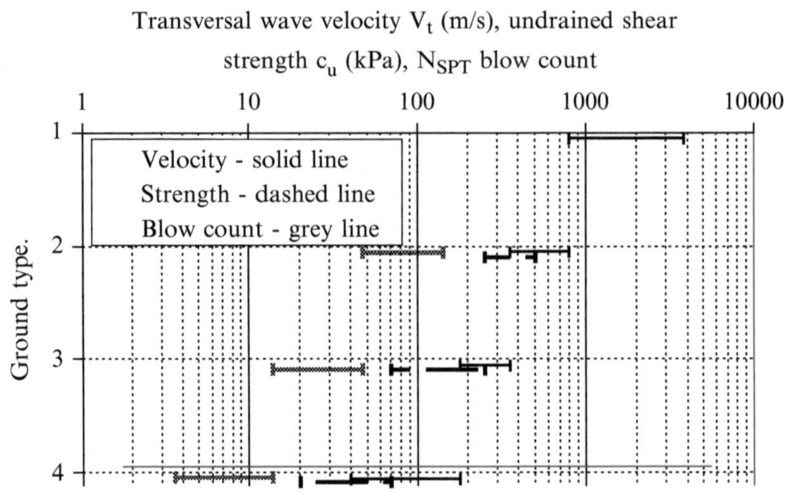

Ground types:
1 - A: Rock (<5m thick top soil)
2 - B: Very dense coarse or very stiff fine grained soil tens of metres thick
3 - C: Dense or medium dense coarse or stiff fine grained soil tens to hundreds metres thick
4 - D: Loose to medium dense coarse or soft to firm fine grained soil

Fig. 2.8 Typical ranges of transversal wave velocities V_t, standard penetration test blow counts N_{SPT} and undrained shear strength c_u for various ground types (Adopted from EN 1998-1, 2004)

2.2 Laboratory Tests

2.2.1 Soil Compressibility, Swellability and Collapsibility in Oedometer

The test is used traditionally for determination of the magnitude and rate of consolidation of soil specimens that are restrained laterally and drained axially while subjected to incrementally applied controlled-stress loading (e.g. BS 1377-5; ASTM D2435/D2435M-11; ISO 17892-5). The load is usually doubled from previous load and test specimens are submerged in water bath except when measuring swellability and collapsibility of soil when water can be added after application of specified load. An example of determination of soil collapsibility and swellability potential is shown in Fig. 2.9 for a specimen of loess. Two methods for determination of time of 100 % of primary consolidation are sketched in Fig. 2.10 after Casagrande and Fadum (1940) and Taylor (1942). Casagrande (1936) proposed a method for determination of preconsolidation pressure p_c as sketched in Fig. 2.11. The values defined from an oedometer tests are:

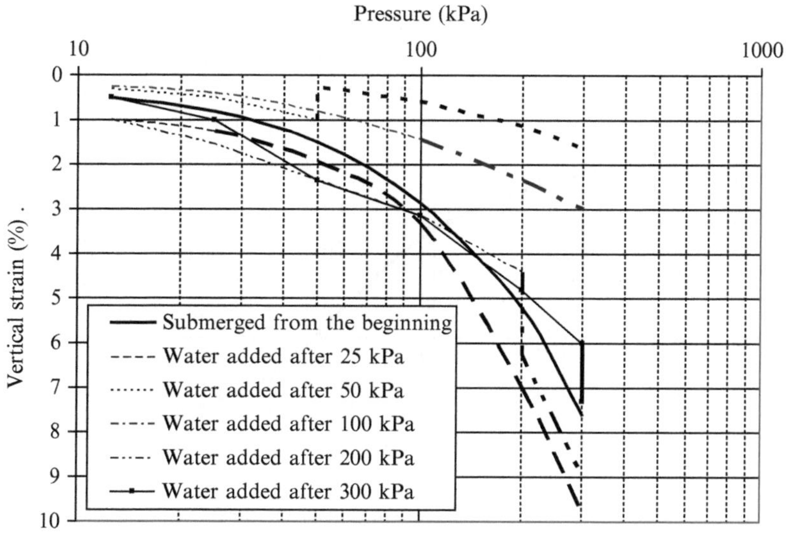

Fig. 2.9 An example of determination of potentials for collapse and swell of specimens of loess in oedometers; swellable for pressures <100 kPa and collapsible for pressure >100 kPa

- **Coefficient of volume compressibility m_v**

$$m_v = \frac{e_1 - e_2}{1 + e_1} \cdot \frac{1}{p_2 - p_1} \tag{2.46}$$

e_1 is the void ratio of a specimen at pressure p_1
e_2 is the void ratio of a specimen at pressure p_2

- **Coefficient of compressibility a_v**

$$a_v = \frac{e_1 - e_2}{p_2 - p_1} \tag{2.47}$$

- **Compression index C_c**

$$C_c = \frac{e_1 - e_2}{\log_{10}(p_2/p_1)} \tag{2.48}$$

- **Compression ratio C_{ce}**

$$C_{ce} = \frac{e_1 - e_2}{1 + e_1} \cdot \frac{1}{\log_{10}(p_2/p_1)} \tag{2.49}$$

- **Recompression index and recompression ratio C_r and C_{re}** for recompression stage of oedometer test similar to Eqs. (2.48) and (2.49)

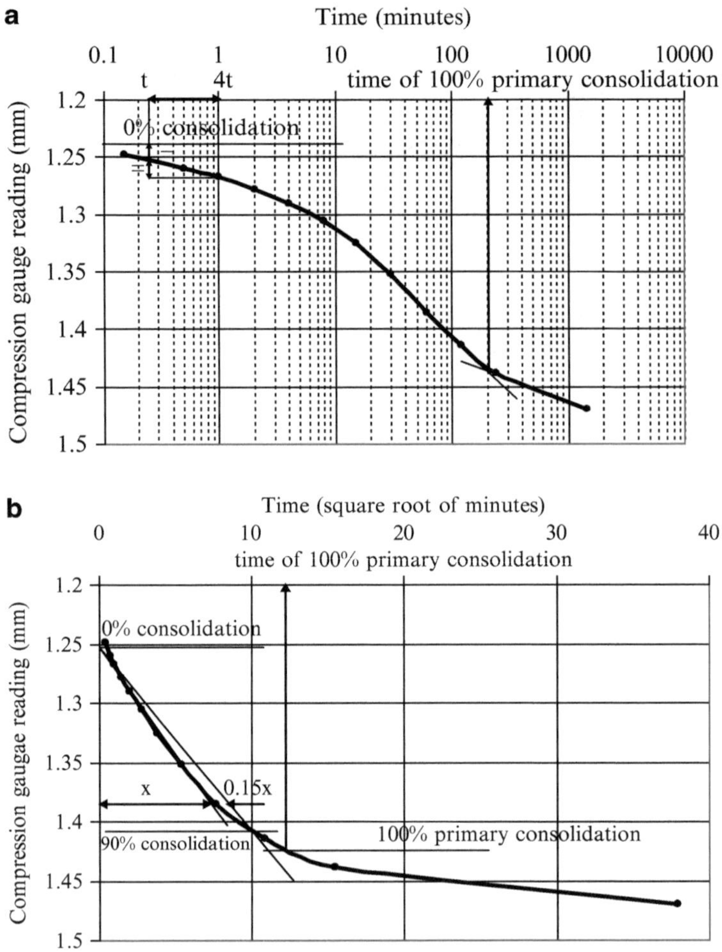

Fig. 2.10 Methods to determine the time of 100 % of primary consolidation in an oedometer test, (**a**) logarithm of time according to Casagrande and Fadum (1940), (**b**) square root of time according to Taylor (1942)

- **Secondary compression index C_α**

$$C_\alpha = \frac{\Delta e}{\log_{10}\left({}^{t_2}\!/_{t_1}\right)} \tag{2.50}$$

Δe is the change in void ratio during secondary compression time $t_2 - t_1$

- **Secondary compression ratio $C\alpha e$**

$$C_{\alpha e} = \frac{C_\alpha}{1 + e_p} \tag{2.51}$$

e_p is the void ratio at the start of secondary compression stage

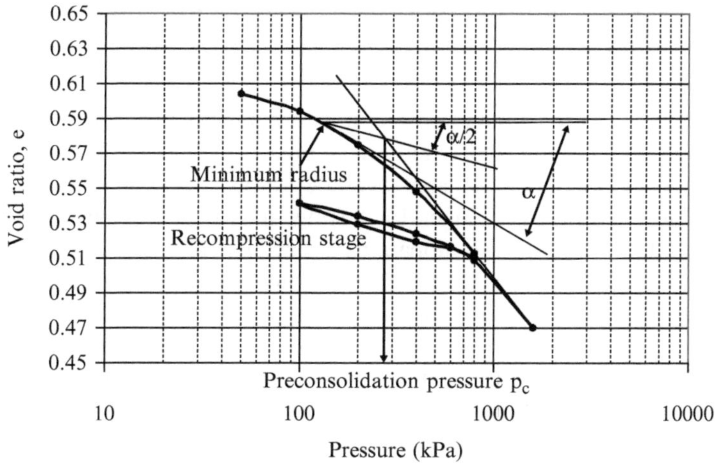

Fig. 2.11 Casagrande (1936) method for determination of preconsolidation pressure p_c using oedometer test results according to Casagrande (1936)

- **Coefficient of consolidation c_v**

$$c_v = \frac{T_v \cdot d^2}{t} = \frac{k}{m_v \cdot \gamma_w} \tag{2.52}$$

T_v is the time factor; $T_{50} = 0.197$, $T_{90} = 0.848$ for top and bottom drainage
d is the maximum length of the drainage path (=1/2 layer or specimen thickness for drainage at the top and bottom)
t is consolidation time (for 50, 90 %, etc. consolidation)
k is the coefficient of permeability
γ_w is the unit weight of water

2.2.2 *Swellability of Rock*

Weathered rock containing high clay content is prone to swelling during further weathering wetting and drying cycles. Suggested method for determining swelling index is published by ISRM (1999).

2.2.3 Soil Static and Cyclic Shear Strength and Stiffness by Simple Shear Apparatus

While direct shear test is standardized it is not favoured because of the stress concentrations at the edges of rigid shear boxes and non-uniform axial stress distribution over the specimen area. Roscoe (1953) developed the first simple shear apparatus with hinged edges for testing in static conditions. Peacock and Seed (1968) first adopted this apparatus to cyclic conditions. The Norwegian Geotechnical Institute uses a wire reinforced membrane while the Swedish Geotechnical Institute uses a series of stacked rings around a cylindrical specimen. When shared in these devices, the specimen does not maintain plane strain condition (Finn 1985). In Roscoe apparatus, uniform simple shear conditions can be induced in most parts of a specimen particularly at small strain. Example plots of shear strain increase (causing shear stiffness decrease because the stiffness is the ratio between applied shear stress and achieved shear strain) and shear strength decrease in cyclic condition are shown in Fig. 2.12 and of cyclic stress ratio dependence on a number of cycles in Fig. 2.13.

2.2.4 Soil Static and Cyclic Shear Strength and Stiffness by Triaxial Apparatus

The test is standardized, e.g.

- **for static conditions**: by BS 1377-7 (total stress), BS 1377-8 (effective stress – total less water pressure), ASTM D2850 (unconsolidated, undrained condition), ASTM D4767 (consolidated undrained condition with excess pore water pressure measurement), ISO 17892-8 (unconsolidated undrained condition), ISO 17892-9 (consolidated undrained or drained)
- **for cyclic conditions**: ASTM D3999 (modulus and damping), ASTM D5311 (shear strength in load controlled condition)

Almost fully water saturated cylindrical specimen (with the height to diameter ratio of at least 2:1 to minimize constraining end effects and the diameter of preferably greater than 100 mm for better involvement of fissures and joints) wrapped into a rubber membrane, surrounded by water under pressure, porous stones placed at the top and bottom is subjected to axial stress increase (sometimes decrease) at a constant rate with/without measurements of volume decrease or pressure increase of pore water within the specimen during application of the axial stress. The measurement of small local strain at the specimen side is not standardized; it can cover the strain range from about 5×10^{-3} to 10^{-5}. Figure 2.14

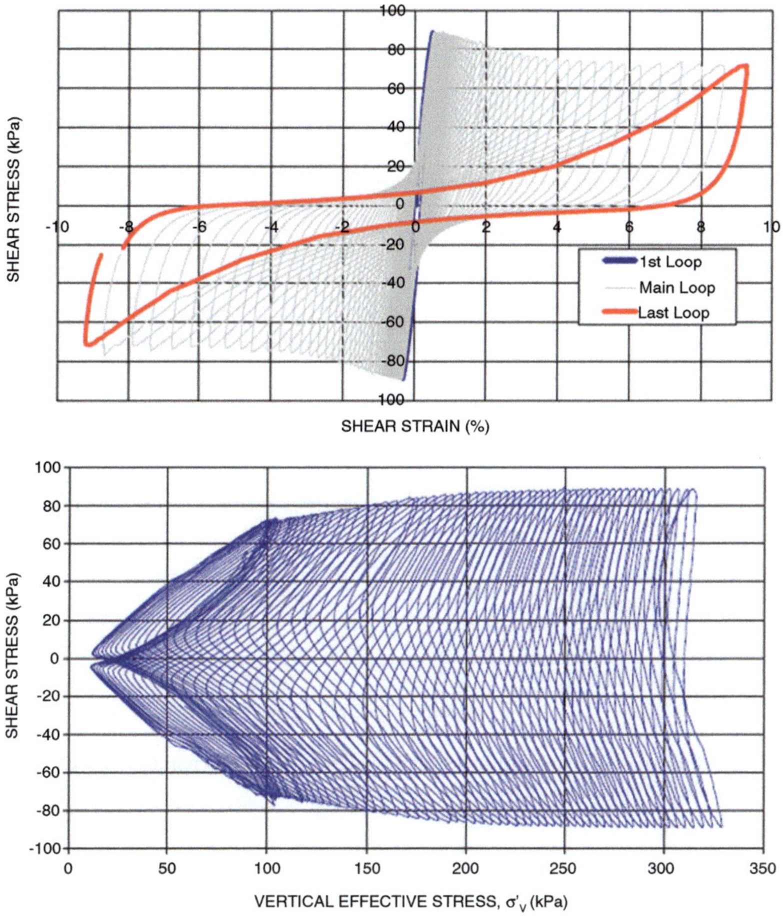

Fig. 2.12 Example of shear strain increase (*top*) and shear strength decrease (*bottom*) with build up of excess pore water pressure in a specimen of silt as the number of cycles increase during stress controlled simple shear test

shows an example of the use of local strain measurement. The cost of a cyclic triaxial test is about ten times greater than the cost of a static triaxial test. Typical cell pressures, rates of axial displacement increase and the results obtained from static tests are:

- **Unconsolidated undrained (UU)** test in terms of total stresses is usually performed with the cell pressure equal to estimated mean total stress at the sample depth, the rate of axial displacement application per minute is about 1 % of the specimen height, typical results are sketched in Fig. 2.15a. Undrained shear strength c_u

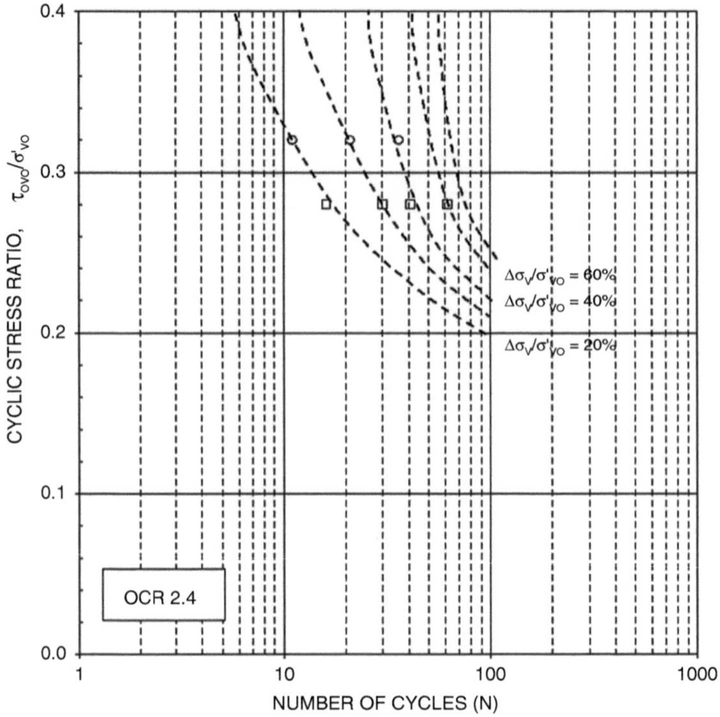

Fig. 2.13 Example of cyclic stress ratio decrease with the increase in number of cycles applied to a specimen of silt during a stress controlled simple shear test

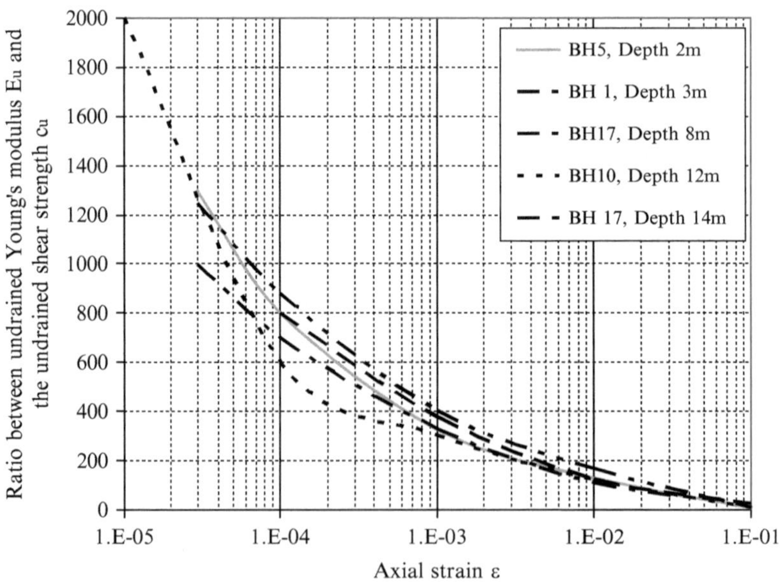

Fig. 2.14 An example of the use of local strain measurements for determination of the ratio between undrained Yong's modules E_u and undrained shear strength c_u of specimens of loess in triaxial apparatus

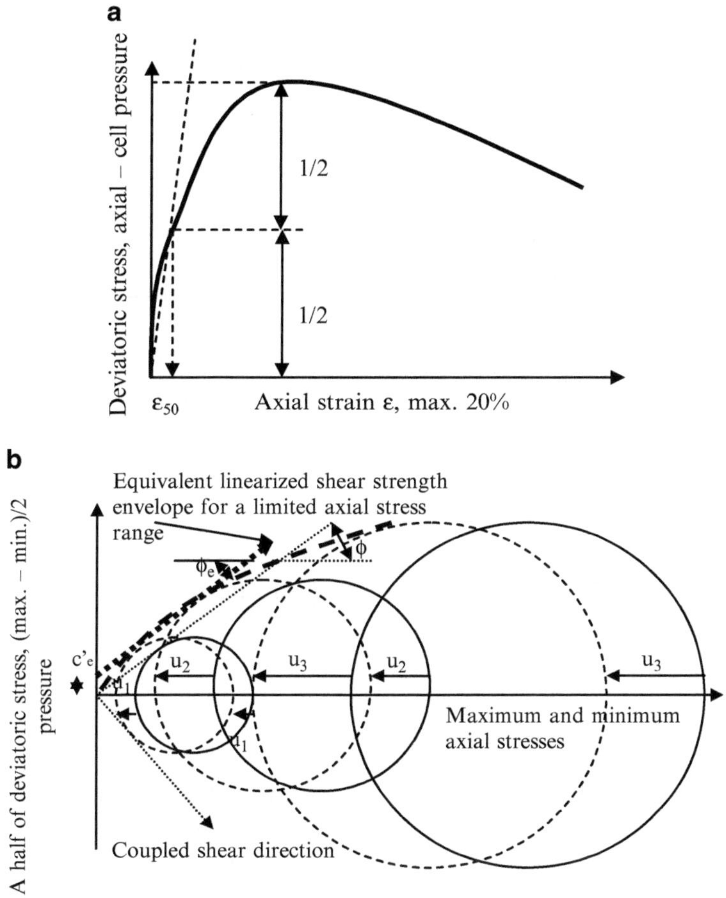

Fig. 2.15 Examples of triaxial test results, (**a**) unconsolidated undrained in terms of total stresses, (**b**) consolidated undrained in terms of total (*solid line*) and effective (*dashed line*) maximum axial stresses and corresponding minimal stress (cell pressure) for three specimens (excess pore water pressures at failure $u_{1,2,3}$), ϕ is ground friction angle (secant for the true non-linear failure envelope)

$$c_u = \frac{\sigma_{\max.axial} - \sigma_{cell}}{2} \tag{2.53}$$

$\sigma_{max.axial}$ is the maximum axial stress
σ_{cell} is the water pressure around a specimen

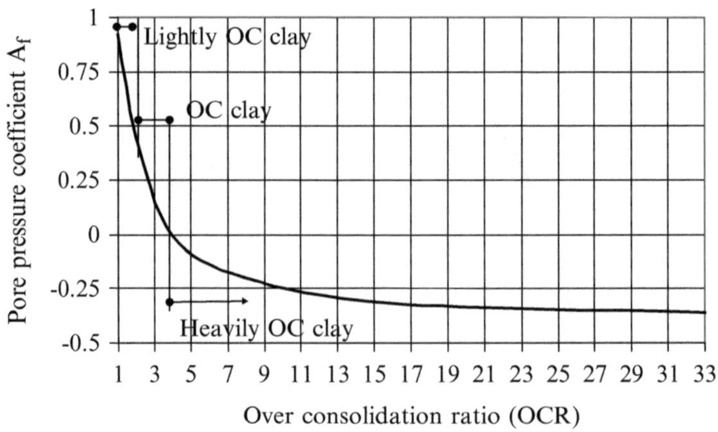

Fig. 2.16 Values of pore pressure coefficient A_f at failure versus the over consolidation ratio (OCR)

Secant modulus E_s in terms of total (undrained) stresses

$$E_s = \frac{\sigma_{\max.axial} - \sigma_{cell}}{2 \cdot \varepsilon_{50}} \tag{2.54}$$

ε_{50} is the axial strain at 50 % of maximum deviatoric stress $\sigma_{max.axial} - \sigma_{cell}$

- **Consolidated undrained (CU)** with measurement of excess pore water pressure during application of the axial displacement is usually performed on three specimens with the cell pressures range of interest (e.g. for soil slopes 50, 100, 200 kPa, for foundations 100, 200, 400 kPa), the rate of axial displacement application per minute to enable pore water pressure equalization within the specimen because of the pore water pressure measurements at the ends e.g. about 0.1 % of the specimen height, typical results sketched in Fig. 2.15b.

  Skempton (1954) provided expression for the increase of excess pore water pressure Δu as a function of the incremental maximum $\Delta\sigma_1$ and minimum $\Delta\sigma_3$ pressures in axisymmetric conditions that exist under loaded areas.

$$\Delta u = B(\Delta\sigma_3 + A \cdot (\Delta\sigma_1 - \Delta\sigma_3)) \tag{2.55}$$

  In a triaxial test in which σ_3 remains constant $\Delta\sigma_3 = 0$. The values of the coefficients A_f at failure and B are shown in Figs. 2.16 and 2.17. In static conditions, greater angle of friction of sand is measured in plane strain condition than in triaxial condition depending on porosity (Cornforth 1964); the difference is about 5° for porosity of 34 % and about 1° for porosity of 42 %.
- The effects of most influential factors on the cyclic shear strength of coarse grained soil measured in triaxial apparatus are listed in Table 2.10 from Townsend

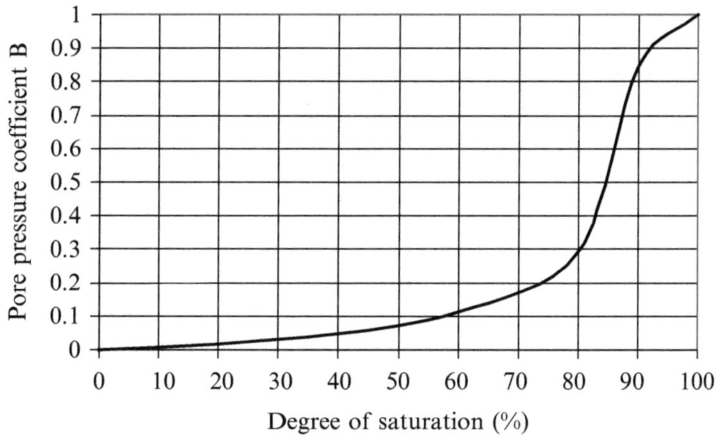

Fig. 2.17 Pore pressure coefficient B versus degree of saturation

Table 2.10 Effects of most influential factors on the cyclic strength of cohesional soil in triaxial apparatus

Factor	Effect
Specimen preparation method	Weakest specimens formed by pluviation through air, while strongest ones formed by vibrating in moist condition. Difference in stress ratio at failure can be 110 %
Reconstituted versus intact	Intact specimens stronger than reconstituted. Strength decrease range from 0 to 100 %
Confining stress	Cyclic strength is directly proportional to confining stress within small range of pressure. Cyclic stress ratio decreases with increasing confining pressure
Loading waveform	Strength increases from rectangular wave shape, via triangular to sine. Sine wave causes approximately 30 % greater strength than rectangular. Irregular wave form can be replaced by equivalent harmonic wave
Frequency	Slower loading frequencies have slightly higher strength. For a range from 1 to 60 cycles per minute, the effect is 10 %. Water presence may affect results at 5 cycles per second
Specimen size	300 mm diameter specimen exhibit approximately 10 % weaker strength than 70 mm diameter specimen
Relative density	Exponential shear strength increase with linear increase in density
Particle size and degradation	Sand with average diameter D_{50} of approximately 0.1 mm has least resistance to cyclic loading. As D_{50} increases from 0.1 to 30 mm, shear strength increases 60 %. As D_{50} decreases from 0.1 mm to silt and clay sizes, a rapid increase in strength is observed. Well graded soil weaker than uniformly graded soil
Pre-straining	Previous cyclic load greatly increases shear strength during current cyclic load
Over consolidation	Over consolidation increased shear strength depending on amount of fines (particles less than about 5 μm)
Anisotropy	Shear strength is increased by increased anisotropy. Method of data presentation influences the effect; isotropic consolidation may not always provide conservative results

Table 2.11 Factors affecting cyclic strength of normally consolidated clay

Factor	Change in factor	Change in undrained shear strength
Cyclic stress	Increase	Decrease approximately linearly with the logarithm of number of cycles
Number of stress cycles	Increase	Decrease
Initial shear stress	Increase	Decrease
Direction of principal stress	90° rotation	Decrease
Shape of cyclic stress	From square to sine	Decrease
Frequency of cyclic stress	From 2 to 1 cycles per second	Decrease
Stiffness of soil	Increase	Increase
Stress state	From triaxial to simple shear	Negligible

(1978) and of the factors affecting cyclic shear strength of fine grained soil are summarised in Table 2.11 from McLelland Engineers (1977). For testing of liquefaction potential, the results from cyclic triaxial tests should be multiplied by a factor (about 0.7 according to data summarised by Kramer 1996).

2.2.5 Axial Strength of Rock Cylinders in Unconfined Condition

This is an important parameter for estimation of rock mass shear strength. The test is standardized by ISRM (1979). A fresh rock sample of minimum NX (54 mm) size with the height to diameter ratio of between 2.5 and 3 must have sufficiently parallel ends in order to avoid undesirable stress concentration and secondary effects (failure in bending instead in axial compression). Rock classification based on unconfined compressive strength (UCS) is given in Table 2.12 from ISO 146891-1 (2003).

2.2.6 Abrasivity of Rock Fill (Los Angeles Test)

The test is important for rock fill subjected to cyclic and dynamic loads such as rip-rap for slope protection from wave impact, fill under vibrating machinery, ballast under train tracks, aggregates for concrete. The test is standardized by

Table 2.12 Rock strength classification based on unconfined compressive strength (UCS)

Type	Field identification	UCS (MPa)
Extremely weak	Indented by thumbnail	0.6–1.0
Very weak	Crumbles under firm blow with point of geological hammer. Peeled by a pocket knife	1–5
Weak	Peeled by a pocket knife with difficulty	5–25
Medium strong	Cannot be scraped with pocket knife. Can be fractured with a single firm blow of geological hammer	25–50
Strong	Requires more than one blow of geological hammer to fracture	50–100
Very strong	Requires many blows of geological hammer to fracture	100–250
Extremely strong	Can only be chipped with geological hammer	>250

ASTM C131 and C535 (chemical weathering C88 for aggregates only). Typical ranges of Los Angeles abrasion test are:

- Hard grains, abrasion 10–15 %
- Medium hard grains, abrasion 15–25 %
- Soft grains, abrasion >25 %

Wilson and Marsal (1979) plotted graphs of the upper bound, average and lower bound deformation modulus M of rock fill versus grain brakeage B_g (%). The B_g is the sum of positive differences between the initial weights retained on sieves after a grain size distribution test and the final retained weights on the sieves after the testing for abrasiveness. B_g in percent is the ratio between the sum of positive differences of the weights and the total weight of the sample subjected to sieve analysis. The fitted functions of modulus M (MPa) are for a range $5\ \% < B_g < 40\ \%$:

$$
\begin{aligned}
M &= \frac{686.6}{B_g^{0.774}}, \quad \textit{the upper bound} \\
M &= \frac{306.1}{B_g^{0.663}}, \quad \textit{the average} \\
M &= \frac{235.1}{B_g^{0.846}}, \quad \textit{the lower bound}
\end{aligned} \tag{2.56}
$$

2.2.7 Transversal Wave Velocity by Bender Elements

When transversal wave velocity has not been measured in situ or for checking of the measured values, the velocity is measured in laboratory using bender element,

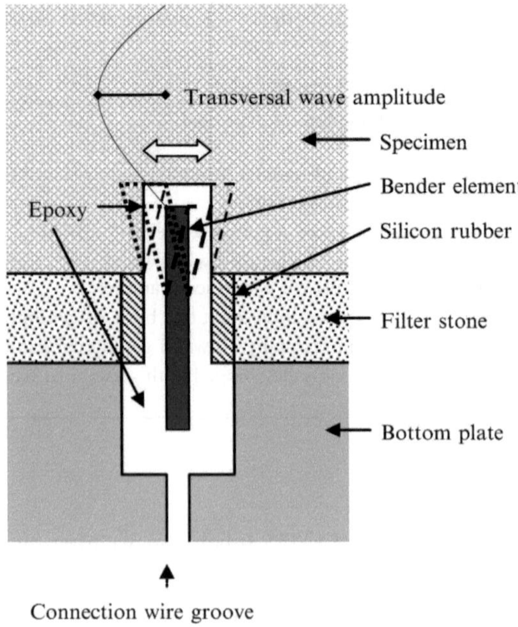

Fig. 2.18 Cross section through a bottom mounted bender element (in a triaxial apparatus). The protruding part is a few millimetres long

Fig. 2.18, in the last 40 years since its description by Shirley (1978). Arulnathan et al. (1998) listed the following potential errors in, and methods of interpreting, the results of cantilever-type, piezoceramic bender element tests for measuring the transversal wave velocity of laboratory soil specimens. Interpretations based on the first direct arrival in the output signal are often masked by near-field effects and may be difficult to define reliably. Interpretations based on characteristic points or cross-correlation between the input and output signals are shown to be theoretically incorrect in most cases because of:

- the effects of wave interference at the boundaries
- the phase lag between the physical wave forms and the measured electrical signals
- non-one-dimensional wave travel and near-field effects. Interpretations based on the second arrival in the output signal are theoretically subject to errors from non-one-dimensional wave travel and near-field effects.

Leong et al. (2009) showed that the performance of the bender–extender elements test can be improved by considering the following conditions:

- the digital oscilloscope used to record the bender–extender element signals should have a high analogue to digital (A/D) conversion resolution
- the size of the bender–extender elements plays an important role in the strength and quality of the receiver signal, especially for compression waves

- using a wave path length to wave length ratio of 3.33 enables a more reliable determination of shear wave velocity

2.2.8 Soil Stiffness and Damping at Small to Large Strain in Cyclic Condition by Resonant Column Apparatus

The test is standardized, ASTM D4015-07. Cylindrical specimens with a minimum diameter of 33 mm and the length to diameter ratio of between 2 and 7 are subjected to axial or tensional vibration of variable frequency until resonance occurs in the strain range from 10^{-6} to 10^{-3}. The vibration apparatus and specimen may be enclosed in a triaxial chamber and subjected to an all-around pressure and axial load. When applied axial stress is greater than the confining stress the specimen length to diameter ratio shall not be greater than 3 to avoid its buckling. In addition, the specimen may be subjected to other controlled conditions (for example, pore-water pressure, degree of saturation, temperature). Skoglund et al. (1976) compared the results obtained by six different investigators and found that the differences in shear and axial modules ranged from minus 19 % to plus 32 % of the average value.

2.2.9 Chemical Testing of Soil and Ground Water

EN 1997-2 (2007) provides recommendations for routine chemical testing in laboratories such as:

- **Organic content**, which can reduce bearing capacity, increase compressibility, increase swelling and shrinkage potential (test BS 1377 (1990) – part 3, ASTM D2974-07a; ISO 10694)
- **Carbonate content**, which can increase degree of cementation of non-calcareous minerals and indicate degree of crushability of calcareous sand offshore (test BS 1377 (1990) – part 3, ASTM 4373-02, ISO 10693). Purely calcareous sand has >80 % of $CaCO_3$.
- **Sulphate content**, which can have detrimental effect on steel and concrete particularly when water soluble instead of hydrochloric acid-soluble (referred to as the total sulphate content) (test BS 1377 (1990) – part 3, ASTM C1580; ISO 11048)
- **Chloride content**, which can have affect on concrete, steel, other materials and soil (test BS 1377 (1990) – part 3, ASTM D1411)
- **pH value**, to assess the possibility of excessive acidity (pH $\ll$7) or alkalinity (pH $\gg$7) (test BS 1377 (1990) – part 3, ASTM D4972, G51-95; ISO 10390)

Table 2.13 An example of classification of sulfate aggressiveness to concrete in natural ground (Adopted from BRE Special Digest 2005)

Design sulfate class	2:1 water/soil extract (SO_4 mg/l)	Groundwater (SO_4 mg/l)	Total potential sulfate (SO_4 %)[a]
DS-1	<500	<400	<0.24
DS-2	500–1,500	400–1,400	0.24–0.6
DS-3	1,600–3,000	1,500–3,000	0.7–1.2
DS-4	3,100–6,000	3,100–6,000	1.3–2.4
DS-5	>6,000	>6,000	>2.4

Notes: [a]Applies only to locations where concrete will be exposed to sulfate ions (SO_4) which may be product of oxidation of sulfides (e.g. pyrite)

Table 2.14 An example of classification of sulfate and magnesium aggressiveness to concrete in fills and at industrial sites (Adopted from BRE Special Digest 2005)

Design sulfate class	2:1 water/soil extract (SO_4 mg/l)	2:1 water/soil extract (Mg mg/l)	Groundwater (SO_4 mg/l)	Groundwater (Mg mg/l)	Total potential sulfate (SO_4 %)[a]
DS-1	<500	–	<400	–	<0.24
DS-2	500–1,500	–	400–1,400	–	0.24–0.6
DS-3	1,600–3,000	–	1,500–3,000	–	0.7–1.2
DS-4	3,100–6,000	<1,200	3,100–6,000	<1,000	1.3–2.4
DS-4m	3,100–6,000	>1,200[b]	3,100–6,000	>1,000[b]	1.3–2.4
DS-5	>6,000	<1,200	>6,000	<1,000	>2.4
DS-5m	>6,000	>1,200[b]	>6,000	>1,000[b]	>2.4

Notes: [a]Applies only to locations where concrete will be exposed to sulfate ions (SO_4) which may be product of oxidation of sulfides (e.g. pyrite)
[b] The limit does not apply to brackish groundwater (chloride content >12,000 mg/l and <17,000 mg/l). Sea water (chloride about 18,000 mg/l) and stronger brines are not covered by the table

Examples of classification of aggressiveness to concrete of sulphates and sulphate and magnesium reach soil and ground water are given in Tables 2.13 and 2.14 based on BRE Special Digest 1 (2005).

2.3 Summary

A summary of the field test types considered and engineering ground properties from the tests results is given in Table 2.15 and for the laboratory test types considered in Table 2.16.

Table 2.15 Summary of field test types considered and the engineering ground properties from the test results

Type	Relative density, D_r	Angle of friction, ϕ	Max. axial stiffness modulus, E_{max}	Max. shear modulus, G_{max}	Coeff. of volume compress., m_v	Trans. wave veloc., V_t	Mass compr. strength, σ	Ult. end bearing stress, q_u	Unconf. compr. strength, UCS	Consistency	Undrained shear strength, c_u	Drained constrained modulus, M_o	Over consolidation ratio, OCR	Water permeability, k	Poisson's ratio
SPT	Coarse soil	Coarse soil	Soil, chalk, weak rock	–	Coarse soil	Soil	Chalk	Chalk	Weak rock	Fine soil	Fine soil	–	–	–	–
CPT	Coarse soil	Coarse soil	Coarse soil	Coarse soil	–	–	–	–	–	Fine soil	Fine soil	Soil	Fine soil	–	–
Piezometer	–	–	–	–	–	–	–	–	–	–	–	–	–	Coarse soil	–
Packer test	–	–	–	–	–	–	–	–	–	–	–	–	–	Rock	–
Plate load	–	–	Soil	–	–	–	–	–	–	–	–	–	–	–	–
CBR	–	–	Soil	–	–	–	–	–	–	–	–	–	–	–	–
Pressuremeter	–	–	Soil, rock	–	–	–	–	–	–	–	–	–	–	–	–
Geophysical	–	–	Soil, rock	Soil, rock	–	Soil, rock	–	—	–	–	–	–	–	–	Soil, rock

Table 2.16 Summary of laboratory test types considered and the engineering ground properties from the test results

Type	Coeff. of volume compress., m_v	Coeff. of compr., a_v	Compr. index and ratio, C_c and C_{ce}	Recompr. index and ratio, C_r and C_{re}	Second. compr. index and ratio, C_α and $C_{\alpha e}$	Coeff. of consolidation, c_v	Swellability	Cyclic shear strength and stiffness	Static undrained shear strength, c_u	Static undrained secant modulus, E_s	Static drained shear strength and equivalent linearized ϕ'_e and c'_e	Unconf. compress. strength, UCS	Abrasivity of rock fill	Deformation modulus, M	Transv. wave vel., V_t	Shear modulus G and damping ratio ξ dependence on shear strain
Oedometer	Fine soil	Fine soil	Fine soil	Fine soil	Fine soil	Fine soil	–	–	–	–	–	–	–	–	–	–
ISRM	–	–	–	–	–	–	Rock	–	–	–	–	–	–	–	–	–
Simple shear	–	–	–	–	–	–	–	Soil	–	–	–	–	–	—	–	–
Triaxial	–	–	–	–	–	–	–	–	Fine soil	Fine soil	Soil	—	–	–	–	–
ISRM	–	–	–	–	–	—	–	–	–	–	–	Rock	–	–	–	–
Los Angeles	–	–	–	–	–	–	–	–	–	–	—	–	Rock fill	Rock fill	–	–
Bender element	–	–	–	–	–	–	–	–	–	–	–	–	–	–	Fine soil	–
Resonant column	–	–	–	–	–	–	–	–	–	–	–	–	–	–	–	Soil

References

Almeida MSS, Jamiolkowski M, Peterson RV (1992) Preliminary results of CPT tests in calcareous Quiou sand. In: Proceedings of the international symposium on calibration chamber testing, Potsdam, New York, 1991. Elsevier, pp 41–53

Ambraseys NN (1988) Engineering seismology. Earthq Eng Struct Dyn 17:1–105

Arulnathan R, Boulanger RW, Riemer MF (1998) Analysis of bender element test. Geotech Test J 21(2):120–131

ASTM C131-06 Standard test method for resistance to degradation of small-size coarse aggregate by abrasion and impact in the Los Angeles machine. American Society for Testing and Materials, Philadelphia, PA

ASTM C1580-09e1 Standard test method for water-soluble sulfate in soil. American Society for Testing and Materials, Philadelphia, PA

ASTM C535-09 Standard test method for resistance to degradation of large-size coarse aggregate by abrasion and impact in the Los Angeles machine. American Society for Testing and Materials, Philadelphia, PA

ASTM C88-05 Standard test method for soundness of aggregates by use of sodium sulfate or magnesium sulfate. American Society for Testing and Materials, Philadelphia, PA

ASTM D1195/D1195M-09 Standard test method for repetitive static plate load tests of soils and flexible pavement components, for use in evaluation and design of airport and highway pavements. American Society for Testing and Materials, Philadelphia, PA

ASTM D1411-09 Standard test methods for water-soluble chlorides present as admixtures in graded aggregate road mixes. American Society for Testing and Materials, Philadelphia, PA

ASTM D1586-11 Standard test method for standard penetration test (SPT) and split-barrel sampling of soils. American Society for Testing and Materials, Philadelphia, PA

ASTM D1883-07e2 Standard test method for CBR (California Bearing Ratio) of laboratory-compacted soils. American Society for Testing and Materials, Philadelphia, PA

ASTM D2435/D2435M-11 Standard test methods for one-dimensional consolidation properties of soils using incremental loading. American Society for Testing and Materials, Philadelphia, PA

ASTM D2850-03a Standard test method for unconsolidated-undrained triaxial compression test on cohesive soils. American Society for Testing and Materials, Philadelphia, PA

ASTM D2974-07a Standard test methods for moisture, ash, and organic matter of peat and other organic soils. American Society for Testing and Materials, Philadelphia, PA

ASTM D3999-11 Standard test methods for the determination of the modulus and damping properties of soils using the cyclic triaxial apparatus. American Society for Testing and Materials, Philadelphia, PA

ASTM D4015-07 Standard test methods for modulus and damping of soils by resonant-column method. American Society for Testing and Materials, Philadelphia, PA

ASTM D4373-02 Standard test method for rapid determination of carbonate content of soils. American Society for Testing and Materials, Philadelphia, PA

ASTM D4428/D4428M-07 Standard test methods for crosshole seismic testing. American Society for Testing and Materials, Philadelphia, PA

ASTM D4429-09a Standard test method for CBR (California Bearing Ratio) of soils in place. American Society for Testing and Materials, Philadelphia, PA

ASTM D4719-07 Standard test methods for prebored pressuremeter testing in soils. American Society for Testing and Materials, Philadelphia, PA

ASTM D4767-11 Standard test method for consolidated undrained triaxial compression test for cohesive soils. American Society for Testing and Materials, Philadelphia, PA

ASTM D4972-01 Standard test method for pH of soils. American Society for Testing and Materials, Philadelphia, PA.

ASTM D5311-11 Standard test method for load controlled cyclic triaxial strength of soil. American Society for Testing and Materials, Philadelphia, PA

ASTM D5777-00 Standard guide for using the seismic refraction method for subsurface investigation. American Society for Testing and Materials, Philadelphia, PA

ASTM D5778-12 Standard test method for electronic friction cone and piezocone penetration testing of soils. American Society for Testing and Materials, Philadelphia, PA

ASTM D6432-11 Standard guide for using the surface ground penetrating radar method for subsurface investigation. American Society for Testing and Materials, Philadelphia, PA

ASTM D6639-01 Standard guide for using the frequency domain electromagnetic method for subsurface investigations. American Society for Testing and Materials, Philadelphia, PA

ASTM D7128-05 Standard guide for using the seismic-reflection method for shallow subsurface investigation. American Society for Testing and Materials, Philadelphia, PA

ASTM G51-95(2005) Standard test method for measuring pH of soil for use in corrosion testing. American Society for Testing and Materials, Philadelphia, PA

ASTM G57-06 Standard test method for field measurement of soil resistivity using the Wenner four electrode method. American Society for Testing and Materials, Philadelphia, PA

Baldi G, Bellotti R, Ghionna V, Jamiolkowski M, Pasqualini E (1986) Interpretation of CPTs and CPTUs; 2nd part: drained penetration of sands. In: Proceedings of the 4th international geotechnical seminar, Singapore, pp 143–156

Baldi G, Bellotti R, Ghionna VN, Jamiolkowski M, Lo Presti DFC (1989) Modulus of sands from CPTs and DMTs. In: Proceedings of the 12th international conference on soil mechanics and foundation engineering, Rio de Janeiro, Balkema, Rotterdam, vol 1, pp 165–170

Bergdahl U, Ottosson E, Malborg BS (1993) Plattgrundlagging. AB Svensk Byggtjanst, Stockholm

BRE Special Digest 1 (2005) Concrete in aggressive ground. Building Research Establishment, Garston

BS 1377-3 (1990) Methods of test for soils for civil engineering purposes. Chemical and electro-chemical tests. British Standards Institution, London

BS 1377-5 (1990) Methods of test for soils for civil engineering purposes. Compressibility, permeability and durability tests. British Standards Institution, London

BS 1377-7 (1990) Methods of test for soils for civil engineering purposes. Shear strength tests (total stress). British Standards Institution, London

BS 1377-8 (1990) Methods of test for soils for civil engineering purposes. Shear strength tests (effective stress). British Standards Institution, London

BS 1377-9 (1990) Methods of test for soils for civil engineering purposes. In-situ tests. British Standards Institution, London

BS 5493 (1977) Code of practice for protective coating of iron and steel structures against corrosion. British Standards Institution, London

BS 5930 (1999) + A2(2010) Code of practice for site investigations. British Standards Institution, London

Burland JB, Burbidge MC (1985) Settlement of foundations on sand and gravel. In: Proceedings of the institution of civil engineers, UK, Part 1, vol 78, pp 1325–1371

Casagrande A (1936) The determination of the pre-consolidation load and its practical significance. In: Proceedings of the 1st international conference on soil mechanics and foundation engineering, Cambridge, UK 3

Casagrande A, Fadum RE (1940) Notes on soil testing for engineering purposes. Harvard University Graduate School of Engineering, Publication No 8

Cearns PJ, McKenzie A (1989) Application of dynamic cone penetrometer testing in East Anglia. In: Proceedings of the geotechnology conference Penetration Testing in the UK organized by the Institution of Civil Engineers, Birmingham, UK, pp 123–127

Clayton CRI (1995) The standard penetration test (SPT): methods and use. Construction Industry Research Information Association report No 143, UK

Cornforth DH (1964) Some experiments on the influence of strain conditions on the strength of sand. Geotechnique 16(2):143–167

DIN 4094-1 (2002) Baugrund – Felduntersuchungen – Teil 1: Drucksondierungen. Deutsche Industries Norm

EN 1997-2 (2007) Eurocode 7: geotechnical design – Part 2: Ground investigation and testing. European Committee for Standardization, Brussels

EN 1998-1 (2004) Eurocode 8: design of structures for earthquake resistance – Part 1: General rules, seismic actions and rules for buildings. European Committee for Standardization, Brussels

EN 1998-5 (2004) Eurocode 8: design of structures for earthquake resistance – Part 5: Foundations, retaining structures and geotechnical aspects. European Committee for Standardization, Brussels

Finn WDL (1985) Aspects of constant volume cyclic simple shear. In: Khosla V (ed) Advances in the art of testing soils under cyclic condition. In: Proceedings of geotechnical engineering division of ASCE convention in Detroit, Michigan, pp 74–98

Hatanaka M, Uchida A (1996) Empirical correlation between penetration resistance and internal friction angle of sandy soils. Soils Found 36(4):1–10

Hvorslev J (1951) Time lag and soil permeability in ground water observations. US Waterways Experimental Station Bulletin 36, Vicksburg, US Army Corps of Engineers

ISO 10390 (2005) Soil quality – determination of pH. International Organization for Standardization, Geneva, Switzerland

ISO 10693 (1995) Soil quality – determination of carbonate content – volumetric method. International Organization for Standardization, Geneva, Switzerland

ISO 10694 (1995) Soil quality – determination of organic and total carbon after dry combustion (elementary analysis). International Organization for Standardization, Geneva, Switzerland

ISO 11048:1995 Soil quality – determination of water-soluble and acid-soluble sulfate. International Organization for Standardization, Geneva, Switzerland

ISO 146891-1 (2003) Geotechnical investigation and testing – identification and classification of rock – Part 1: Identification and description. International Organization for Standardization, Geneva, Switzerland

ISO 22476-1 (2011) Geotechnical investigation and testing. Field testing. Part 1. Electrical cone and piezocone penetration tests. International Organization for Standardization, Geneva, Switzerland

ISO 22476-3 (2008) Geotechnical investigation and testing – field testing – Part 3: Standard penetration test. International Organization for Standardization, Geneva, Switzerland

ISO 22476-12 (2009) Geotechnical investigation and testing – field testing – Part 12: Mechanical cone penetration test (CPTM). International Organization for Standardization, Geneva, Switzerland

ISO 22476–13 (future) Geotechnical investigation and testing – field testing – Part B: Plate loading test. International Organization for Standardization, Geneva, Switzerland

ISO/TS 17892-5 (2004) Geotechnical investigation and testing – laboratory testing of soil – Part 5: Incremental loading oedometer test. International Organization for Standardization, Geneva, Switzerland

ISO/TS 17892-8 (2007) Geotechnical investigation and testing – laboratory testing of soil – Part 8: Unconsolidated undrained triaxial test. International Organization for Standardization, Geneva, Switzerland

ISO/TS 17892-9 (2007) Geotechnical investigation and testing – laboratory testing of soil – Part 9: Consolidated triaxial compression tests on water-saturated soils. International Organization for Standardization, Geneva, Switzerland

ISRM (1979) Suggested method for determining the uniaxial compressive strength and deformability of rock materials. Parts 1 and 2. In: Ulusay R, Hudson JA (eds) The complete ISRM suggested methods for rock characterization, testing and monitoring: 1974–2006. The commission on testing methods. International Society for Rock Mechanics, Lisbon

ISRM (1999) Suggested methods for laboratory testing of swelling rocks. Parts 1 to 4. In: Ulusay R, Hudson JA (eds) The complete ISRM suggested methods for rock characterization, testing and monitoring: 1974–2006. The commission on testing methods. International Society for Rock Mechanics, Lisbon

Janbu N (1963) Soil compressibility as determined by oedometer and triaxial tests. In: Proceedings of the European conference on soil mechanics and foundation engineering, Wiesbaden, vol 1, pp 19–25

Japan Road Association (2003) Specifications for highway bridges, Part V: Seismic design, English version translated by PWRI, Japan

Kramer SL (1996) Geotechnical earthquake engineering. Prentice Hall, Englewood Cliffs

Kulhaway FH, Mayne PH (1990) Manual on estimating soil properties for foundation design. Electric Power Research Institute, EPRI, Palo Alto

Leong EC, Cahyadi J, Rahardjo H (2009) Measuring shear and compression wave velocities of soil using bender-extender elements. Can Geotech J 46(7):792–812

Lord JA, Clayton CRI, Mortimore RN (2002) Engineering in chalk. Construction Industry Research Information Association report No C574, UK

Lugeon M (1933) Barrages et Geologie. Dunod, Paris

Lunne T, Christophersen HP (1983) Interpretation of cone penetrometer data for offshore sands. In: Proceedings of the offshore technology conference, Richardson, TX, paper No. 4464

Lunne T, Robertson PK, Powell JJM (1997) Cone penetration testing in geotechnical practice. Spoon Pres, New York

Marsland A, Powell JJM (1988) Investigation of cone penetration tests in British clays carried out by the Building research establishment 1962–1987. In: Proceedings of the geotechnology conference: Penetration Testing in the UK, Birmingham. Thomas Telford, London, pp 209–214

McDowell PW, Barker RD, Butcher AP, Culshaw MG, Jackson PD, McCann DM, Skip BO, Matthews SL, Arthur JCR (2002) Geophysics in engineering investigations. Construction Industry Research Information Association report C562, UK

McLelland Engineers (1977) Extract from technical report. In: Poulos HG (1988) marine geotechnics. Unwin Hyman, Boston, MA

Meight AC (1987) Cone penetration testing methods and interpretation. Construction Industry Research and Information Association report, UK

Peacock WH, Seed HB (1968) Sand liquefaction under cyclic loading simple shear condition. J Soil Mech Found Div ASCE 94(SM3):689–708

Peck RB, Hanson WE, Thornburn TH (1953) Foundation engineering. Wiley, New York

Penetration Testing in the UK (1989) Proceedings of the geotechnology conference organized by the Institution of Civil Engineers in Birmingham in 6–8 July 1988. Thomas Telford, UK

Powell JJM, Quarterman RST (1988) The interpretation of cone penetration tests in clays, with particular reference to rate effects. In: Proceedings of the international symposium on Penetration Testing, ISPT-1, Orlando. Balkema, Rotterdam, vol 2, pp 903–910

Powell WS, Potter JF, Mayhew HC, Numm ME (1984) The structural design of bituminous roads. The Transport and Road Research Laboratory report LR1132, UK

Power PT (1982) The use of electric cone penetrometer in the determination of the engineering properties of chalk. In: Proceedings of the 2nd European symposium on Penetration Testing, ESOPT-II, Amsterdam. Balkema, Rotterdam, vol 2, pp 769–774

Rix GJ, Stokoe KH (1992) Correlation of initial tangent modulus and cone resistance. In: Proceedings of the international symposium on calibration chamber testing, Potsdam, New York, 1991. Elsevier, pp 351–362

Robertson PK (1990) Soil classification using the cone penetration test. Can Geotech J 27(1):151–158

Robertson PK, Campanella RG (1983) Interpretation of cone penetration tests, Part I: Sand. Can Geotech J 20(4):718–733

Robertson PK, Campanella RG, Gillespie D, Greig J (1986) Use of piezometer cone data. In: Proceedings of the ASCE specialty conference In Situ'86: use of in situ tests in geotechnical engineering, Blacksburg 1263–1280, American Society of Civil Engineers (ASCE)

Roscoe KH (1953) An apparatus for the application of simple shear to soil samples. In: Proceedings of the 3rd international conference on soil mechanics, Zurich, vol 1, pp 186–191

Sanglerat G (1972) The penetrometer and soil exploration. Elsevier, Amsterdam

Schmertmann JH (1970) Static cone to compute settlement over sand. J Soil Mech Found Div ASCE 96(SM3):1011–1043

Schmertmann JH (1975) Measurement of in situ shear strength. In: Proceedings of specialty conference on in situ measurement of soil properties, ASCE. North Carolina State University, Raleigh, NC, vol 2, pp 57–138

Seed HB, Tokimatsu K, Harder LF, Chung RM (1985) Influence of SPT procedures in soil liquefaction resistance evaluations. J Geotech Eng 111(12):1425–1445

Senneset K, Sandven R, Lunne T, By T, Amundsen T (1988) Piezocone test in silty soil. In: Proceedings of the international symposium on Penetration Testing, ISOPT-1, Orlando. Balkema, Rotterdam, vol 2, pp 955–966

Shirley DJ (1978) An improved shear wave transducer. J Acoust Soc Am 63(5):1643–1645

Skempton AW (1954) The pore pressure coefficients A and B. Geotechnique 4(4):143–147

Skempton AW (1986) Standard penetration test procedures and the effects in sands of overburden pressure, relative density, particle size, ageing and overconsolidation. Geotechnique 3(3):425–447

Skoglund GR, Marcuson WF 3rd, Cunny RW (1976) Evaluation of resonant column test devices. J Geotech Div ASCE 11:1147–1188

Stenzel G, Melzer KJ (1978) Soil investigations by penetration testing according to DIN 4094 (in German). Tiefbau 20(S):155–160, 240–244

Stroud MA (1989) The standard penetration test – its application and interpretation. In: Proceedings of the geotechnology conference Penetration Testing in the UK organized by the Institution of Civil Engineers, Birmingham, UK, pp 29–49

Stroud MA, Butler FG (1975) The standard penetration test and the engineering properties of glacial materials. In: Proceedings of symposium on engineering behaviour of glacial materials, University of Birmingham, UK, pp 124–135

Taylor DW (1942) Research on consolidation of clays. Massachusetts Institute of Technology, Publication No 82

Terzaghi K, Peck RB (1948) Soil mechanics in engineering practice, 1st edn. Wiley, New York

Terzaghi K, Peck RB (1974) Soil mechanics in engineering practice, 2nd edn. Wiley, New York

Townsend FC (1978) A review of factors affecting cyclic triaxial tests. Special Technical Publication 654, ASTM, pp 356–358

Wilson SD, Marsal RJ (1979) Current trends in design and construction of embankment dams. Report prepared for International Commission on Large Dams (ICOLD) and Geotechnical Division of ASCE. American Society of Civil Engineers, New York

Chapter 3
Geo-properties

Abstract This chapter contains information on classification (index), durability and engineering properties of soil (coarse and fine grained with particle diameters less than 0.06 mm (BS) or 0.074 mm) (ASTM) as well as of rock.

Soil properties considered:

- Particle size distribution – grading (coarse grained soil)
- Plasticity and consistency (fine grained soil)
- Over consolidation (fine grained soil)
- Unit weight and other basic parameters
- Undrained shear strength in static and cyclic condition (fine grained soil)
- Angle of friction in static (coarse and fine grained soil) and cyclic conditions (coarse grained soil)
- Stiffness in static and cyclic condition
- Water permeability
- Consolidation (fine grained soil)
- Content of carbonates, chlorides, sulphates and organic matter

Soil properties concerning collapse, swelling, erosion, liquefaction, frost susceptibility and contamination are considered in Chap. 4

Rock properties considered:

- Identification
- Rock mass rating
- Quality number of rock mass
- Shear strength
- Stiffness
- Water permeability

Rock properties concerning solution, swelling and tectonic stresses are considered in Chap. 4

M. Srbulov, *Practical Guide to Geo-Engineering: With Equations, Tables, Graphs and Check Lists*, Geotechnical, Geological and Earthquake Engineering 29, DOI 10.1007/978-94-017-8638-6_3,

3.1 Soil Properties

3.1.1 Particle Size Distribution – Grading (Coarse Grained Soil)

The results of particle size distribution (PSD) test described in BS 1377-2; ASTM D422, D6913-04, D5519-07; ISO 11277; ISO 17892-4 are used for classification of coarse grained soil. Figure 3.1 shows **boundaries of coarse grained soil types** according to BS 5930 and Fig. 3.2 according to ASTM D2487.

Besides PSD, coarse grained soil is described according to its colour, density (based on N_{SPT}), inclusions (e.g. pieces of wood, nodules of calcium carbonates, pockets of sand, seams of silt) and geological age (e.g. Holocene – about 12,000 till present, Pleistocene – from about 2.5 million years ago to 12,000 years ago, Quaternary – in the last about 2.5 million years, Tertiary – from about 2.5 million years ago to 65 million years ago, and older).

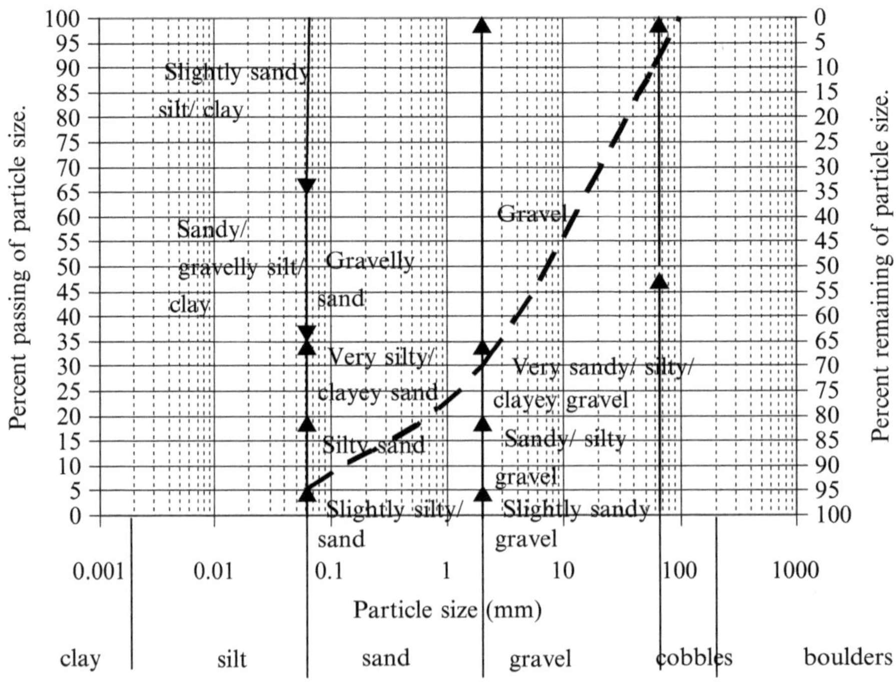

Fig. 3.1 Boundaries between different coarse grained soil types according to BS 5930 (1999). *Dashed line* is an example of slightly silty, very sandy GRAVEL. Cobbles size up to 100 mm

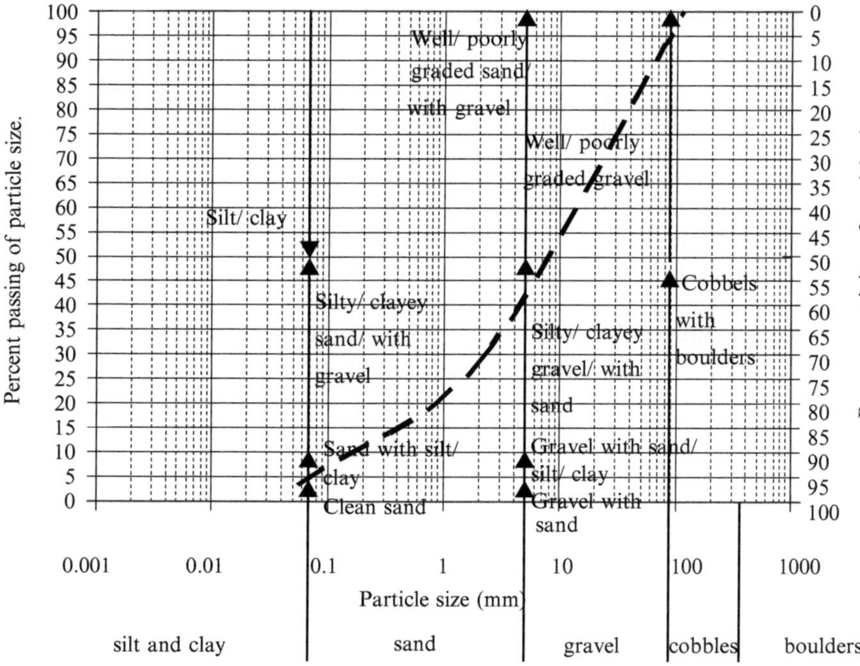

Fig. 3.2 Boundaries between different coarse grained soil types according to ASTM D2487. *Dashed line* is an example of well graded silty GRAVEL with sand

Hydrometer test (BS 1377-2; ASTM D422; ISO 11277) can be used for further determination of particle sizes of fine grained soil, which is sticky when wet and which is capable of retaining shape at zero confining stress.

Additional criteria for the classification according ASTM D2487 are the **coefficient of uniformity C_u and curvature C_{cr}.**

$$C_u = \frac{D_{60}}{D_{10}} \tag{3.1}$$

$$C_{cr} = \frac{{D_{30}}^2}{D_{60} \cdot D_{10}} \tag{3.2}$$

D_{60} is the opening that 60 % of particle sizes pass
D_{10} is the opening that 10 % of particle sizes pass
D_{30} is the opening that 30 % of particle sizes pass

Well graded gravel has $C_u > 4$ while sand >6 and C_c between 1 and 3 otherwise it is poorly graded, silty or clayey.

3.1.2 Plasticity and Consistency (Fine Grained Soil)

Plasticity is the ability of a material to be shaped without fracturing. Of five limits identified by Atterberg (1911) only three have been used in soil mechanic:

- **The shrinkage limit** corresponds to the moisture content at which further drying of soil does not cause reduction in its volume
- **The plastic limit (PL)** corresponds to the minimum amount of water necessary to maintain the flexibility of the electro-chemical bonds between clay particles and pore water
- **The liquid limit (LL)** corresponds to the amount of water at which clay particles are far enough apart to reduce the electro-chemical attraction forces to almost zero

Other parameters used for definition of soil plasticity and consistency are:

- **Plasticity index *PI***

$$PI = LL - PL \tag{3.3}$$

Skempton (1953) defined **clay activity A** as

$$A = \frac{PI}{\%\, clay} \tag{3.4}$$

- **Consistency index I_c**

$$I_c = \frac{LL - w}{PI} \tag{3.5}$$

w is actual moisture content. The states of soil consistency are given in Table 3.1

- **Liquidity index *LI***

$$LI = \frac{w - PL}{PI} = 1 - I_c \tag{3.6}$$

w is actual moisture content, i.e. the ratio between weight of water and weight of solids within a soil volume.

Tests to define *PL* and *LL* are described in BS 1377-2; ASTM D4318; ISO 17892-12 and for moisture content *w* in BS 1377-2; ASTM D2216; ISO 17892-1, 16586. Shrinkage limit tests are described in BS 1377-2; ASTM D4943; ISO 17892.

Classification charts for fine grained soil based on its *LL* and *PI* are shown in Figs. 3.3 and 3.4. An approximate moisture content at the shrinkage limit can be determined according to Casagrande (quoted by Carter and Bentley 1991) by

Table 3.1 States of soil consistency based on consistency index I_c

I_c	State
<0	Liquid
0–0.25	Very soft (plastic)
0.25–0.5	Soft (plastic)
0.5–0.75	Medium stiff (plastic)
0.75–1	Stiff (plastic)
1–1.25	Very stiff (semi solid)
>1.25	Hard (solid)

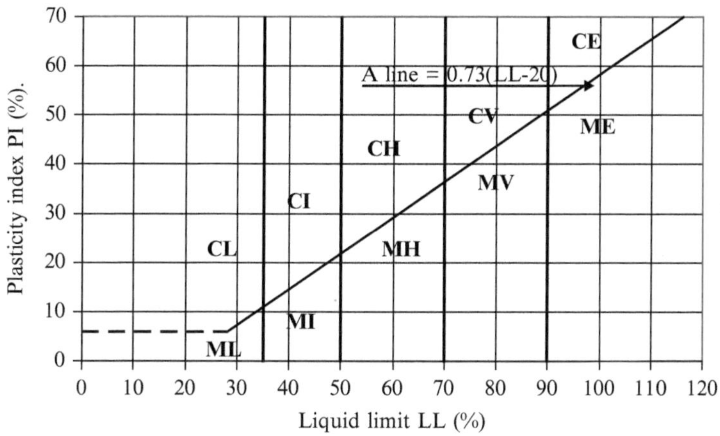

Fig. 3.3 Classification chart of fine grained soil and fines of coarse grained soil according to BS 1377-2 Legend: *C* clay, *M* silt, *L* low plasticity, *I* intermediate plasticity, *H* high plasticity, *V* very high plasticity, *E* extremely high plasticity

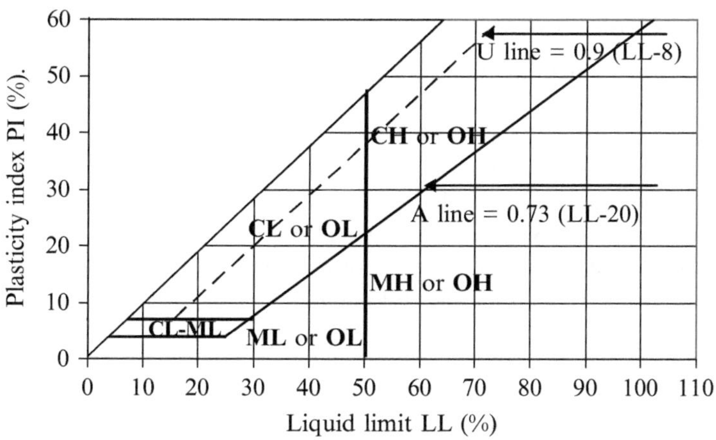

Fig. 3.4 Classification chart of fine grained soil and fines of coarse grained soil according to ASTM D4318 Legend: *C* clay, *M* silt, *O* organic soil, *L* low plasticity, *H* high plasticity

extending the U and A lines in Fig. 3.4 to their intersection at LL = −43.5 %, PI = −46.4 %, connecting the intersection with a line to the point of soil LL and PI and reading the value of LL at PI = 0 as the shrinkage limit.

Besides plasticity, fine grained soil is described according to its colour, consistency, inclusions (e.g. pieces of wood, nodules of calcium carbonates, pockets of sand, seams of silt) and geological age (e.g. Holocene – about 12,000 till present, Pleistocene – from about 2.5 million years ago to 12,000 years ago, Quaternary – in the last about 2.5 million years, Tertiary – from about 2.5 million years ago to 65 million years ago, and older).

3.1.3 Over Consolidation (Fine Grained Soil)

True over consolidation occurs as a result of removal of ground layers by erosion or melting of ice sheet in the post-glacial period. The over consolidation pressure that exited in the past can be estimated based on Fig. 2.11. Over consolidation ratio (OCR) is the ratio between previous and existing over burden effective stress. Apparent over consolidation is caused by:

- Desiccation (drying)
- Cementation (by calcium carbonates, iron and aluminium oxides)
- Creep (secondary compression)

3.1.4 Unit Weight and Other Basic Parameters

Tests to determine soil unit weight are described in BS 1377-2, BS 1377-4, BS 5930; ASTM D1556, D2167, D2937, D4253, D4254, D4564, D7263, ISO 17892-2. Other basic parameters are:

- **Porosity *n*** is defined as the ratio of the volume of voids to the total volume of soil.
- **Void ratio *e*** is defined as the ratio of the volume of voids to the volume of solids within the total volume of soil.
- **Degree of saturation S_r** is defined as the ratio of the volume of water and the volume of voids within the total volume of soil.
- **Air content n_a** is defined as the ratio of the volume of air and the total volume.

Correlations between the basic soil parameters are:

$$n = \frac{e}{1+e} = \frac{w \cdot G_s}{1 + w \cdot G_s} = w \cdot \frac{\gamma_d}{\gamma_w} \tag{3.7}$$

$$e = \frac{n}{1-n} = \frac{G_s \cdot \gamma_w}{\gamma_d} - 1 \tag{3.8}$$

Table 3.2 Typical ranges of void ratio and dry unit weight of coarse grained soil (Adopted from Das 1985)

	Void ratio e		Dry unit weight γ_d (kN/m^3)	
Soil type	Maximum	Minimum	Minimum	Maximum
Gravel	0.6	0.3	16	20
Coarse sand	0.75	0.35	15	19
Fine sand	0.85	0.4	14	19
Gravely sand	0.7	0.2	15	22
Silty sand	1	0.4	13	19
Silty sand and gravel	0.85	0.15	14	23

w is moisture content
G_s is the specific gravity of soil solids (2.65–2.68 for gravel and sand, 2.66–2.7 for silt, 2.68–2.8 for clay)
γ_w is unit weight of water
γ_d is dry unit weight of soil.

$$S_r = \frac{w \cdot G_s}{e} \tag{3.9}$$

$$n_a = n \cdot (1 - S_r) = \frac{e - w \cdot G_s}{1 + e} \tag{3.10}$$

$$\gamma = \gamma_d \cdot (1 + w) = \frac{G_s \cdot \gamma_w \cdot (1 + w)}{1 + e} = G_s \cdot \gamma_w \cdot (1 - n) \cdot (1 + w) \tag{3.11}$$

In saturated condition $S_r = 1$.

$$\begin{aligned} \gamma_s &= \frac{(G_s + e) \cdot \gamma_w}{1 + e} = \left[G_s - n \cdot (G_s - 1)\right] \cdot \gamma_w \\ w_s &= \frac{\gamma_w}{\gamma_d} - \frac{1}{G_s} = \frac{n}{G_s \cdot (1 - n)} = \frac{\gamma_s}{\gamma_d} - 1 = n \cdot \frac{\gamma_w}{\gamma_d} \end{aligned} \tag{3.12}$$

Submerged soil unit weight γ'

$$\gamma' = \gamma_s - \gamma_w \tag{3.13}$$

Relative density D_r of coarse grained soil

$$D_r = \frac{e_{\max} - e}{e_{\max} - e_{\min}} = \frac{\gamma_{d,\max}}{\gamma_d} \cdot \frac{\gamma_d - \gamma_{d,\min}}{\gamma_{d,\max} - \gamma_{d,\min}} \tag{3.14}$$

Typical ranges of void ratio and dry unit weight of coarse grained soil are given in Table 3.2.

Table 3.3 Classification of clay sensitivity based on S_t

S_t	Sensitivity
~1	Insensitive
1–2	Low sensitivity
2–4	Medium sensitivity
4–8	Sensitive
8–16	Extra sensitive
>16	Quick

3.1.5 Undrained Shear Strength in Static and Cyclic Condition (Fine Grained Soil)

- For **normally consolidated clay in static condition**, Skempton (1957) proposed the ratio between **undrained shear strength** $\boldsymbol{c_u}$ and effective overburden pressure σ'_v (total less water pressure)

$$\frac{c_u}{\sigma'_v} = 0.11 + 0.0037 \cdot PI \tag{3.15}$$

PI is the plasticity index in percent

Undisturbed natural clay exhibit reduction in undrained shear strength when remoulded. The ratio between undisturbed and remoulded undrained shear strength is termed **clay sensitivity** $\boldsymbol{S_t}$.

$$S_t = \frac{c_u}{c_{u,remoulded}} \tag{3.16}$$

Classification of clay sensitivity based on S_t is given in Table 3.3. It should be noted that the quick clay in Scandinavia and Canada when disturbed can flow fast like heavy fluid for long distances. Causes of clay sensitivity are summarized by Mitchell and Houston (1969). Wroth and Wood (1978) proposed the following expression for the **undrained shear strength of remoulded (compacted) clay** in kPa

$$c_{u,remoulded} = 170 \cdot e^{-4.6 \cdot LI} \tag{3.17}$$

LI is the liquidity index

Skempton and Northey (1952) data for the correlation between liquidity index *LI* and clay sensitivity S_t can be fitted by a simple expression

$$S_t \tilde{\ } 0.8 \cdot e^{2.6 \cdot LI} \tag{3.18}$$

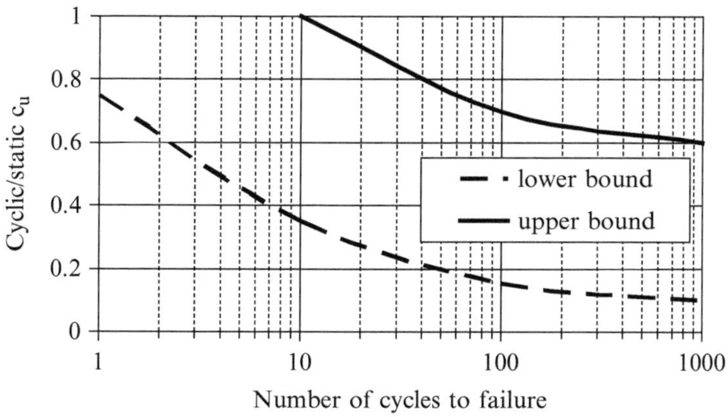

Fig. 3.5 Ratios between cyclic and static undrained shear strength of mainly undisturbed but also some compacted and remoulded clay

- For **over consolidated clay in static condition**, the ratio between averaged **undrained shear strength c_u** and effective overburden pressure σ'_v (total less water pressure)

$$\frac{c_u}{\sigma'_v} = (0.11 + 0.0037 \cdot PI) \cdot OCR^{0.8} \tag{3.19}$$

PI is the plasticity index in percent
OCR is over consolidation ratio

The exponent 0.8 is based on Ladd and Foot (1974).

- The **residual shear strength** at large strain (displacement) and fast rates of shear in undrained condition has been tested in ring shear apparatus. In general, clay of high plasticity develops a significant decrease of the peak shear strength as shown in the following subsection for drained conditions at slow shear rates. Bray et al (2004) suggest that the residual shear strength $c_{u,r}$ is determined as follows:

for $w/LL \geq 0.85$ and $PI \leq 12\ \%$, $c_{u,res} = c_{u,remoulded}$
for $w/LL \geq 0.8$ and $12\ \% < PI < 20\ \%$, $c_{u,res} = 0.85\ c_u$
for $w/LL < 0.8$ and $PI \geq 20\ \%$, $c_{u,res} = c_u$
w is moisture content
LL is liquid limit
PI is plasticity index

- In **cyclic condition**, the boundaries of the ratios between the cyclic and static undrained shear strength depend on the number of cycles to failure are shown in Fig. 3.5 based on data by Lee and Focht (1976). In Fig. 3.5, the lower bound strength ratios are obtained for the remoulded clay from Anchorage – Alaska, and undisturbed clay from North Sea while the ratios greater than one for the number of cycles less than 10 are caused by the rate (viscous) and dilatancy effects.

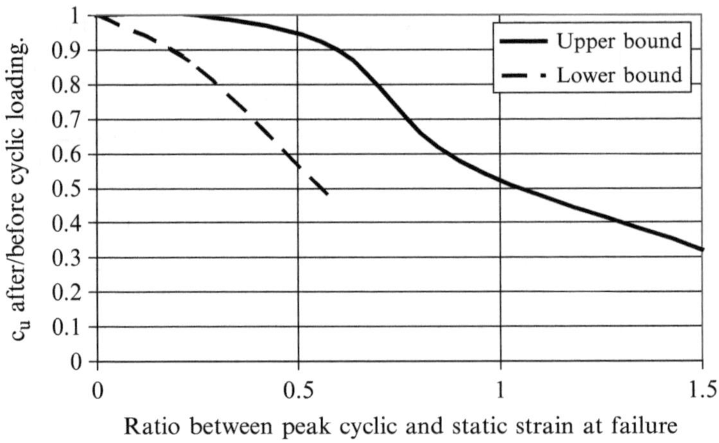

Fig. 3.6 Boundaries of the ratio between undrained shear strengths c_u after and before cyclic loading depending on the peak cyclic to static failure strain ratio

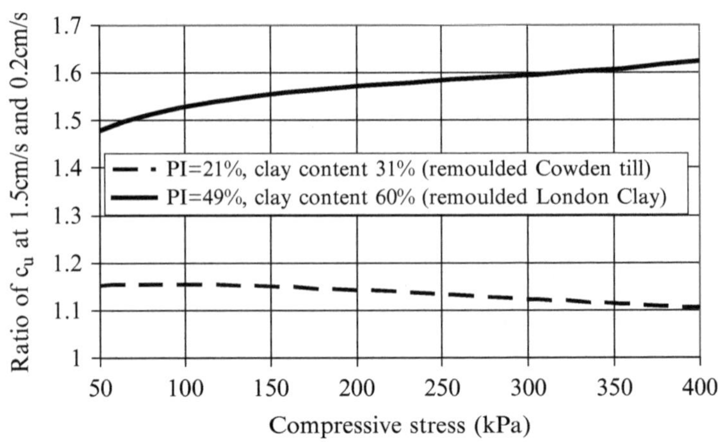

Fig. 3.7 Rate of shear effect on the shear strength of two soil when the rate of shear increases from 0.2 to 1.5 cm/s (undrained conditions)

The amount of strength decrease at a number of cycles is dependent on the cyclic strain. Figure 3.6 shows the boundaries of the ratios between the undrained shear strengths after and before cyclic loading versus the ratio between the peak cyclic strain and the strain at failure before cyclic loading according to data by Thiers and Seed (1969).

- The **rate of strain effect** on undrained shear strength increase is shown in Fig. 3.7 from data by Parathiras (1995) for the shear strength obtained using ring shear apparatus and remoulded soil. Shear strength increase was minimal beyond the rate of shear of 2 cm/s (CPT penetration rate). Parathiras (1995) indicates that the shear strength at fast shear rates could be 150–250 % greater than the residual

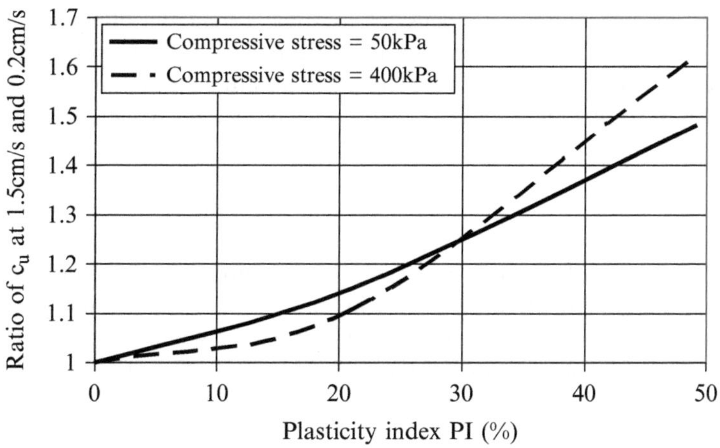

Fig. 3.8 Rate of shear effect on the shear strength of remoulded soil with different plasticity indices when the rate of shear increased from 0.2 to 1.5 cm/s (undrained conditions)

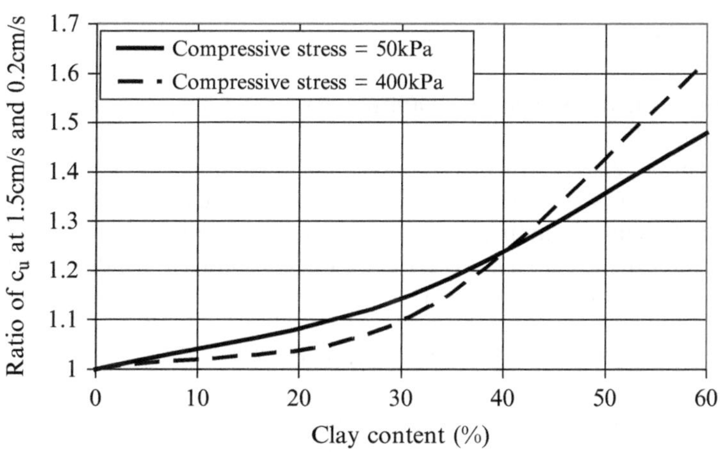

Fig. 3.9 Rate of shear effect on the shear strength of remoulded soil with different clay contents when the rate of shear increased from 0.2 to 1.5 cm/s (undrained conditions)

shear strength at slow rate (1.5×10^{-4} cm/s). Figures 3.7, 3.8 and 3.9 indicate much smaller effect of compressive stress in comparison with the plasticity index *PI* and clay content effects on the undrained shear strength of fine grained soil at fast shear rates. The cyclic to static shear strength ratio of 1 for PI = 0 and clay content = 0 are plotted based on data for sand by Hungr and Morgenstern (1984) who found slight influence of rate of shear on the shear strength of sand. Tika et al. (1996) also reported on the shear strength at fast shear rates but their results for the residual strength were affected significantly by uncontrolled opening of the gap between the upper and lower shear rings and soil loss through the gap.

The rate of shear effect compensates to some extent the effect of increasing number of cycles on shear strength decrease except for very soft and sensitive clay. Kulhaway and Mayne (1990) suggested a formula for taking into account of the effect of axial strain rate $d\varepsilon/dt$ (%/h) on clay undrained shear strength c_u:

$$\frac{c_u}{c_u at1\%/h} = 1 + 0.1 \cdot \log_{10}\frac{d\varepsilon}{dt} \tag{3.20}$$

CPT test rate of 2 cm/s corresponds to approximately 36,000 %/h for a specimen height of 20 cm used in triaxial testing for a 100 mm diameter of specimen, and consequently the multiplying factor would be about 1.45. This explains why N_{kt} factors used for the calculation of c_u from CPT q_t in Eq. (2.32) is greater than 9 that would be normally used in static condition.

3.1.6 Angle of Friction in Static (Fine and Coarse Grained Soil) and Cyclic Conditions (Coarse Grained Soil)

- For **fine grained soil in static condition**, a range of maximum angles of friction is shown in Fig. 3.10 based on data by Kenny (1959). The angles of residual friction at large strain (displacements) are shown in Fig. 3.11 based on data by Lupini et al. (1981) and in Fig. 3.12 based on data by Stark et al. (2005). The **coefficient of earth pressure at rest K_o** is connected with the friction angle according to Jaky (1944) for normally consolidated condition (OCR = 1) and Mayne and Kulhaway (1982) for over consolidated condition (OCR > 1)

$$K_o' = (1 - \sin\phi) \cdot OCR^{\sin\phi} \tag{3.21}$$

Besides plasticity index and clay content, the friction angle of fine grained soil is also affected by existing effective compressive stress σ'. A simple expression for the **secant friction angle ϕ** based on a non-linear shear strength relationship shown in Fig. 3.13 is

$$\phi = \phi_{b,r} + \frac{\Delta\phi}{1 + \frac{\sigma'}{p_n}} \tag{3.22}$$

$\phi_{b,r}$ is the basic residual friction angle
$\Delta\phi$ is the maximum angle difference
p_n is the median angle normal stress

The values of $\phi_{b,r}$, $\Delta\phi$, p_n for different clay are given in Table 3.4. Some angle of residual friction dependences on effective compressive stress are plotted in Fig. 3.14.

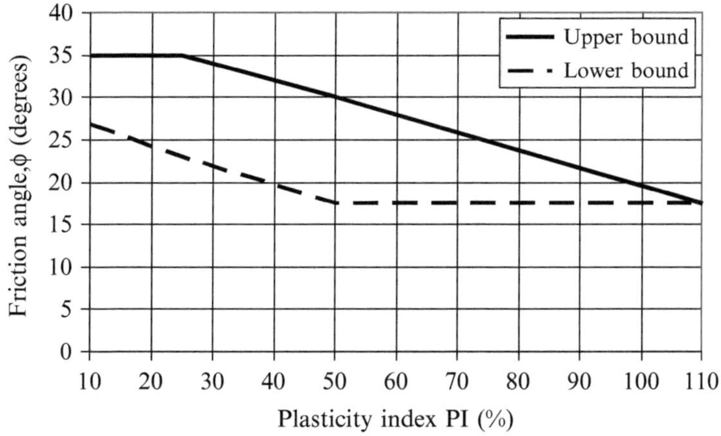

Fig. 3.10 Ranges of maximum angles of friction depending on plasticity index of normally consolidated clay

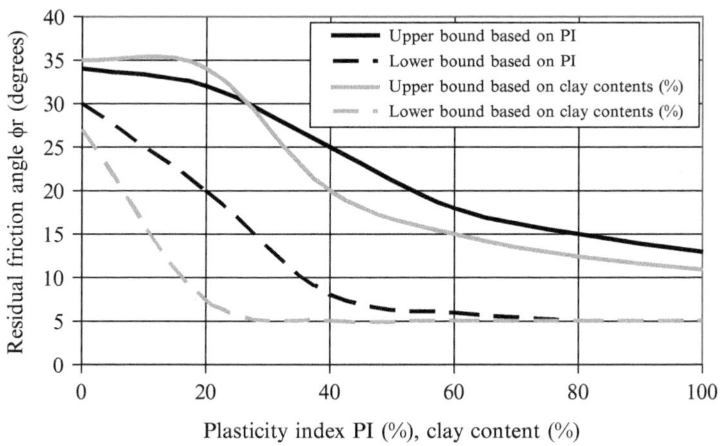

Fig. 3.11 Ranges of residual friction angles depending on plasticity index and clay content

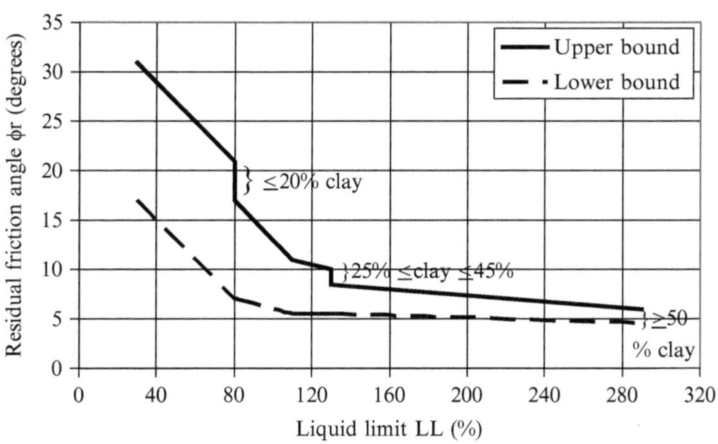

Fig. 3.12 Ranges of residual friction angles depending on liquid limit

Fig. 3.13 Non-linear shear strength relationship with the parameters $\phi_{b,r}$, $\Delta\phi$, p_n, ϕ is ground friction angle (secant). Equivalent constant friction angle ϕ_e for the range of effective compressive stresses from σ_1 to σ_2

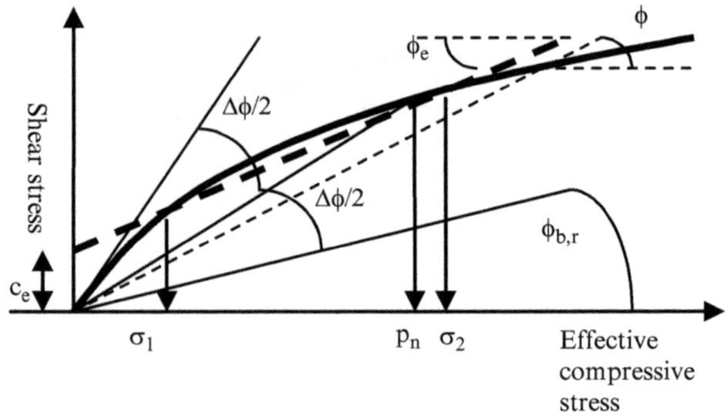

Table 3.4 Shear strength parameters for non-linear failure envelopes of fine grained soil

Description	$\phi_{b,r}$ (degrees)	$\Delta\phi$ (degrees)	p_n (kPa)	Reference
London Clay – landslides, average	8.0	9.0	120.0	Maksimovic (1989a)
	10.3	6.0	50.0	
Lias clay – landslides, average	6.5	13.5	50.0	
	4.8	10.9	133.3	
Atherfield clay – Sevenoaks bypass	7.0	8.9	130.0	
Weald clay – ring shear test	8.2	11.8	55.7	
Amuay clay, PI = 15 %	17.9	15.6	31.3	
Amuay clay, PI = 30 %	10.9	15.6	31.3	
Amuay clay, PI = 50 %	5.4	15.6	31.3	
Kaolinite	10.6	13.4	55.0	
Hydrous mica	19.6	13.5	100.0	
Na montmorillonite	2.1	4.3	110.0	
Altamira bentonitic tuff	3.3	16.1	157.0	Maksimovic (1995)
compacted London Clay	16.3	48.1	28.2	Maksimovic (1989b)

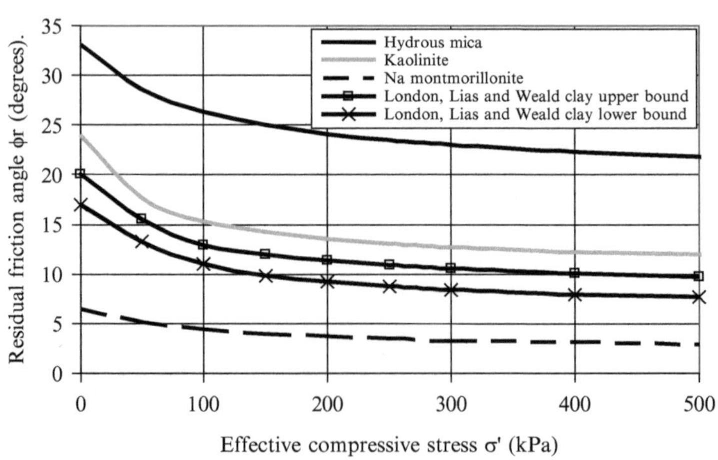

Fig. 3.14 Residual friction angles for some minerals and clay

Table 3.5 Shear strength parameters for non-linear failure envelopes of coarse grained soil from Maksimovic (1996a)

Description	$\phi_{b,r}$ (degrees)	$\Delta\phi$ (degrees)	p_n (kPa)
Sand from limestone	35.8	22.9	320
Crushed anthracite	32.2	16.5	107
Aluminium oxide – $D_r = 49\ \%$	31.7	10.5	4,709
Granitic gneiss – dense	39.5	21.4	460
Crushed basalt – max. size 152.4 mm	34.0	19.6	920
Crushed basalt – $D_r = 97$ to 100 %, max size 76.2 mm	32.7	25.7	2,176
Diorite – dense, from El Infernillo dam	32.2	18.9	544
Argilite – max. size 152.4 mm from Pyramid dam	33.6	18.9	870
Silicified conglomerate – El Infernillo dam	30.0	17.8	1,674
Conglomerate – Malpasso dam	32.2	19.1	1,402
Gravel – max. size of cobbles 152.4 mm from Oroville dam	32.7	15.8	3,614
Sand and gravel – Pinzandaran dam	36.4	19.4	5,94
Molsand – $D_r = 83\ \%$	34.3	11.5	223
Very dense sand	37.8	16.5	454
Very dense silt	37.2	19.1	1,772

Equivalent constant friction angle ϕ_e and cohesion c_e within axial stress range $\sigma_1 < \sigma_2$ (Figs. 2.15b and 3.13) are

$$\begin{aligned} \phi_e &= \arctan \frac{\sigma_2 \tan \phi_{\sigma_2} - \sigma_1 \tan \phi_{\sigma_1}}{\sigma_2 - \sigma_1} \\ c_e &= \sigma_1 \left(\tan \phi_{\sigma_1} - \tan \phi_e \right) \end{aligned} \tag{3.23}$$

- For **coarse grained soil in static condition,** Bjerrum et al. (1961) proposed that the **peak friction angle ϕ** of fine sand can be correlated with the initial sand porosity n_s. A best fit functional relationship based on data by Bjerrum et al. (1961) is

$$\phi = 12 + \sqrt{27^2 - \left[\frac{27}{11.5} \cdot (n_s - 36) \right]^2}, 36\% \le n_s \le 47.5\% \tag{3.24}$$

ϕ is in degrees
n_s is the initial sand porosity in percent

The shear strength of coarse grained soil is very dependent on existing effective compressive stress σ'. Dependence of the secant friction angle ϕ on σ' is described by Eq. (3.22). Table 3.5 contains the values of $\phi_{b,r}$,$\Delta\phi$, p_n for different soil types. The boundaries of peak friction angle dependence on effective compressive stress are plotted in Fig. 3.15.

- For **coarse grained soil in cyclic condition** the apparent friction angles during earthquakes are shown in Figs. 2.2 to 2.4 when excess pore water pressure

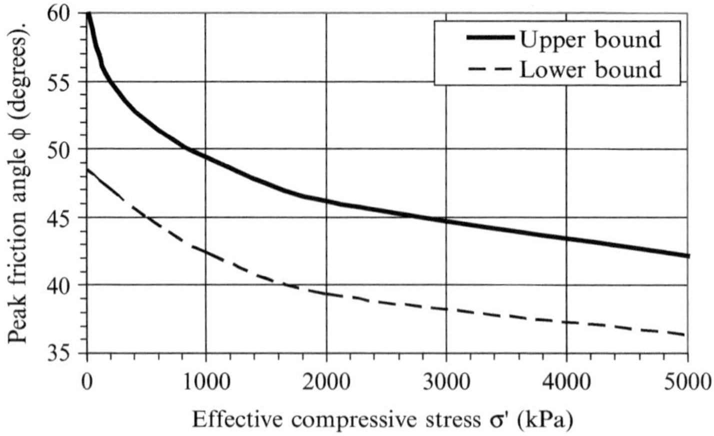

Fig. 3.15 Peak friction angles for some coarse grained soil and rock fill

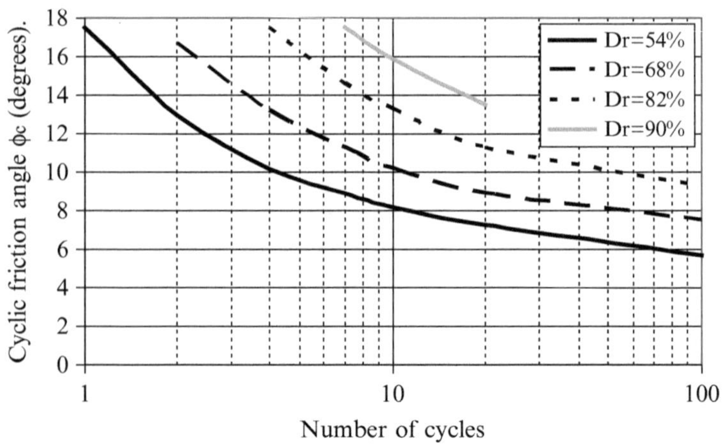

Fig. 3.16 Cyclic friction angles for initial liquefaction depending on the relative density D_r

build-up has not been considered explicitly. For other cyclic conditions such as those caused by waves and machinery, the results of strain controlled cyclic shear tests similar to those described in Sect. 2.2.3 can be used. However, undisturbed samples of coarse grained soil are rarely recovered and the tests need to be performed on disturbed samples. As a first approximation when the results of specific cyclic tests are not available it can be assumed that the dependence of friction angle on the number of cycles is similar to the values shown in Fig. 3.16 based on the results of tests by De Alba et al. (1975).

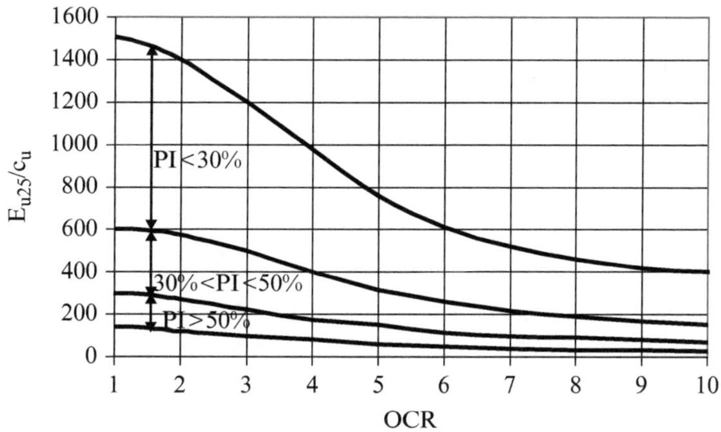

Fig. 3.17 Ratio between undrained modulus E_{u25} at 25 % of the undrained shear strength c_u versus over consolidation ratio (OCR) for different plasticity indices (PI)

3.1.7 Stiffness in Static and Cyclic Conditions

- For **fine grained soil in static condition**, Duncan and Buchigani (1975) provided a graph of the ranges of normalised axial modulus $\boldsymbol{E_u}$ **in undrained condition** and over consolidation ratio OCR, Fig. 3.17.

  Skempton (1944) proposed the relationship between **compression index $\boldsymbol{C_c}$** and liquid limit *LL* (%)

$$C_c = 0.007 \cdot (LL - 10) \tag{3.25}$$

  Terzaghi and Peck (1967) proposed similar relationship with the reliability range of ±30 %

$$C_c = 0.009 \cdot (LL - 10) \tag{3.26}$$

  $LL < 100$ %
  $S_t < 4$ is clay sensitivity

  Wroth and Wood (1978) provided relationship for **remoulded (compacted) clay**

$$C_c = 0.5 \cdot PI \cdot G_s \tag{3.27}$$

  PI is plasticity index
  G_s is the specific gravity of soil grains (~2.65).

  Roscoe et al. (1958) proposed typical values of **recompression index $\boldsymbol{C_r}$** (in over consolidated state) range from 0.015 to 0.35 and often assumed to be 5–10 % of C_c.

Secondary compression index C_α is usually assumed to be related to C_c, with the ratio of C_α/C_c typically in the range 0.025 to 0.006 for inorganic soil and 0.035 to 0.085 for organic soil. Mesri (1973) plotted correlation between **secondary compression ratio $C_{\alpha e} = C_\alpha/(1 + e_p)$** and moisture content w (%), where e_p is the void ratio at the start of linear portion of the e-log pressure curve. From Mesri's (1973) data it follows that on average

$$C_{\alpha e} = 0.0001 \cdot w \tag{3.28}$$

w is moisture content in the range 30 % < w < 600 %
The variability range of $C_{\alpha e}$ is from about 3 to about 0.5 of the average value.

Maximum shear stiffness modulus G_{max} at small strain ($<10^{-6}$)

$$G_{\max} = \frac{625}{0.3 + 0.7 \cdot e^2} \cdot OCR^{-5 \cdot 10^{-5} \cdot PI + 0.0096 \cdot PI + 0.0014} \cdot \sqrt{p_a \cdot \sigma'_m} \tag{3.29}$$

e is void ratio, the function in the denominator is according to Hardin (1978)
OCR is over consolidation ratio
PI is plasticity index (%), the relationship in the exponent is fitted from data by Hardin and Drnevich (1972)
p_a is atmospheric pressure in the unit of σ'_m
σ'_m is the mean principal effective stress

Zhang et al. (2005) provided the following relationship for the ratio of **secant shear modulus G** and G_{max} based on a statistical analysis of resonant column and torsional shear test results from 122 specimens

$$\frac{G}{G_{\max}} = \frac{1}{1 + \left(\frac{\gamma}{\gamma_r}\right)^\alpha}$$

$$\gamma_r = \gamma_{r1} \cdot \left(\frac{\sigma'_m}{p_a}\right)^k \tag{3.30}$$

$$\sigma'_m = \sigma'_v \cdot \frac{1 + 2 \cdot K_o}{3}$$

γ is actual shear strain
γ_r is the reference shear strain from Table 3.6
$\alpha, \kappa, \gamma_{r1}$ are the coefficients from Table 3.6
σ'_m is the mean principal effective stress
σ'_v is the effective vertical stress
K_o is the coefficient of earth pressure at rest

Table 3.6 Recommended mean values of α,k,γ_{r1} and ξ_{min} at $\sigma'_m = 100$ kPa (Zhang et al. 2005, with permission from ASCE)

		Plasticity index, PI (%)					
Geologic age	Parameter	0	15	30	50	100	150
Quaternary	α	0.83	0.87	0.90	0.94	1.04	1.15[a]
	k	0.316	0.255	0.207	0.156	0.077	0.038[a]
	γ_{r1}	0.075	0.092	0.108	0.13	0.186	0.241[a]
	ξ_{min}	0.82	0.94	1.06	1.23	1.63	2.04[a]
Tertiary and older	α	1.03	1.04	1.05	1.07	1.11[a]	–
	k	0.316	0.268	0.227	0.182	0.105[a]	–
	γ_{r1}	0.031	0.037	0.43	0.051	0.072[a]	–
	ξ_{min}	0.82	0.94	1.06	1.23	1.63[a]	–
Residual/saprolite soil	α	0.79	0.86	0.92	1.01[a]	–	–
	k	0.42	0.212	0.107	0.043[a]	–	–
	γ_{r1}	0.039	0.053	0.067	0.086[a]	–	–
	ξ_{min}	0.82[b]	0.94[b]	1.06[b]	1.23[b]	–	–

[a]Tentative value; extrapolated from the range of available data
[b]Tentative value; no small-strain torsional shear damping measurements available
– Little or no data available

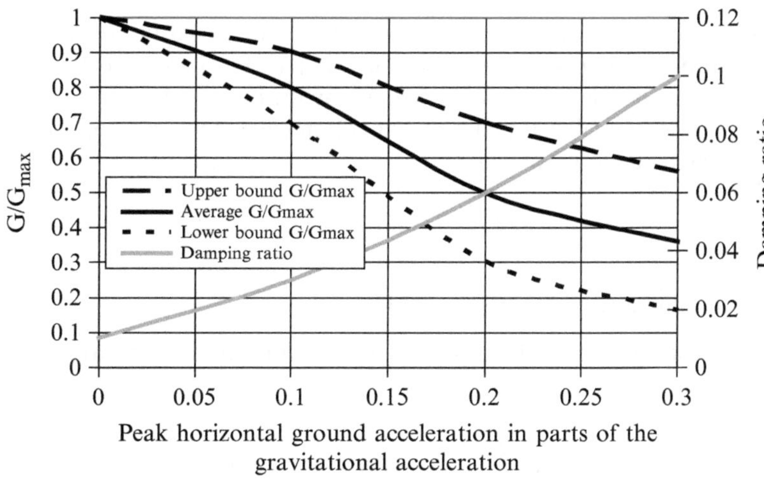

Fig. 3.18 Secant shear modulus ratio and damping ratio dependence on the peak horizontal ground acceleration

Because of inter dependence of shear strain $\gamma = \tau/G$ and shear modulus G with application of shear stress τ, EN 1998-5 (2004) tabulated the values of the **shear modulus ratio G/G_{max} and damping ratio ξ** for different peak horizontal ground accelerations. The values are shown in Fig. 3.18.

At small strain and for small strain increments, the following relationship between the **axial modulus E** and shear modulus G exists

$$E = 2 \cdot (1 + \nu) \cdot G \tag{3.31}$$

ν is Poisson ratio

Park and Hashash (2008) summarised the results of a number of tests performed at shear strain range from 8×10^{-6} to 1.2×10^{-4} and showed that the increase in shear modulus obtained at the cyclic frequency of 0.5 Hz was less than 10 % for the frequency increase to 50 Hz indicating a small effect on stiffness of the increase in rate of strain.

- For **coarse grained soil in static and cyclic condition**, Seed and Idriss (1970) provided the following relationship for the **maximum shear stiffness modulus** $\boldsymbol{G_{max}}$ (lb/ft^2) **of sand** at small strain ($<10^{-6}$)

$$\begin{aligned} G_{\max} &= 1000 \cdot 123.1 \cdot \exp(-1.44 \cdot e) \cdot \sqrt{\sigma'_m} \\ G_{\max} &= 1000 \cdot (0.586 + 16.47 \cdot D_r) \cdot \sqrt{\sigma'_m} \end{aligned} \tag{3.32}$$

e is void ratio, $0.4 < e < 0.9$, the exp function is fitted from the table by Seed and Idriss (1970)
D_r is relative density (%), 30 % $< D_r <$ 90 %, the functional dependence of G_{max} on D_r is fitted from the table by Seed and Idriss (1970)
σ'_m is the mean principal effective stress (lb/ft^2); lb/ft^2 ~ 0.0479 kPa, 1 kPa ~ 20.885 lb/ft^2

- For **gravely cobble deposits**, Lin et al. (2000) suggested the relationship for the **maximum shear stiffness modulus** $\boldsymbol{G_{max}}$ (MPa) based on the results of measurements of transversal wave velocity by the down hole method and large-scale dynamic triaxial tests and resonant-column tests performed on the samples from Taichung area of Taiwan

$$G_{\max} = 305 \cdot e^{0.0025 \cdot \sigma'_3} \tag{3.33}$$

σ'_3 is confining effective pressure in kPa

- For **fine and coarse grained soil in cyclic condition, damping ratio** ξ is according to Zhang et al. (2005)

$$\xi = 10.6 \cdot \left(\frac{G}{G_{\max}}\right)^2 - 31.6 \cdot \frac{G}{G_{\max}} + 21.0 + \xi_{\min} \tag{3.34}$$

G/G_{max} is given in Eq. (3.30)
ξ_{min} is given in Table 3.6

Park and Hashash (2008) summarised the results of a number of tests performed at shear strain range from 10^{-5} to 10^{-4} and showed that the rate of strain increase had a significant effect on the increase in damping ratio. Figure 3.19 shows

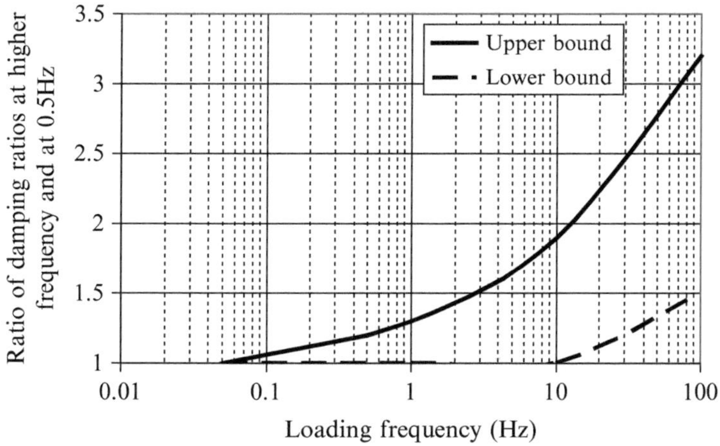

Fig. 3.19 Effect of the increase of loading frequency on damping ratio in the strain range from 10^{-5} to 10^{-4} and for soil plasticity index PI range from 20 to 30 %

dependence of normalised damping ratio with variation in loading frequency based on resonant column and torsional shear test results for undisturbed cohesive soil (PI = 20–30 %) from Treasure Island in California.

- **California Bearing Ratio (CBR)** used for pavement design depend on soil plasticity and moisture content. Agarwal and Ghanekar (1970) obtained the best fit relationship for *CBR* (%) fine grained soil as

$$CBR = 21 - 16 \cdot \log_{10}(OMC) + 0.07LL \tag{3.35}$$

OMC is the optimum moisture content (%) to achieve the maximum density
LL is liquid limit (%)
The standard deviation is 1.8

De Graft-Johnson and Bhatia (1969) plotted a relationship between suitability index *SI* and soaked CBR (%), from which the best fit average value is

$$\begin{aligned} CBR_{average} &= 36.5 \cdot SI - 4, 0.75 \leq SI \leq 4 \\ SI &= \frac{PP}{LL \cdot \log_{10} PI} \end{aligned} \tag{3.36}$$

PP is the percent of grains passing 2.4 mm opening sieve
LL is liquid limit (%)
PI is plasticity index (%)

De Graft-Johnson et al. (1972) plotted a relationship between CBR (%) soaked for 48 h and the ratio between maximum dry density $\gamma_{d,max}$ (kg/m^3) and plasticity index PI (%), from which the best fit average value is

$$CBR_{soaked} = 29.2 \cdot \ln_e \left(\frac{\gamma_{d,\max}}{PI} \right) - 84.2, \quad 70 \leq \frac{\gamma_{d,\max}}{PI} \leq 600 \tag{3.37}$$

Black (1962) plotted a correction factor c of CBR for effective degree of saturation S_r (%), which best fit is

$$c = 2 \cdot 10^{-5} \cdot S_r^{2.33}, 50\,\% \leq S_r \leq 90\,\% \tag{3.38}$$

3.1.8 Water Permeability

An average velocity of water permeability of soil is frequently described as a product of coefficient of permeability k and hydraulic gradient i according to Darcy (1856).

- For **fine grained soil**, coefficient of water permeability k can be obtained from Eq. (2.52).
- For **coarse grained soil** Hazen (1911) gave an empirical formula for coefficient of water permeability k (cm/s)

$$k = 100 \cdot D_{10}^2 \tag{3.39}$$

D_{10} (cm) is the effective size of soil grains so that 10 % of soil by weight has smaller sizes. 0.1 mm $< D_{10} <$ 3 mm, The coefficient of uniformity $C_u < 5$. The value of k changes linearly with the change in the ratio $e^3 (1 + e)^{-1}$, e is void ratio.

Coefficient of water permeability of a soil mass depends also on the presence of seams and cracks and is usually anisotropic (different in different direction).

3.1.9 Consolidation (Fine Grained Soil)

Approximate correlations between coefficient of consolidation c_v (cm^2/s) and liquid limit LL (%) are according to US Navy DM-7.01 (1986) graphs

- For **undisturbed soil in compression**

$$c_v = \frac{108}{LL^{2.84}}, 30\ \% \leq LL \leq 160\ \% \tag{3.40}$$

- For **undisturbed soil in recompression**

$$c_v \geq \frac{5452}{LL^{3.55}}, 30\ \% \leq LL \leq 120\ \% \tag{3.41}$$

- For **remoulded (compacted) soil**

$$c_v \leq \frac{1.2}{LL^{2.04}}, 30\ \% \leq LL \leq 120\ \% \tag{3.42}$$

3.2 Rock Properties

3.2.1 Identification

- Visual identification is based on examination of rock masses and samples with observations of decomposition and discontinuities (ISO 14689-1)
- Weathering classification is related to geological processes (ISO 14689-1)
- Discontinuities (bedding planes, joints, fissures, cleavage and faults) are quantified concerning pattern, spacing, inclination and infill if any (ISO 14689-1)
- Rock quality designation (RQD), total core recovery (TCR), and solid core recovery (SCR) are stated in borehole logs and determined according to ISO 22475-1.

TCR (%) is the ratio of core recovered (solid and non intact) to length of core run.
SCR (%) is the ratio of solid core recovered to length of core run.
RQD (%) is the ratio of solid core pieces longer than 100 mm to length of core run (Deere 1964).

Table 3.7 contains a description of rock classification used by Hoek and Brown (1980). Table 3.7 refers to the following classification systems:

CSIR – the South African's Council for Scientific and Industrial Research rating (Bieniawski 1974) that is renamed into rock mass rating (RMR) described in Sect. 3.2.2.
NGI – the Norwegian Geotechnical Institute rating (Barton et al. 1974) that is renamed into rock quality number (Q) described in Sect. 3.2.3.

Table 3.7 Basic rock types and states used by Hoek and Brown (1980) except RQD ranges from Deer (1964)

Main rock types	Typical representatives	Rock quality types	Weathering	Joint spacing	CSIR rating	NGI rating	RQD (%)
Carbonate rock with well developed crystal cleavage	Dolomite Limestone Marble	Very good	Tightly interlocking undisturbed rock with unweathered joints	±3 m	85	100	90–100
Lithified argillaceous rock	Mudstone Siltsone Shale Slate	Good	Fresh to slightly weathered joints, slightly disturbed	1–3 m	65	10	75–90
Arenaceous rock with strong crystals and poorly developed crystal cleavage	Sandstone Quartzite	Fair	Several sets of moderately weathered joints	0.3–1 m	44	1.0	50–75
Fine grained polyminerallic igneous crystalline rock	Andesite Dolerite Diabase Rhyolite	Poor	Numerous weathered joints with some gouge filling	30–500 m	23	0.1	25–50
Coarse grained polymineralic igneous and metamorphic crystalline rock	Amphibolite Gabbro Gneiss Granite Norite Quartz-diorite	Very poor	Numerous heavily weathered joints with gouge filling	<50 mm	3	0.01	<25

3.2.2 Rock Mass Rating

Rock mass rating (RMR) between zero and 100 is obtained by summation of the ratings of the following five individual parameters and adjusting this total by taking into account the joint orientations (e.g. Hoek and Brown 1980; Stacey and Page 1986).

- Unconfined compressive strength of rock pieces (UCS)
- Rock quality designation (RQD)
- Joint spacing
- Joint roughness and separation
- Groundwater

The following parameters have not been considered by RMR:

- Confining stress
- Number of joint sets
- Durability

Table 3.8 contains individual ratings. Table 3.9 contains the rating adjustment for joint orientation.

3.2.3 Quality Number of Rock Mass

The quality number Q between 0.001 and 1,000 is based on three aspects:

- Rock block size (RQD/J_n)
- Joint shear strength (J_r/J_a)
- Confining stress (J_w/SFR)

$$Q = \frac{RQD}{J_n} \cdot \frac{J_r}{J_a} \cdot \frac{J_w}{SRF} \tag{3.43}$$

RQD the rock quality designation
J_n is the joint set number
J_r is the joint alteration number
J_w is the joint water reduction factor
SRF is the stress reduction factor

The Q system does not consider the following parameters:

- Rock strength explicitly but implicitly in SRF
- Joint orientation

Table 3.8 Individual ratings for RMR

UCS (MPa)	Rating	RQD (%)	Rating	Joint spacing	Rating	Joint condition	Rating	Ground water	Rating
>200	15	90–100	20	>3 m	30	very rough, not continuous, not separated, hard wall	25	inflow per 10 m tunnel length = 0 or ratio between joint water pressure and joint principal stress = 0 or general conditions = dry	10
100–200	12	75–90	17	1–3 m	25	slightly rough, separation <1 mm, hard wall	20	inflow per 10 m tunnel length <25 l/min or ratio between joint water pressure and joint principal stress = 0 to 0.2 or general conditions = moist	7
50–100	7	50–75	13	0.3–1 m	20	Slightly rough separation <1 mm soft wall	12	inflow per 10 m tunnel length = 25 to 125 l/min or ratio between joint water pressure and joint principal stress = 0.2 to 0.5 or general conditions = moderate water pressure	4
25–50	4	25–50	8	50–300 mm	10	Slickensided or gouge < 5 mm thick or joint width 1–5 mm continuous	6	inflow per 10 m tunnel length >125 l/min or ratio between joint water pressure and joint principal stress >0.5 or general conditions = severe water problems	0
10–25	2	<25	3	<50 mm	5	soft gouge >5 mm thick or joint width >5 mm continuous	0		
3–10	1								
1–3	0								

Table 3.9 RMR adjustment for joint orientation

Strike and dip orientations of joints		Very favourable	Favourable	Fair	Unfavourable	Very unfavourable
Rating	Tunnels	0	−2	−5	−10	−12
	Foundations	0	−2	−7	−15	−25
	Slopes	0	−5	−25	−50	−60

Values of the individual parameters are provided in textbooks (e.g. Hoek and Brown 1980; Stacey and Page 1986). Bieniawski (1976) found the following relationship between RMR and Q number:

$$RMR = 9 \cdot \ln_e Q + 44 \tag{3.44}$$

3.2.4 Shear Strength

The shear strength τ of a **rock mass** is defined as:

$$\tau = A \cdot UCS \cdot \left(\frac{\sigma}{UCS} + s_1\right)^B \tag{3.45}$$

A, s_1, B are the coefficient in Table 3.9 from Hoek and Brown (1980)
UCS is unconfined compressive strength of rock pieces
σ is the axial compressive stress acting on the plane in which τ is calculated

For **triaxial condition** of stresses in a rock mass, the relationship between the maximum σ_1 and the minimum σ_3 stresses is:

$$\sigma_1 = \sigma_3 + \sqrt{m \cdot \sigma_3 \cdot UCS + s \cdot UCS^2} \tag{3.46}$$

σ_1 is the maximum (principal) stress
σ_3 is the minimum (confining) stress
m,s are the coefficients in Table 3.10 from Hoek and Brown (1980)
UCS is unconfined compressive strength of rock pieces

A plot of stresses in triaxial compression is shown in Fig. 2.15b.

Based on the results of in situ measurements of stresses in Earth's crust, Hoek and Brown (1980) suggested that **the ratio of the average horizontal to vertical stresses** is

Table 3.10 Rock mass coefficients from Hoek and Brown (1980)

Types	Dolomite Limestone Marble	Mudstone Siltsone Shale Slate	Sandstone Quartzite	Andesite Dolerite Diabase Rhyolite	Amphibolites Gabbro Gneiss Granite Norite Quartz-diorite
Intact peaces	A = 0.816	A = 0.918	A = 1.044	A = 1.086	A = 1.22
	s_1 = 0.14	s_1 = 0.099	s_1 = 0.067	s_1 = 0.059	s_1 = 0.04
	B = 0.658	B = 0.677	B = 0.692	B = 0.696	B = 0.705
	m = 7.0	m = 10.0	m = 15.0	m = 17.0	m = 25.0
	s = 1.0	s = 1.0	s = 1.0	s = 1.0	s = 1.0
Tightly interlocking undisturbed rock with unweathered joints	A = 0.651	A = 0.739	A = 0.848	A = 0.883	A = 0.998
	s_1 = 0.028	s_1 = 0.02	s_1 = 0.013	s_1 = 0.012	s_1 = 0.008
	B = 0.679	B = 0.692	B = 0.702	B = 0.705	B = 0.712
	m = 3.5	m = 5.0	m = 7.5	m = 8.5	m = 12.5
	s = 0.1	s = 0.1	s = 0.1	s = 0.1	s = 0.1
Fresh to slightly weathered joints, slightly disturbed	A = 0.369	A = 0.427	A = 0.501	A = 0.525	A = 0.603
	s_1 = 0.006	s_1 = 0.004	s_1 = 0.003	s_1 = 0.002	s_1 = 0.002
	B = 0.669	B = 0.683	B = 0.695	B = 0.698	B = 0.707
	m = 0.7	m = 1.0	m = 1.5	m = 1.7	m = 2.5
	s = 0.004	s = 0.004	s = 0.004	s = 0.004	s = 0.004
Several sets of moderately weathered joints	A = 0.198	A = 0.234	A = 0.28	A = 0.295	A = 0.346
	s_1 = 0.0007	s_1 = 0.0005	s_1 = 0.0003	s_1 = 0.0003	s_1 = 0.0002
	B = 0.662	B = 0.675	B = 0.688	B = 0.691	B = 0.7
	m = 0.14	m = 0.2	m = 0.3	m = 0.34	m = 0.5
	s = 0.0001	s = 0.0001	s = 0.0001	s = 0.0001	s = 0.0001
Numerous weathered joints with some gouge filling	A = 0.115	A = 0.129	A = 0.162	A = 0.172	A = 0.203
	s_1 = 0.0002	s_1 = 0.0002	s_1 = 0.0001	s_1 = 0.0001	s_1 = 0.0001
	B = 0.646	B = 0.655	B = 0.672	B = 0.676	B = 0.686
	m = 0.04	m = 0.05	m = 0.08	m = 0.09	m = 0.13
	s = 0.00001	s = 0.00001	s = 0.00001	s = 0.00001	s = 0.00001
Numerous heavily weathered joints with gouge filling	A = 0.042	A = 0.05	A = 0.061	A = 0.065	A = 0.078
	s_1 = 0	s_1 = 0	s_1 = 0	s_1 = 0	s_1 = 0
	B = 0.534	B = 0.539	B = 0.546	B = 0.548	B = 0.556
	m = 0.007	m = 0.01	m = 0.015	m = 0.017	m = 0.025
	s = 0	s = 0	s = 0	s = 0	s = 0

Note: Table 3.7 contains more detailed description of rock types and condition

Table 3.11 Shear strength parameters for non-linear failure envelopes of continuous rock joints from Maksimovic (1996b)

Description	$\phi_{b,r}$ (degrees)	$\Delta\phi$ (degrees)	p_n (kPa)
Mudstone	24.8	9.7	1,811
Sandstone	31.0	19.0	1,922
Sandstone	31.5	28.8	2,277
Foliation in micaschist	25.5	39.2	623
Drammen granite (tentative)	29.0	27.2	1,450

$$\frac{100}{z} + 0.3 < \frac{\sigma_{hor}}{\sigma_{vert}} < \frac{1500}{z} + 0.5 \tag{3.47}$$

z is depth (m) below surface

For $z < 500$ m, the horizontal stresses σ_{hor} are significantly greater than the vertical stresses $\sigma_{vert} = \gamma\ z$, γ is the unit weight of rock. Sheorey (1994) proposed the following equation for the ratio between horizontal and vertical in situ stresses in rock mass:

$$\frac{\sigma_{hor}}{\sigma_{vert}} = 0.25 + 7 \cdot E_h \cdot \left(0.001 + \frac{1}{z}\right) \tag{3.48}$$

E_h is horizontal deformation modulus (GPa)
z is depth below surface (m)

For continuous **joints governing the shear strength** of a rock mass, the angle of friction is defined in Eq. (3.22). The coefficients for rock joints are given in Table 3.11 from Maksimovic (1996b).

3.2.5 *Stiffness*

Figure 3.20 shows upper and lower boundaries of rock mass stiffness modulus obtained from in situ measurements and data plotted by Hoek and Diederichs (2006). For weaker rock, the scale in GPa is too large and therefore rock

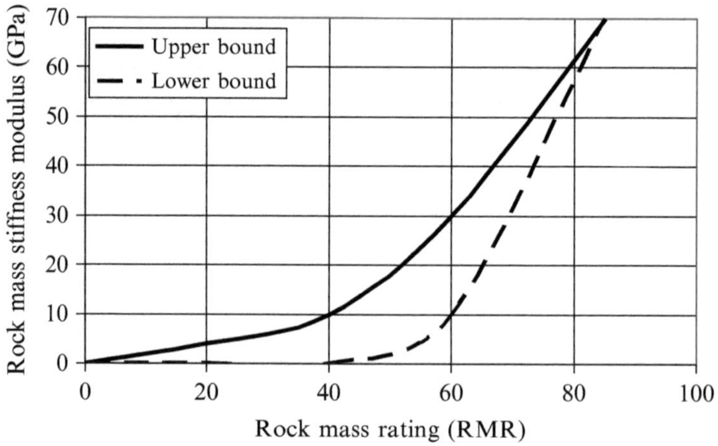

Fig. 3.20 Rock mass modulus limits based on data from in situ measurements

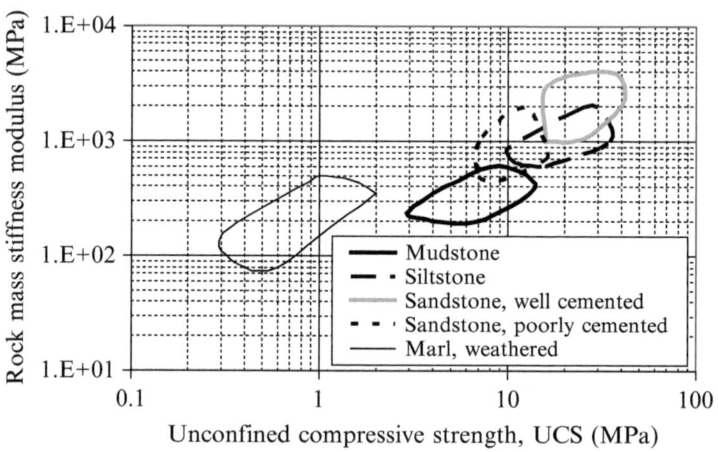

Fig. 3.21 Rock mass stiffness modulus limits for weaker rock from Trias age

mass stiffness modulus ranges from data by Hobbs (1974) are shown in Fig. 3.21 in MPa. Figure 3.22 shows small strain Young modulus and unconfined compressive strength of chalk from data by Matthews and Clayton (1993).

3.2.6 Water Permeability

For a set of parallel planar fractures, the permeability of rock mass according to Snow (1968) is:

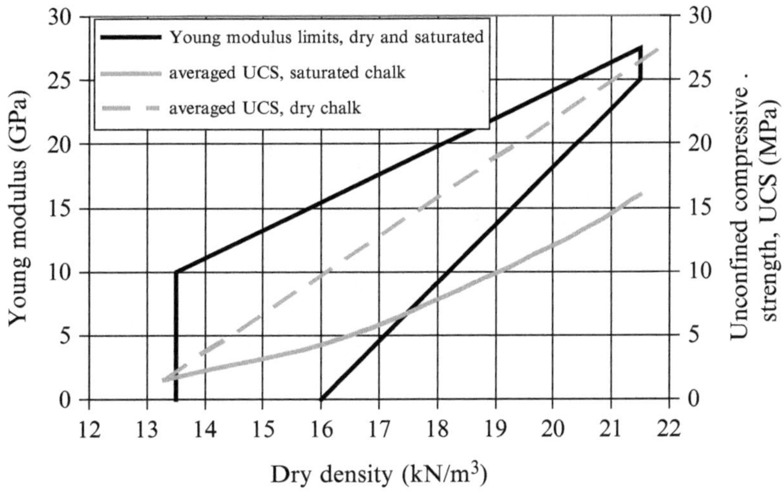

Fig. 3.22 Young modulus and unconfined compressive strength of chalk peace

$$p = \frac{w^3}{12 \cdot S} \tag{3.49}$$

p is rock mass permeability in m^2. 1 darcy $= 10^{-12}$ $m^2 \sim 10^{-5}$ m/s for water in usual condition
w is fracture aperture (m)
S is fracture spacing (m)

3.3 Summary

Necessary ground properties and tests for foundation and retaining wall design in soil and rock are listed in Table 3.12. For other geo-structures such as slopes and tunnels, most of the ground parameters listed in Table 3.12 are required with addition of the results of field measurements, which main types are described in Chap. 6.

Table 3.12 Necessary ground properties and test types for large scale foundations and retaining walls design in soil and rock

State	Condition	Ground type	Property	Test type
Classification and durability	Static & cyclic	Soil	Grain size distribution; content of organic matter, sulphates, chlorides and carbonates	Classification
		Silt, clay	Liquid and plastic limits (LL, PL)	Classification
		Rock	Type and state	Field description of rock type, joints, weathering, TCR, SCR, RQD
		All	Electrical resistivity for earthening	Geophysical electrical resistivity
Ultimate limit state (ULS) (strength condition)	Static & cyclic	All	Unit weight (γ) and moisture content (w)	Classification
		Sand	Angle of friction (ϕ)	SPT or CPT, cyclic simple shear in laboratory
		Clay	Undrained condition shear strength (c_u)	UU triaxial
			Drained condition equivalent cohesion (c'_e), and friction angle (ϕ'_e)	CU triaxial with pore pressure measurements
		Rock	Compressive strength of rock pieces, number of joints per m′, freshness	UCS, joint spacing, weathering amount from rock cores
Serviceability limit state (SLS) (deformation condition)	Static	Soil	Young modulus (E_{max}) at small strain	SPT or CPT in sand, local measurement of small strain in triaxial apparatus on clay specimens
		Clay	Coefficient of volume compressibility (m_v), secondary compression index (C_α), coefficient of consolidation (c_v)	Oedometer
		Rock	Young modulus (E)	UCS, joint spacing and condition from rock cores; ground water level
	Cyclic	Sand and clay	Shear modulus (G) dependence on shear strain	Bender element & resonant column test
		Rock	Shear stiffness (G_{max}) at small strain	Geophysical wave refraction

Note: Contamination tests not included

References

Agarwal KB, Ghanekar KD (1970) Prediction of CBR from plasticity characteristics of soils. In: Proceedings of the 2nd South-east Asian conference on soil engineering, Singapore, pp 571–576

ASTM D1556-07 Standard test method for density and unit weight of soil in place by the sand-cone method. American Society for Testing and Materials, Philadelphia, PA

ASTM D2167-08 Standard test method for density and unit weight of soil in place by the rubber balloon method. American Society for Testing and Materials, Philadelphia, PA

ASTM D2216-10 Standard test methods for laboratory determination of water (moisture) content of soil and rock by mass. American Society for Testing and Materials, Philadelphia, PA

ASTM D2487-11 Standard practice for classification of soils for engineering purposes (Unified Soil Classification System). American Society for Testing and Materials, Philadelphia, PA

ASTM D2937-10 Standard test method for density of soil in place by the drive-cylinder method. American Society for Testing and Materials, Philadelphia, PA

ASTM D422-63(2007) Standard test method for particle-size analysis of soils. American Society for Testing and Materials, Philadelphia, PA

ASTM D4253-00(2006) Standard test methods for maximum index density and unit weight of soils using a vibratory table. American Society for Testing and Materials, Philadelphia, PA

ASTM D4254-00(2006)e1 Standard test methods for minimum index density and unit weight of soils and calculation of relative density. American Society for Testing and Materials, Philadelphia, PA

ASTM D4318-10 Standard test methods for liquid limit, plastic limit, and plasticity index of soils. American Society for Testing and Materials, Philadelphia, PA

ASTM D4564-08e1 Standard test method for density and unit weight of soil in place by the sleeve method. American Society for Testing and Materials, Philadelphia, PA

ASTM D4943-08 Standard test method for shrinkage factors of soils by the wax method. American Society for Testing and Materials, Philadelphia, PA

ASTM D5519-07 Standard test methods for particle size analysis of natural and man-made riprap materials. American Society for Testing and Materials, Philadelphia, USA

ASTM D6913-04(2009) Standard test methods for particle-size distribution (gradation) of soils using sieve analysis. American Society for Testing and Materials, Philadelphia, PA

ASTM D7263 – Test methods for laboratory determination of density (unit weight) of soil specimens. American Society for Testing and Materials, Philadelphia, PA

Atterberg A (1911) Lerornas forhallande till vatten, deras plasticitetsgranser och plastiitetsgrader (The behavior of clay with eater, their limits of plasticity and their degrees of plasticity). Kungliga Lantbruksakademiens Handlingar och Tidskrift 50(2):132–158

Barton N, Lien R, Lunde J (1974) Engineering classification of rock masses for the design of tunnel support. Rock Mech 6(4):189–236

Bieniawski ZT (1974) Geomechanics classification of rock masses and its application in tunnelling. In: Proceedings of the 3rd international congress on rock mechanics, ISRM, Denver 11A:27–32

Bieniawski ZT (1976) Rock mass classification in rock engineering. In: Proceedings of the symposium on exploration for rock engineering, Johannesburg, Balkema 1:97–106

Bjerrum L, Krinstad S, Kummeneje O (1961) The shear strength of fine sand, In: Proceedings of the 5th international conference on soil mechanics and foundation engineering, Paris 1:29–37

Black WPM (1962) A method for estimating CBR of cohesive soil from plasticity data. Geotechnique 12:271–272

Bray JD, Sancio RB, Durgunoglu T, Onalp A, Youd TL, Stewart JP, Seed RB, Cetin OK, Bol E, Baturay MB, Christensen C, Karadayilar T (2004) Subsurface characterization at ground failure sites in Adapazari, Turkey. J Geotech Geoenviron Eng ASCE 130(7):673–685

BS 1377-2 (1990) Methods of test for soils for civil engineering purposes – part 2: classification tests. British Standards Institution, UK

BS 1377-4 (1990) Soils for civil engineering purposes. Compaction-related tests (AMD 8259) (AMD 13925). British Standards Institution, UK
BS 5930 (1999) + A2(2010) Code of practice for site investigations. British Standards Institution, UK
Carter M, Bentley SP (1991) Correlations of soil properties. Pentech Press, London
Darcy H (1856) Les fontaines publiques de la ville de Dijon. Dalmont, Paris
Das BM (1985) Advanced soil mechanics. McGraw-Hill, New York
De Alba P, Chan CK, Seed HB (1975) Determination of soil liquefaction characteristics by large-scale laboratory tests. Earthquake Engineering Research Centre report no EERC 75–14, College of Engineering, University of California, Berkeley, CA
De Graft-Johnson JWS, Bhatia HS (1969) The engineering characteristics of the lateritic gravels of Ghana. In: Proceedings of the 7th international conference on soil mechanics and foundation engineering, Mexico 2:13–43
De Graft-Johnson JWS, Bhatia HS, Yaboa SL (1972) Influence of geology and physical properties on strength characteristics of lateritic gravels for road pavements. Highway Res Board Rec 405:87–104
Deere DU (1964) Technical description of rock cores for engineer purposes. Rock Mech Eng Geol 1(1):17–22
Duncan JM, Buchigani AL (1975) An engineering manual for settlement studies. Department of Civil Engineering, University of California, Berkeley
EN 1998–5 (2004) Eurocode 8: design of structures for earthquake resistance – Part 5: Foundations, retaining structures and geotechnical aspects. European Committee for Standardization. Brussels
Hardin BO (1978) The nature of stress-strain behaviour of soils. In: Proceedings of the conference on earthquake engineering and soil dynamics, ASCE, Pasadena, CA 1:3–89
Hardin BO, Drnevich VP (1972) Shear modulus and damping in soils: design equations and curves. Journal of the soil mechanics and foundation division, ASCE 98(SM7_:667–692)
Hazen A (1911) Discussion of "Dams on sand foundation". Trans ASCE 73:199
Hobbs NB (1974) Factors affecting the prediction of settlement of structures on rock with particular reference to the Chalk and Trias. In: Settlement of structures, proceedings of the conference organised by the British Geotechnical Society at Cambridge, Pentech Press, London, pp 579–654
Hoek E, Brown ET (1980) Underground excavations in rock. Institution of Mining and Metallurgy, London
Hoek E, Diederichs MS (2006) Empirical estimation of rock mass modulus. Int J Rock Mech Mining Sci 43:203–215
Hungr O, Morgenstern NR (1984) High velocity ring shear test on sand. Geotechnique 34:415–421
ISO 11277 (2009) Soil quality. Determination of particle size distribution in mineral soil material. Method by sieving and sedimentation. International Organization for Standardization, Geneva, Switzerland
ISO 14689-1 (2003) Geotechnical investigation and testing – Identification and classification of rock – Part 1: Identification and description. International Organization for Standardization, Geneva, Switzerland
ISO 16586 (2003) Soil quality – Determination of soil water content as a volume fraction on the basis of known dry bulk density – Gravimetric method. International Organization for Standardization, Geneva, Switzerland
ISO 22475-1 (2006) Geotechnical investigation and testing – sampling methods and groundwater measurements – Part 1: Technical principles for execution. International Organization for Standardization, Geneva, Switzerland
ISO/TS 17892-1 (2004) Geotechnical investigation and testing – laboratory testing of soil – Part 1: Determination of water content. International Organization for Standardization, Geneva, Switzerland

ISO/TS 17892-12 (2004) Geotechnical investigation and testing – laboratory testing of soil – Part 12: Determination of Atterberg limits. International Organization for Standardization, Geneva, Switzerland

ISO/TS 17892-2 (2004) Geotechnical investigation and testing – laboratory testing of soil – Part 2: Determination of density of fine grained soil. International Organization for Standardization, Geneva, Switzerland

ISO/TS 17892-4 (2004) Geotechnical investigation and testing laboratory testing of soil – Part 4: Determination of particle size distribution. International Organization for Standardization, Geneva, Switzerland

Jaky J (1944) The coefficient of earth pressure at rest. J Soc Hungarian Arch Eng 355–358

Kenney TC (1959) Discussion. J Soil Mech Div ASCE 85(3):67–79

Kulhaway GH, Mayne PW (1990) Manual of estimating soil properties for foundation design. Geotechnical Engineering Group, Cornell University, Ithaca

Ladd CC, Foot R (1974) New design procedures for stability of soft clays. J Geotech Eng Div ASCE 100(GT7):763–786

Lee KL, Focht JA (1976) Strength of clay subjected to cyclic loading. Marine Georesour Geotech 1(3):165–168

Lin SY, Lin PS, Luo H-S, Juang CH (2000) Shear modulus and damping ratio characteristics of gravely deposits. Can Geotech J 37:638–651

Lupini JF, Skinner AE, Vaughan PR (1981) The drained residual strength of cohesive soils. Geotechnique 31(2):181–213

Maksimovic M (1989a) On the residual shearing strength of clays. Geotechnique 39(2):347–351

Maksimovic M (1989b) Nonlinear failure envelope for soils. J Geotech Eng ASCE 115 (4):581–586

Maksimovic M (1995) Discussion. J Geotech Eng ASCE 121(9):670–672

Maksimovic M (1996a) A family of nonlinear failure envelopes for non-cemented soils and rock discontinuities. Elect J Geotech Eng 1:2002–2022

Maksimovic M (1996b) The shear strength components of a rough rock joint. Int J Rock Mech Mining Sci Geomech Abstr 33(8):769–783

Matthews MC, Clayton CRI (1993) Influence of intact porosity on the engineering properties of a weak rock. In: Anagnostopoulos A, Schlosser F, Kalteziotis N, Frank R (eds) Geotechnical engineering of hard soil – soft rocks, vol 1. AA Balkema, Rotterdam, pp 639–702

Mayne PW, Kulhaway FH (1982) K_o – OCR relationships in soil. J Geotech Eng Div ASCE 108:851–872

Mesri G (1973) The coefficient of secondary compression. J Soil Mech Found Div ASCE 103:417–430

Mitchell JK, Houston WN (1969) Causes of clay sensitivity. J Soil Mech Found Div ASCE 95 (3):845–869

Parathiras A (1995) Rate of displacement effects on fast residual strength. In: Ishihara K (ed) Proceedings of the 1st international conference on earthquake geotechnical engineering, Tokyo 1:233–237

Park D, Hashash MA (2008) Rate-dependent soil behaviour in seismic site response analyses. Can Geotech J 45:454–469

Roscoe KH, Schofield AN, Wroth CP (1958) On the yielding of soils. Geotechnique 8:2–52

Seed HB, Idriss IM (1970) Soil moduli and damping factors for dynamic response analyses. Report EERC 70-10, Earthquake Engineering Research Centre, University of California, Berkeley

Sheorey PR (1994) A theory for in situ stresses in isotropic and transversely isotropic rock. Int J Rock Mech Min Sci Geomech Abstr 31(1):23–34

Skempton AW (1944) Notes on the compressibility of clays. Quat J Geol Soc 100:119–135

Skempton AW (1953) The colloidal activity of clay. In: Proceedings of the 3rd international conference on soil mechanics and foundation engineering, Zurich 1:57–61

Skempton AW (1957) Discussion: the planning and design of new Hong Kong airport. Proc Inst Civil Eng 7:305–307

Skempton AW, Northey R (1952) The sensitivity of clay. Geotechnique 3:30–53
Snow DT (1968) Rock fracture spacing, openings and porosities. J Soil Mech Found Eng ASCE 94:73–91
Stacey TR, Page CH (1986) Practical handbook for underground rock mechanics. Trans Tech Publications, Series in Rock and Soil mechanics, vol 12
Stark TD, Choi H, McCone S (2005) Drained shear strength parameters for analysis of landslides. J Geotech Geoenviron Eng ASCE 131(5):575–588
Terzaghi K, Peck RB (1967) Soil mechanics in engineering practice. Wiley, London
Thiers GR, Seed HB (1969) Strength and stress-strain characteristics of clays subjected to seismic loading conditions. In: Vibration effects on soils and foundations, Special Technical Publication 450, ASTM, Philadelphia, pp 3–56
Tika TE, Vaughan PR, Lemos LJLJ (1996) Fast shearing of pre-existing shear zones in soil. Geotechnique 46(2):197–233
US Navy DM-7.01 (1986) Design manual: soil mechanics. Navy facilities engineering command, NAVFAC, U.S. Naval Publications and Forms Centre, Virginia
Wroth CP, Wood CM (1978) The correlation of index properties with some basic engineering properties of soils. Can Geotech J 15:137–145
Zhang J, Andrus RD, Juang CH (2005) Normalized shear modulus and material damping ratio relationships. J Geotech Geoenviron Eng ASCE 131:453–464

Chapter 4
Geo-hazards

Abstract This chapter contains descriptions, extents, identifications and remediation measures for the following geo-hazards:

- hydraulic failure
- erosion
- liquefaction
- rock dissolving and caves
- collapse of soil structure
- subsidence of ground surface
- heave of soil and rock
- slope instability
- contamination
- vibration
- earthquakes
- volcanoes
- frozen ground
- unexploded ordnance (UXO)
- gasses underground
- rock burst

Ground investigation contractors use cable and pipe detection tools and hand excavation of trial pits to 1 m depth to avoid hazards from high voltage cables and gas pipes. Biological hazards (poisonous animals and plants and harmful bacteria and viruses) should be subject of environmental assessment reports.

Risk is a product of hazard, structural vulnerability and cost for repairs or replacement.

M. Srbulov, *Practical Guide to Geo-Engineering: With Equations, Tables, Graphs and Check Lists*, Geotechnical, Geological and Earthquake Engineering 29,
DOI 10.1007/978-94-017-8638-6_4,

4.1 Hydraulic Failure

4.1.1 Description

EN 1997-1 (2004) defines four types of hydraulic failures of which two (global) are considered in this section and the other two (local) in the following sections.

- Uplift (buoyancy) of fine grained soil belonging to a low permeability ground layer, buried hollow structures, lightweight embankments over more permeable ground layer subjected to ground water pressure.
- Uplift (boiling) of coarse grained well permeable soil due to upward seepage forces that act against soil submerged weight to cause lifting of soil particles by flowing water. Popular term is so called "quick" sand when objects on top of it sink.

4.1.2 Extent

Several situations that can be expected to cause global ground hydraulic failures are sketched in Fig. 4.1. While the small scale deep excavations are frequent, large scale artesian ground water pressures are less frequent.

Particular problem is great speed at which hydraulic failures can develop.

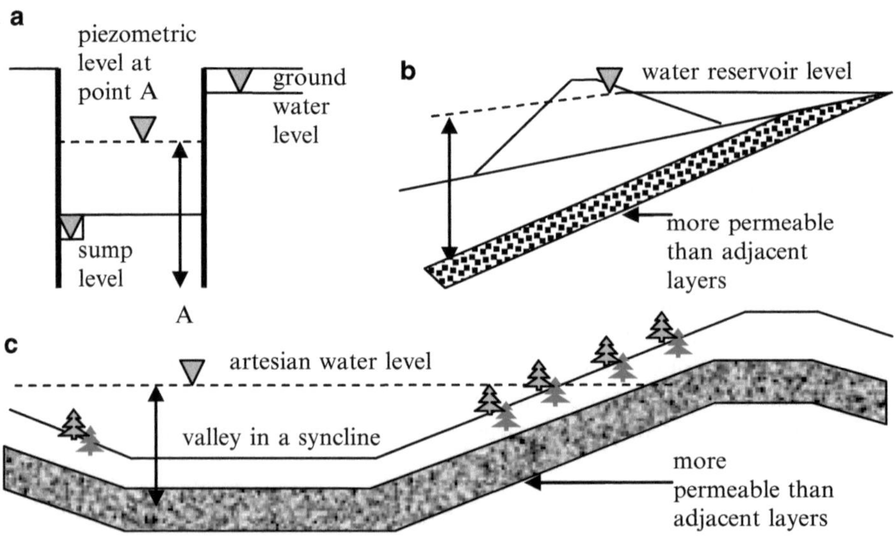

Fig. 4.1 Sketches of cases when buoyancy and boiling can occur (**a**) small scale deep excavation, (**b**) medium scale fill dam on top of ground of low permeability above well permeable ground surfacing in a water reservoir, (**c**) large scale artesian water in the valley formed within the syncline of a geological fold

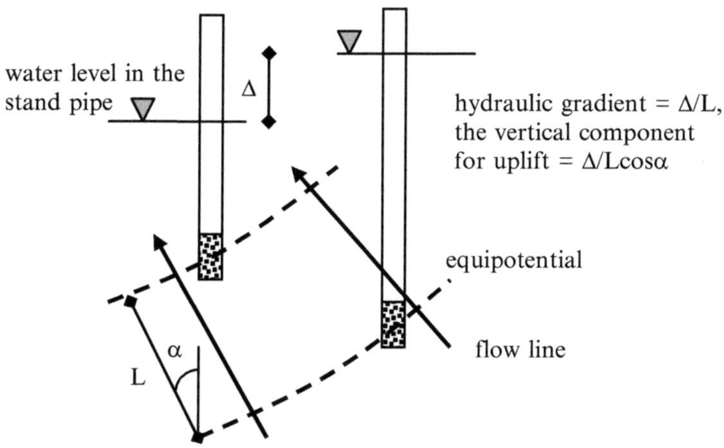

Fig. 4.2 Sketches of water flow lines and equipotentials for definition of a hydraulic gradient and its vertical component for the calculation of uplift force

4.1.3 Identification

The global hydraulic failures will occur when the following conditions exist:

- Uplift (buoyancy) when the water pressure acting at the underside of low permeable soil layer, buried hollow structure or lightweight embankment exceeds the pressure from the total weight of the layer, structure or embankment.
- Uplift (boiling) when upward water flow forces acting against the submerged weight of soil reduce the vertical effective stress to zero. The upward water flow force per unit volume is equal to the product of the unit weight of water and the upward component of hydraulic gradient, as sketched in Fig. 4.2. Besides knowledge of piezometric levels at two places, it is necessary to know, at least approximately, the direction of water flow, which may be estimated based on a sketch of local flow net (e.g. Cadergen 1997; Harr 2003).

Local engineering codes prescribe the required factor of safety against uplift.

4.1.4 Remediation

The measures most commonly adopted to resist failure by uplift are for:

- Uplift (buoyancy): decreasing by drainage of the water pressure acting at the underside of a soil layer with low permeability to water, buried hollow structure or lightweight embankment, increasing the weight of the structure or anchoring the structure into the underlying strata.

- Uplift (boiling): increasing the surcharge on soil, decreasing the hydraulic gradient within soil by using lengthening of seepage path by barriers or interception of the seepage path by relief wells.

More information on ground water control is provided by Preene et al. (2000) for example and in Chap. 6.

4.2 Erosion

4.2.1 Description

The local hydraulic failures are:

- Internal erosion (suffosion) of coarse grained soil particles caused by their transport by water flow within a soil layer, at the interfaces of soil layers with coarser and finer size particles, or at the interface between soil and a structure.
- Concentrated erosion (piping) of the wall of pipe-shaped conduit formed within a coarse grained non-cohesive soil by the arching effect or in a fine grained cohesive but dispersive soil, at the interface between cohesive and non-cohesive soil strata, or at the interface between soil and a structure.
- Surface erosion by rainfall and water waves

First two types of erosion are described in EN 1997-1 (2004).

4.2.2 Extent

- Internal erosion (suffosion) most frequently occurs in loose to medium dense coarse grained soil with gap gradation with the smaller particles passing between the voids that exist between much larger soil particles when subjected to fast ground water flow. Erosion of the surface (i.e. scour, e.g. May et al. 2002; Hoffmans and Verheij 1997) or walls of pipe-shaped conduits in non-cohesive soil is caused by strong water flow, which is able to detach and carry particles long distances without their sedimentation. Figure 4.3 illustrates the conditions for surface erosion and sedimentation process depending on particle size and water flow speed.
- Concentrated erosion (piping) occurs in coarse grained soil at the surface where concentrated flow gradient exists and extents retrogressively inside the soil mass. Change of chemistry of water films around fine grains in clay by removal of sodium cations causes breakage of the existing chemical bonds and loss of cohesion i.e. piping type failures (e.g. Bell and Culshaw 1998).
- Surface erosion by overtopping and internal erosion by seepage from water reservoirs are most frequent cause of the spectacular failures of several fill dams. Surface erosion of fertile soil is subject of agriculture.

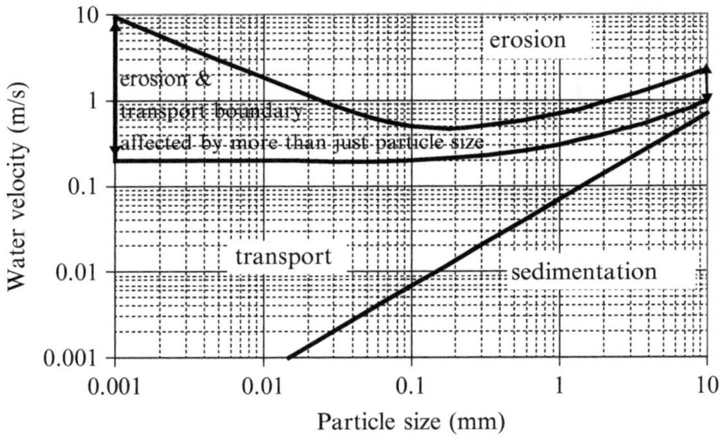

Fig. 4.3 Hjulstrom diagram with modification from Sundborg (1956) and approximate transport/sedimentation boundary (Adopted from Dean 2009)

4.2.3 *Identification*

4.2.3.1 Internal Erosion (Suffosion)

Lane (1935) introduced a term called weighted creep distance L_w

$$L_w = \frac{\Sigma L_{h,i}}{3} + \Sigma L_{v,i} \tag{4.1}$$

$\Sigma L_{h,i}$ is the sum of horizontal distances along the shortest flow path (around a structure or low permeability ground layer)
$\Sigma L_{v,i}$ is the sum of vertical distances along the shortest flow path (around a structure or low permeability ground layer)

The weighted creep ratio *WCR* according to Lane (1935) is:

$$WCR = \frac{L_w}{H_1 - H_2} \tag{4.2}$$

H_1 is the height of free water column at the inflow boundary
H_2 is the height of free water column at the outflow boundary

An example for a varying (inclined) boundary is sketched in Fig. 4.4.

Safe values for the weighted creep ratio are given in Table 4.1.

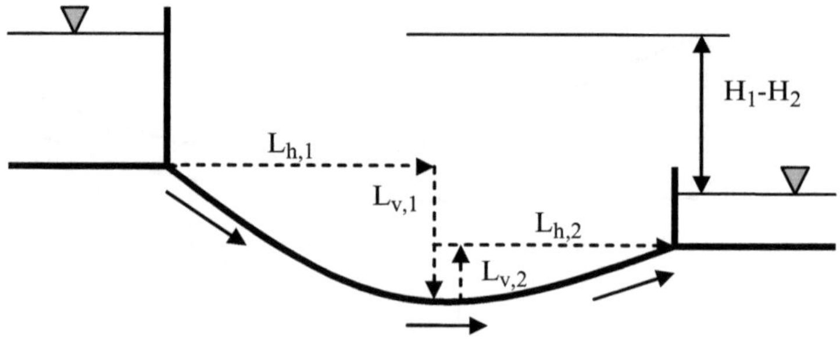

Fig. 4.4 An example of the equivalent horizontal and vertical distances along the shortest flow path, and the water head difference $H_1 - H_2$

Table 4.1 Safe values for the weighted creep ratio (Lane 1935)

Material	Safe weighted creep ratio
Very fine sand or silt	8.5
Fine sand	7.0
Medium sand	6.0
Coarse sand	5.0
Fine gravel	4.0
Coarse gravel	3.0
Soft to medium stiff clay	2.0–3.0
Hard clay	1.8
Hard pan	1.6

4.2.3.2 Concentrated Erosion (Piping)

- Piping in coarse grained non-cohesive soil can be identified as for the surface erosion, Fig. 4.3.
- Piping in fine grained cohesive but dispersive soil can be dependent on the sodium adsorption ratio (SAR)

$$SAR(meq/litre) = \frac{Na}{\sqrt{0.5(Ca + Mg)}} \tag{4.3}$$

A SAR of more than six suggests soil sensitivity to piping. In Australia, Aitcison and Wood (1965) regarded soil in which SAR exceeded 2 as dispersive. Bell and Culshaw (1998) support the limit of 2. The exchangeable sodium percentage (ESP) is:

$$ESP(meq/100\,g) = \frac{exchageble\ sodium}{cation\ exchange\ capacity} \times 100 \tag{4.4}$$

Where 100 g refers to dry clay. A threshold value of 10 % has been recommended for susceptibility to dispersivity of their free salts under seepage of relatively pure water. Soil with ESP greater than 15 % are highly dispersive (e.g. with smectitic and some illites mineral). High values of ESP and high dispersivity are rare in kaolinitic clay. Soil with high cation exchange capacity values and a plasticity index greater than 35 % swell so much that their dispersion is not significant. Dispersive soil occurs in semi-arid regions and low-lying areas with gently rolling topography and smooth relatively flat slopes where the rainfall generated seepage has a high SAR (Bell and Culshaw 1998). BS 1377-5 (1990) describes three tests:

1. The pinhole test use the flow of water under a high hydraulic gradient through a cavity in the soil
2. The crumb test use observation of the behaviour of crumbs of soil in a static dilute sodium hydroxide
3. The dispersion method (double hydrometer test) is based on comparison of the extent of natural dispersion of clay particles with that obtained using standard chemical and mechanical dispersion

ASTM D4647-06e1 exist for the pin hole test and ASTM D4221-11 for double hydrometer test.

4.2.3.3 Surface Erosion

Regular visual inspections of slope surfaces reveal the extent and severity of the erosion.

4.2.4 Remediation

4.2.4.1 Internal Erosion (Suffosion)

- Decreasing the hydraulic gradient within soil by using lengthening of seepage path by barriers or interception of the seepage path by relief wells (Chap. 6).
- Alternatively or in combination by placing granular filters at the surface/interfaces of layers with different grain sizes. Bertram (1940) provided the following criteria to be satisfied by the filter (F) and protected soil (S).

$$\begin{aligned} \frac{D_{15(F)}}{D_{85(S)}} &\leq 4 \text{ to } 5 \\ \frac{D_{15(F)}}{D_{15(S)}} &\geq 4 \text{ to } 5 \end{aligned} \tag{4.5}$$

$D_{15(F)}$ diameter through which 15 % of filter material will pass by weight
$D_{15(S)}$ diameter through which 15 % of soil to be protected will pass by weight
$D_{85(S)}$ diameter through which 85 % of soil to be protected will pass by weight

The U.S. Navy (1971) requires the following conditions to be fulfilled:

$$\begin{aligned} &\frac{D_{15(F)}}{D_{85(S)}} < 5 \\ &\frac{D_{50(F)}}{D_{50(S)}} < 25 \\ &\frac{D_{15(F)}}{D_{15(S)}} < 20 \end{aligned} \tag{4.6}$$

If the uniformity coefficient C_u of the protected soil is less than 1.5, $D_{15(F)}/D_{85(S)}$ may be increased to 6. Also, if C_u of the protected soil is greater than 4, $D_{15(F)}/D_{15(S)}$ may be increased to 40.

$$\frac{D_{15(F)}}{D_{15(S)}} > 4 \tag{4.7}$$

- The maximum particle size of the filter = 76.2 mm (3 in.) to avoid segregation
- The filter should have no more than 5 % by weight of the particles passing a No. 200 sieve (0.074 mm) to avoid internal movement of fines in the filter
- For perforated drainage pipes

$$\begin{aligned} &\frac{D_{85(F)}}{slot\ \ width} > 1.2\ \ to\ \ 1.4 \\ &\frac{D_{85(F)}}{hole\ \ diameter} > 1\ \ to\ \ 1.2 \end{aligned} \tag{4.8}$$

The thickness of granular filters should be not only sufficient to retain the particles of the protected soil and conduct out flow water but be possible to construct with available equipment to prescribed construction tolerances.

Task Force 25 (1991) recommended the following criteria for geotextile protection against erosion:

$$\begin{aligned} &O_{95} < 0.6\,\text{mm},\ for\ \ No.200\ \ sieve\ \ passing \leq 50\,\% \\ &O_{95} < 0.3\,\text{mm}\ \ for\ \ No.200\ \ sieve\ \ passing > 50\,\% \end{aligned} \tag{4.9}$$

O_{95} is the 95 % opening size of the geotextile

The apparent opening size is determined by dry-sieving method (U.S.) i.e. the filtration opening size is determined by wet sieving in Europe and hydrodynamic sieving in Canada. The wet and hydrodynamic sieving is preferable to dry sieving (Koerner 1998).

Luettich et al. (1992) provided flow charts for geotextile filter design for steady-state and dynamic flow conditions.

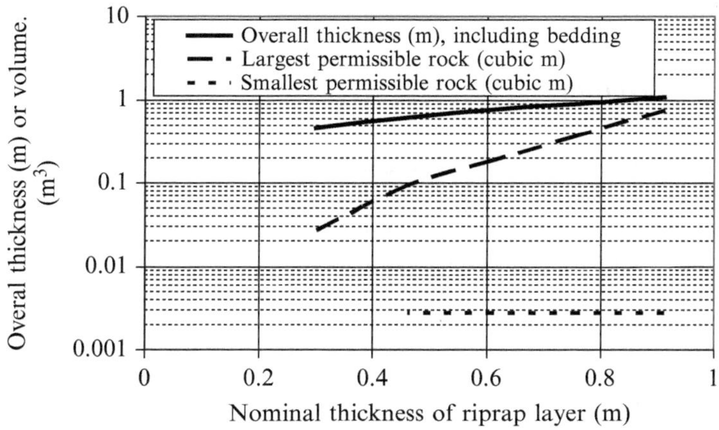

Fig. 4.5 Permissible rock sizes for various thicknesses of riprap (From Table 6 in Earth Manual 1980)

4.2.4.2 Concentrated Erosion (Piping)

- For coarse grained soil: increase its density, mix it to eliminate gap gradation, decrease the hydraulic gradient within soil by using lengthening of seepage path by barriers or interception of the seepage path by relief wells (Chap. 6), use filters as for internal erosion (suffosion).
- For fine grained soil (clay) Sherard et al. (1977) stated that many homogeneous dams without filters, in which dispersive clay has been properly compacted (at 2 % above its optimum moisture content to inhibit shrinkage and cracking), experienced no leaks and failures. Alternatively, hydrated lime (4 %), pulverized fly ash (6 %), gypsum (in soil mixture or in reservoir water) and aluminium sulphate (0.6 %) have been used to treat dispersive clay used in earth dam (Bell and Culshaw 1998). Care must be taken that such mixing does not cause undesirable brittleness (which can lead to development of shrinkage cracks) and swelling pressure.

4.2.4.3 Surface Erosion

It is usually prevented by vegetating slopes whenever feasible or by placing various covers (from geomembranes to rip-rap). More information on geomembranes is provided in Chap. 6. Permissible rock sizes for various thicknesses of rip-rap are shown in Fig. 4.5 from Table 6 in Earth Manual (1980). More detailed approach for sizing of rock armour of slopes is provided in CIRIA C683 (2007).

4.3 Liquefaction

4.3.1 Description

Soil liquefaction occurs when submerged loose and medium dense coarse grained soil is subjected to rapid static and cyclic loading with prevented dissipation of excess pore water pressure generated in response to imposed loading and when extremely sensitive saturated clay is disturbed by excavation, vibration or rapid loading. Clay and fine silt can suffer a loss of strength/stiffness during earthquakes termed ‘cyclic mobility’. Obermeier (1996) listed causes of static liquefaction of coarse grained soil:

- Rapid sedimentation and loading (by placing fill hydraulically and by rapidly moving landslides)
- Artesian pressures
- Slumping
- Chemical weathering
- Periglacial environment

Known causes of liquefaction of coarse grained soil induced by cyclic loading are:

- Earthquakes
- Pile driving
- Soil vibratory compaction (e.g. Ekstrom and Olofsson 1985)
- Blasting for demolition and excavation and for compaction by explosives
- Conduction of geophysical survey (e.g. Hryciw et al. 1990)

Causes of liquefaction of extremely sensitive saturated clay may be:

- Leaching of salty water and its replacement by fresh water from rainfall (in Scandinavia: Rankka et al. 2004, in Japan: Ehgashira and Ohtsubo 1982)
- Loss of natural cementation (in Canada: Geertsema and Torrance 2005; Crawford 1968)

More information on sensitive clay is provided by Mitchell and Houston (1969).

Consequences of soil liquefaction could be minor such as local ground cracking and subsidence to major such as flow type failures of inclined ground and level ground near soil slopes (Olson 2001), sinking of shallow foundations (Liu and Dobry 1997), uplifting of shafts and conduits, pile bending, quay wall settlement and tilting and structural damage or collapse (e.g. Ishihara 1993; Hamada and O’Rourke 1992).

4.3.2 Extent

- The most frequent and extensive is the liquefaction caused by earthquakes. The largest known depth of coarse grained soil liquefaction during earthquakes is 20 m (e.g. EN 1998-5:2004(E)). If liquefaction occurred at larger depths then sills (intrusions) were formed within upper sandy layer out of sight. The ground surface does not liquefy during earthquakes because the shear stress is zero at the surface. Also, when ground water level is at some depth below the ground surface and the thickness of non-liquefied layer is less than 10 m then sand volcanoes can be formed facilitated by the presence of previous desiccation cracks, rotten tree roots and channels burrowed by animals. Liquefaction of sloping ground and level ground adjacent to slopes can cause flow type failures. In flow type failures, non-liquefied top layer can exert lateral pressure on the existing structures equal to the passive resistance.

- Liquefaction of extremely sensitive clay should be limited to marine clay uplifted and exposed to atmosphere at present or in the past. It should be mentioned that the current sea level was about 120 m lower than the present level during glaciations so that present day marine clay could have been located onshore in the past.

4.3.3 Identification

- Saturated clay with the sensitivity (Sect. 3.1.5) greater than 16 is susceptible to liquefaction.
- Bray et al. (2004) have suggested the following classification based on moisture content *w*, liquid limit *LL* and plasticity index *PI*:

  1. w/LL $\geq$0.85 and PI $\leq$0.12 then susceptible to liquefaction or 'cyclic mobility'
  2. w/LL $\geq$0.8 and 0.12$<$PI$<$0.20 then moderately susceptible to liquefaction or 'cyclic mobility'
  3. w/LL $<$0.8 and PI $\geq$0.20 then no liquefaction or 'cyclic mobility' but may undergo significant deformation if cyclic shear stress is greater than static undrained shear strength

- Liquefaction potential of coarse grained soil depends mainly on the particle size distribution, density, and the level of excitation. Liquefaction of pre Pleistocene deposits are rare as are such deposits near the ground surface. Particle size distributions of liquefiable coarse grained soil are shown in Fig. 4.6. Kokusho (2007) compiled available data on liquefied gravely soil, Fig. 4.7. The effects of density and excitation are combined using empirically determined boundaries shown in Fig. 4.8 from Seed et al. (1985) and EN 1998-5:2004(E). $(N_1)_{60}$ in Fig. 4.8 is measured standard penetration tests (SPT) blow count N normalised to an effective overburden pressure of 100 kPa according to Liao and Whitman (1986a) and corrected to an energy ratio of 60 % (the average ratio of the average energy E_m delivered by hammer to the theoretical free-fall energy E_{ff}).

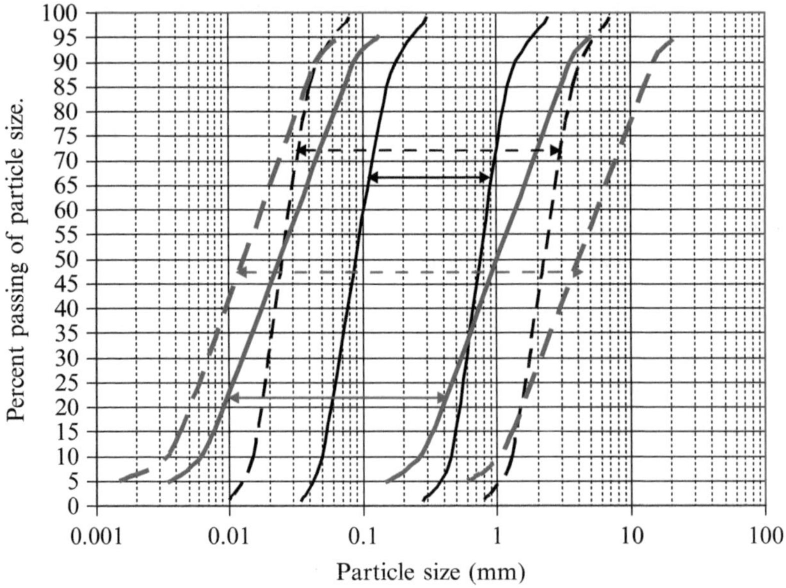

Fig. 4.6 Particle size distribution ranges of coarse grained soil susceptible to liquefaction depending on its density and the amount of excitation (Adapted from design standards for port and harbour structures of Japan, 1971); *black* – soil with low coefficient of uniformity, *grey* – soil with large coefficient of uniformity, *solid line* – high possibility of liquefaction, *dashed line* – possibility of liquefaction

$$(N_1)_{60} = N_{SPT} \cdot \sqrt{\frac{100}{\sigma_v^{'}} \cdot \frac{E_m}{0.6 \cdot E_{ff}}}, \sigma_v^{'} \; is \; in \; \text{kPa}$$
$$0.5 < \sqrt{\frac{100}{\sigma_v^{'}}} < 2 \;\; (\text{EN} \;\; 1998 - 5 : 2004(\text{E})) \tag{4.10}$$
$$N_{SPT}^{'} = 0.75 \div N_{SPT} \;\; at \; depths \leq 3\,\text{m} \;\; (\text{EN} \;\; 1998 - 5 : 2004(\text{E}))$$

The basic cyclic stress ratio (CSR) for earthquake magnitude of 7.5, effective overburden stress of 100 kPa and level ground is:

$$\frac{\tau}{\sigma_v^{'}} = 0.65 \cdot \frac{a_{\max}}{g} \cdot \frac{\sigma_v}{\sigma_v^{'}} \cdot r_d \tag{4.11}$$

τ is shear stress at a depth where the overburden stress is acting
σ'_v and σ_v are the effective and total overburden stress respectively
a_{max} is the peak horizontal acceleration at the ground surface
g is the gravitational acceleration
r_d is stress reduction coefficient with depth

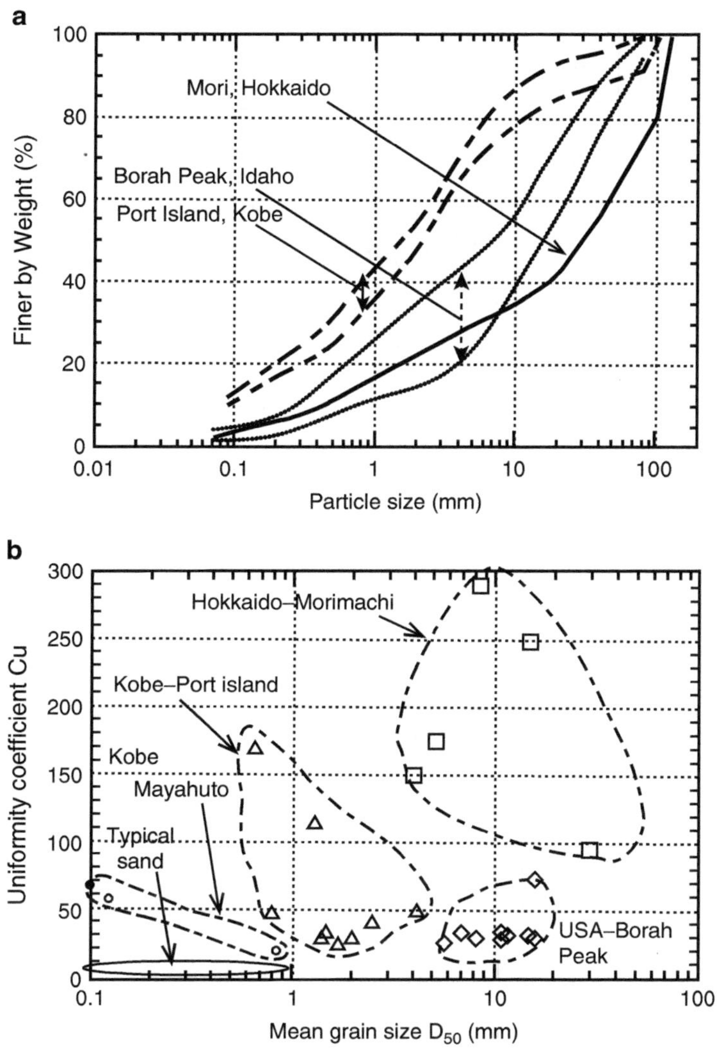

Fig. 4.7 (**a**) Typical particle sizes of liquefied gravely soil, (**b**) mean grain size versus uniformity coefficient (Kokusho 2007)

Liao and Whitman (1986b) proposed an averaged value of r_d as

$$r_d = 1 - 0.00765 \cdot z, \quad z \leq 9.15\,\text{m}$$
$$r_d = 1.174 - 0.0267 \cdot z, \quad 9.15\,\text{m} \leq z \leq 23\,\text{m} \tag{4.12}$$

Seed and Idriss (1971) proposed similar average value as well as a range of r_d for different soil profiles. EN 1998-5:2004(E) considers $r_d = 1$.

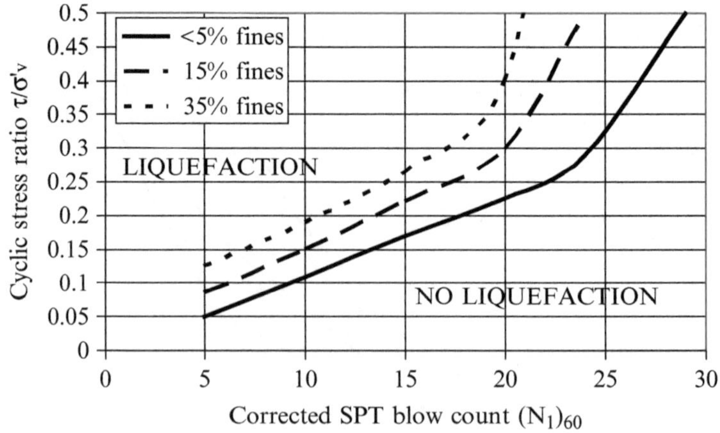

Fig. 4.8 Boundaries between liquefied and non-liquefied soil with different fines contents for earthquake magnitude 7.5

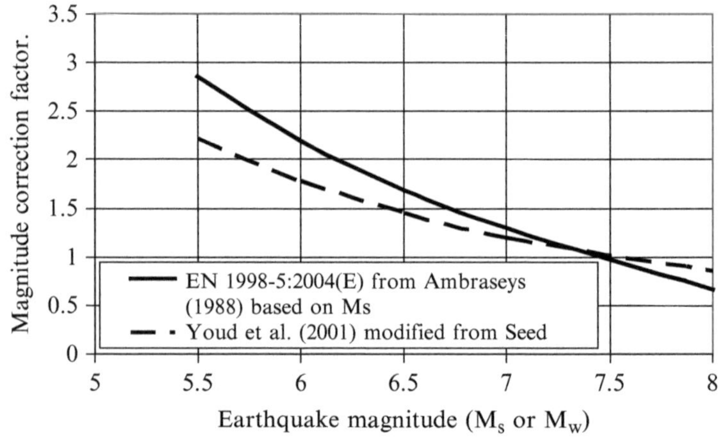

Fig. 4.9 Factor for the cyclic stress ratio division depending on earthquake magnitude (M_s – surface wave magnitude, M_w – moment magnitude)

The scaling coefficients of CSR for different earthquake magnitudes are provided by EN 1998-5:2004(E) from Ambraseys (1988) and by Youd et al. (2001) modified from Idriss (1990), Fig. 4.9.

Youd et al. (2001) proposed similar procedure for the assessment of liquefaction potential based on cone penetration test results and shear wave velocity measurements.

Harder and Boulanger (1997) noted that a wide range of sloping ground correction coefficient of the CSR have been proposed. Youd et al. (2001) recommended that the published correction coefficients for sloping ground should not be used by non-specialists. Srbulov (2010b) proposed that the CSR of sloping ground is determined using the sliding block concept (New-mark 1965) as applied to any slip surface by Ambraseys and Menu (1988). The basic equation is:

$$\frac{\tau}{\sigma'_v} = \frac{a_h}{g} \cdot \cos\alpha \cdot \left(\frac{\tan\varphi}{F_s} \cdot \sin\alpha + \frac{\sum N}{\sum N'} \cdot \cos\varphi \right) \tag{4.13}$$

τ is the shear stress acting on the slip surface of an equivalent sliding block
σ'_v is the effective axial stress acting on the slip surface of an equivalent sliding block
a_h is the horizontal acceleration acting above considered slip surface
α is inclination to the horizontal of the slip surface of an equivalent sliding block
g is the gravitational acceleration
φ is an equivalent friction angle at the base of the sliding block = arctan $[(Fs * \Sigma T)/\Sigma N']$
F_s is the factor of safety of slope stability (=1 when the critical horizontal acceleration is applied) and is usually calculated using the limit equilibrium method
$\Sigma N/\Sigma N'$ is the ratio between total and effective resultant forces acting in perpendicular direction to the slip surface considered from the equivalent sliding block approach. The ratio is proportional to the $\Sigma W/\Sigma W'$ where ΣW and $\Sigma W'$ are the total and effective weight of soil above the slip surface considered
ΣT is the resultant of forces acting along a slip surface considered, Fig. 5.12.

The correction coefficients for the effective stress different from 100 kPa (Eq. 4.10) of SPT averaged blow counts N along the slip surface considered and for the earthquake magnitude different from 7.5 (Fig. 4.9) are applied to Eq. (4.13) when it is used in connection with the graph shown in Fig. 4.8.

4.3.4 Remediation

Kramer (1996) summarise common techniques used:

- Vibro compaction by vibroflot (pendulum) and vibro rod is most effective when the fines content is less than 20 %
- Dynamic compaction by dropping a heavy weight (up 1,500 kN) from height up to 40 m is most effective when the fines content is less than 20 % and to depths less than 12 m.
- Blasting at greater depths is most effective in dry soil and when the fines content is less than 20 %

- Compaction grouting is used when vibration is not allowed and space is limited and is most effective when the fines content is less than 20 %
- Installation of stone columns that increases density, drainability and stiffness is most effective when the fines content is less than 20 %
- Installation of compaction piles that provide reinforcement and densification in combination with increased bearing capacity and limited settlement
- Mixing soil and binder (lime or cement) to depths over 60 m in Japan when the fines content exceeds 20 %
- Jet grouting under high water and air pressure
- Dewatering by lowering of ground water level
- Installation of gravel drains to dissipate excess pore water pressure when the installation of stone columns is not feasible

Other techniques include (PHRI 1997):

- Increase in the lateral capacity of the structures within/next to liquefied soil
- Preloading
- Installation of sand compaction piles

More information on compaction, mixing, grouting and drainage works is provided in Chap. 6.

4.4 Rock Dissolving and Caves

4.4.1 Description

Fast chemical weathering of rock such as:

- Halite (rock salt)
- Gypsum
- Chalk
- Limestone (and dolomite $MgCO_3$)

combined with ground water flow that removed the products of rock dissolving lead to creation and the existence of numerous cavities. Solution features formed by infill of ground cavities with soil are also hazardous to foundations and excavations.

The chemical formulas for rock dissolving are:

$$\begin{aligned}&for\ \ halite:\\&2NaCl + 2H_2O \rightarrow Cl_2 + H_2 + 2NaOH\\&for\ \ gypsum:\\&CaSO_4 \cdot 2H_2O \rightarrow Ca_2 + SO_4 + 2H_2O\\&for\ \ chalk\ \ and\ \ limestone:\\&CaCO_3 + H_2O + CO_2 \rightarrow Ca(HCO_3)_2\end{aligned} \tag{4.14}$$

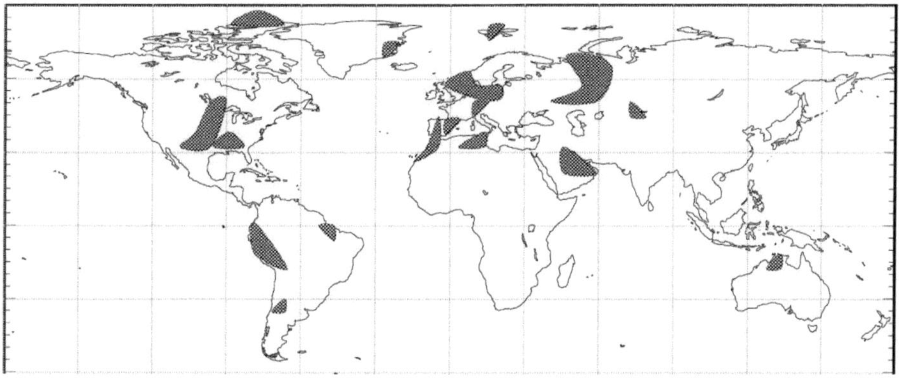

Fig. 4.10 Known major locations onshore of halite and sulphate sediments (gypsum) from various sources

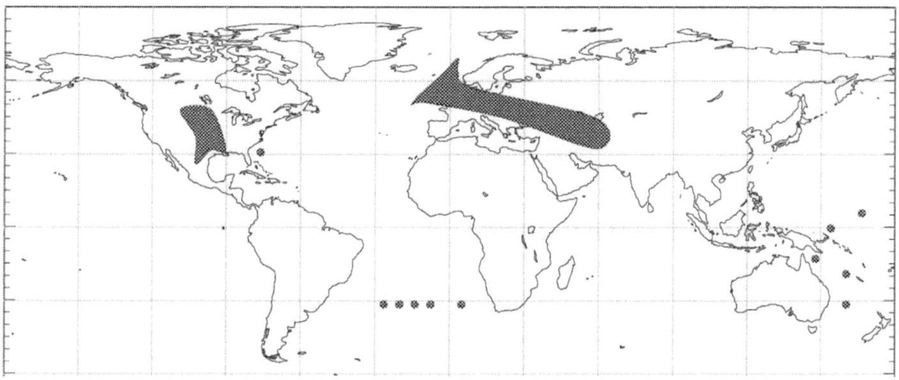

Fig. 4.11 Known major locations of chalk (Adopted from Mortimore 1990)

4.4.2 Extent

Known locations of major deposits of halite, gypsum, chalk and limestone are shown in Figs. 4.10, 4.11, and 4.12 inclusive.

4.4.3 Identification

- The locations of halite, gypsum, chalk and limestone are indicated on local geological maps.
- The locations of cavities are best detected using geophysical cross-hole, electromagnetic and resistivity profiling and ground penetrating radar according to Table 7 in BS 5930:1999+A2:2010. The ASTM designations for these tests are provided in Sect. 2.1.8.

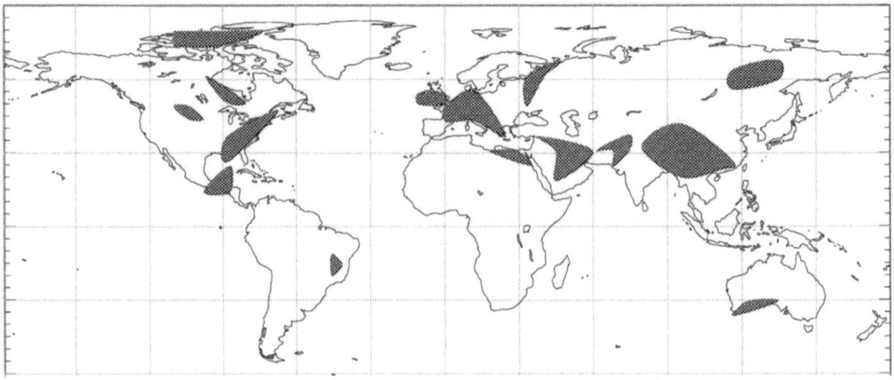

Fig. 4.12 Known major locations of limestone (& dolomite) (Adopted from Ford and Williams 1989)

4.4.4 Remediation

Large cavities are filled with grout and concrete. Smaller cavities can be spanned by geogrid reinforced fills, raft foundations and capping slabs.

4.5 Collapse of Soil Structure

4.5.1 Description

Known collapsible soil types on submergence by water, exposure to strong shaking (by earthquakes, ground vibration due to blasting in construction and mining or due to vibro equipment such as compressors and hammers) or on thawing are:

- Loess (wind blown fine sand and silt particles connected by calcium carbonate and/or moist clay)
- Residual soil in situ (the last stage of rock weathering)
- Compacted soil at moisture content much less than the water content on soil saturation
- Formerly frozen soil after thawing of ice because ice volume is about 9 % greater than the volume of water after thawing

4.5.2 Extent

Known locations of major loess deposits are shown in Fig. 4.13, of residual tropical soil in Fig. 4.14 and of frozen soil in Fig. 4.15.

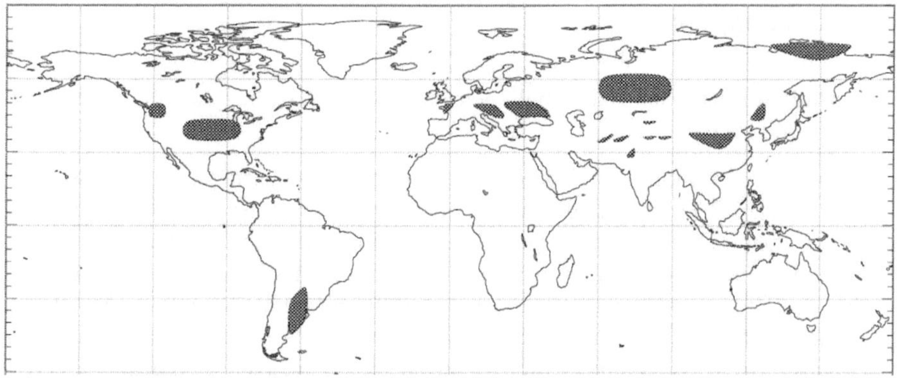

Fig. 4.13 Known locations of major loess deposits (Adopted from http://www.physicalgeography.net/fundamentals/10ah.html)

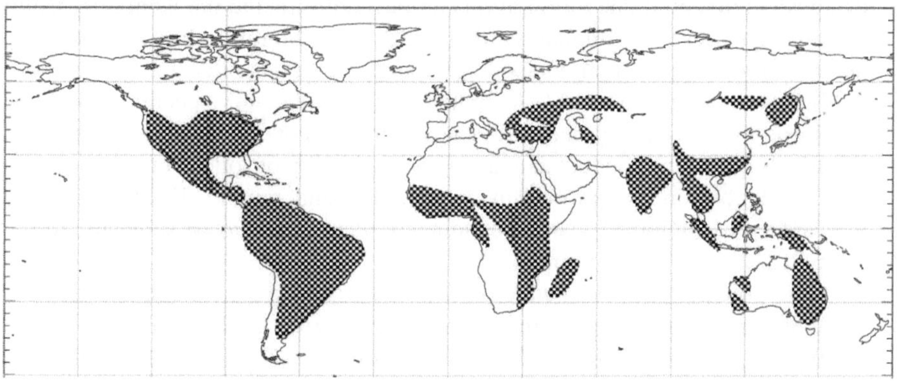

Fig. 4.14 Known major locations of residual tropical soil (Adopted from Fookes 1997)

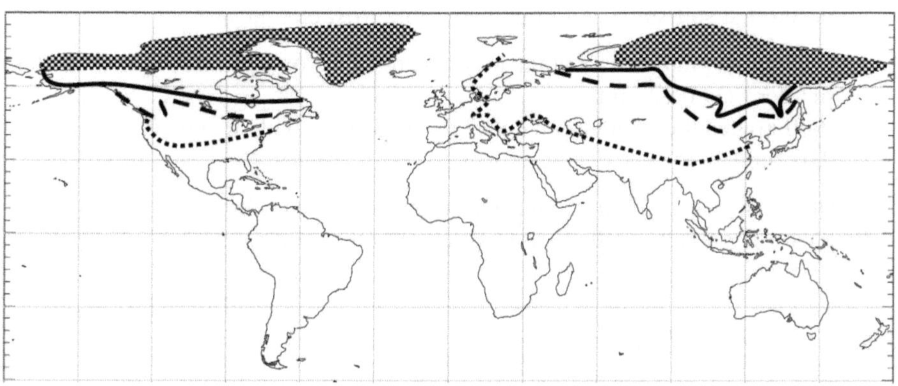

Fig. 4.15 Known locations of frozen ground: *shaded* – continuous permafrost several hundred metres thick, *solid line* – southern border of discontinuous permafrost up to several tens of metres thick, *dashed line* – southern border of sporadic permafrost a few metres thick, *dotted line* – southern limit of substantial frost penetration of a couple of metres (Adopted from Burdick et al. 1978)

4.5.3 Identification

Collapsible soil has usually low bulk density (below 16 kN/m^3) and high void ratio (above 1 and up to 2).

- Collapsible potential on inundation by water can be determined from oedometer tests (Sect. 2.2.1) when undisturbed soil specimens can be obtained. Figure 2.9 shows that collapsible soil can exhibit swelling at small compressive stress. When potential to collapse is measured in oedometer then it should be recognized that partially saturated soil can suck up water from saturated porous stones placed at the end of specimen and collapse quickly so that a part of its swellability or collapsibility would not be recorded if the specimen thickness change has not been monitored from the beginning. When undisturbed soil specimens cannot be obtained then field inundation tests can be conducted to assess soil collapsibility after its submergence in water.
- Collapsible potential due to vibration can be determined in the laboratory using shaking table test (to simulate earthquakes) and vibratory table test (used for determination of maximum index density and unit weight of coarse grained soil: ASTM D4253 and BS 1377-4) when undisturbed soil specimens can be obtained. When undisturbed soil specimens cannot be obtained then field blasting trials can be conducted to assess collapsibility of soil during earthquakes and due to blasting operations in construction and mining. Field tests using relevant vibro equipment can be conducted when soil is to be subjected to the vibrations from vibro equipment.
- Collapse potential due to thawing of formerly frozen ground can be determined in laboratory by performing freezing and thawing tests when undisturbed soil specimens can be obtained. When undisturbed soil specimens cannot be obtained then field investigations by digging trenches to sufficient depth may reveal the depth of highly porous soil or alternatively local knowledge may exist concerning the maximum depth of frost penetration. Casagrande (1932) proposed that soil subjected to unfavourable frost effects are:

    1. Uniformly graded if contain more than 10 % of particles with diameters smaller than 0.02 mm
    2. Well graded if contain more than 3 % of particles with diameters smaller than 0.02 mm.

    The existence of high ground water level facilitate formation of ice lenses by enabling greater capillary rise, which increases soil saturation with water.

4.5.4 Remediation

Koerner (1985) summarises the following methods for soil which collapsibility is caused by water saturation or vibration:

- Excavation and replacement (recompaction when possible), if shallow
- Wetting in situ, if thick

- Densification from the surface by vibratory rollers or by deep dynamic compaction, if shallow
- Deep densification by vibroflot or vibro rod, if deep
- Use of deep foundations (piles or caissons)

More information on geo-structures and works is provided in Chaps. 5 and 6. For remediation of existing shallow foundations, the following measure can be used:

- Addition of segmented pushed-in precast piles under shallow foundation or mini bored and cast in place piles on the sides of shallow foundation
- Compaction grouting or chemical grouting underneath shallow foundation
- Casting of a wide raft between existing strip foundations and their connection to the raft

For soil, which collapsibility is caused by thawing of frozen ground, lowering of foundation depth below the depth of frost penetration is widely practised. This is not practical for roads and other traffic areas for which the use of well compacted coarse grained and well drained sub-base prevents formation of large ice lenses, which thawing would cause the appearance of large voids and uneven road surfaces with pot holes.

4.6 Subsidence of Ground Surface

4.6.1 Description

Decrease in the level of existing ground surface within an area can be caused by:

- Lowering or rising of ground water level and thawing of frozen ground
- Shrinkage of clay on drying or extraction of moisture by tree roots
- Deep excavations and tunnelling
- Collapse of the openings that remained after mining including extraction of gas and oil
- Tectonic fault movement and ground slumping
- Biodegradation of more than 4 % of organic matter

4.6.2 Extent

4.6.2.1 Lowering or Rising of Ground Water Level and Thawing of Frozen Ground

- When ground water level is lowered, the unit weight of soil increases from submerged to moist causing an increase in effective overburden pressure, which causes additional settlement depending on the thickness and stiffness of soil later

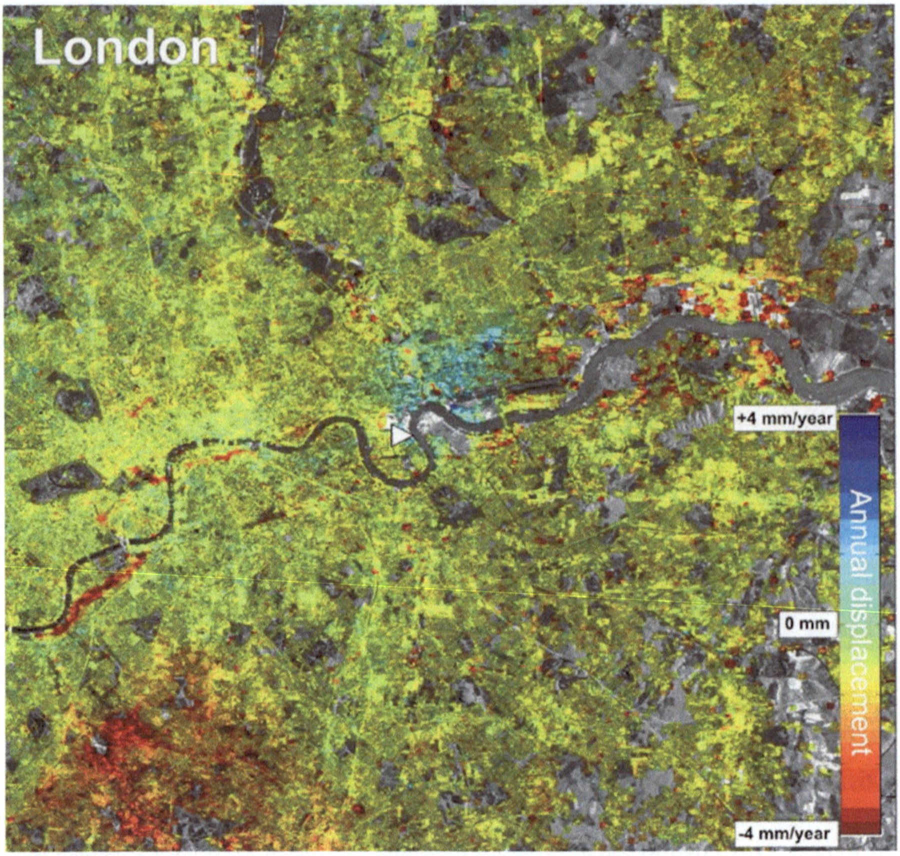

Fig. 4.16 Interpolated permanent scatterer InSAR (PSI) image of 900 km^2 area of London with *red* indicated subsidence rate due to ground water extraction by pumping and *blue* ground heave rate due to ground water rising after the ground water extraction ceased (Courtesy of Nigel Press Associates and Tele-Rilevamento Europa, the Society for Earthquake and Civil Engineering Dynamics Newsletter 20(1): 5–8, 2007)

over bedrock. Pumping of ground water in deep excavation can cause subsidence of structures kilometres away and for this reason water recharge wells are used when necessary.

- When ground water level rises, effective overburden stress decreases causing decrease in soil stiffness when it is dependent on effective stress level and in turn increases in ground subsidence.
- On thawing of frozen ground, the volume of water that remains after the melting of ice lenses decreases about 9 % in comparison with the volume of previous ice. Examples of the scale and the rate of ground surface subsidence are shown in Figs. 4.16 and 4.17.

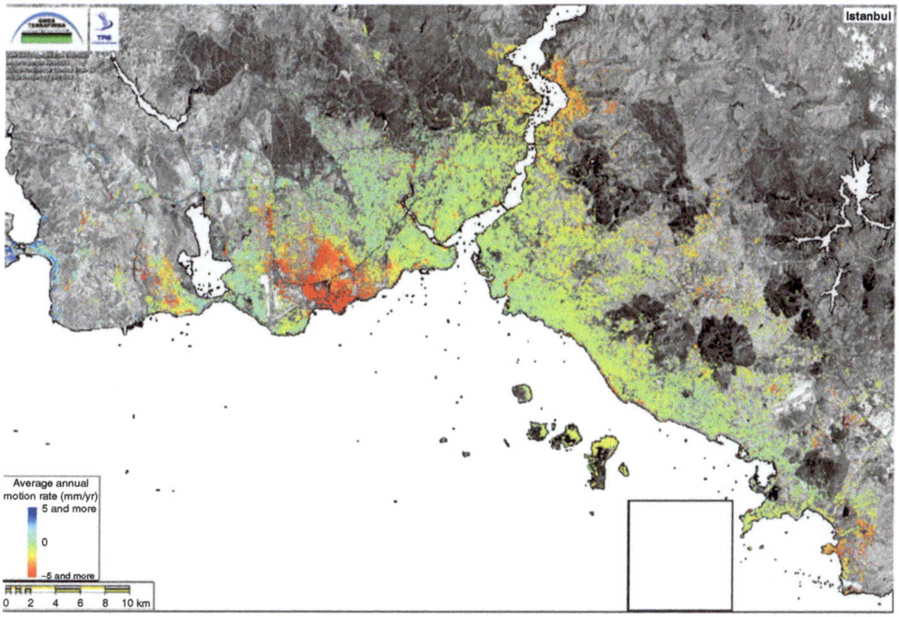

Fig. 4.17 Map of subsidence rate in Istanbul in Turkey (Courtesy of Tele-Rilevamento Europa and Terrafirma, Adopted from the Society for Earthquake and Civil Engineering Dynamics Newsletter 20(1): 5–8, 2007)

4.6.2.2 Shrinking of Clay on Drying or Extraction of Moisture by Tree Roots

- Clay shrinkage on drying can reach several percent so that ground surface subsidence due to clay shrinkage can amount to several centimetres depending on the thickness of clay soil layer undergoing shrinkage, clay plasticity index (PI) and percentage of clay particles sizes (CPS) smaller than 0.002 mm. For PI >35 % and CPS >95 % of all particle sizes by weight, the shrinkage potential is very high. For PI <18 % and CPS <30 % of all particle sizes by weight, the shrinkage potential is low (Driscoll 1984).
- Roots of threes extract moisture from soil by increasing suction. Increased suction causes increase in tensile forces applied by water menisci on soil skeleton and consequently its decrease in volume, which is manifested in vertical direction as subsidence. Table 4.2 lists various tree species, the maximum tree heights and the maximum distances of influence recorded according to Driscoll (1984).

4.6.2.3 Deep Excavations and Tunnelling

Extraction of soil at depths influences soil tendency to fill the void which in turns results in formation of a depression at the ground surface. Mair and Taylor (1997)

Table 4.2 Tree species damage ranking (Adopted from Driscoll 1984)

Ranking	Species	Max. tree height (m)	Max. distance recorded (m)
1	Oak	16–23	30
2	Poplar	25	30
3	Lime	16–24	20
4	Common ash	23	21
5	Plane	25–30	15
6	Willow	15	40
7	Elm	20–25	25
8	Hawthorn	10	11.5
9	Maple/sycamore	17–24	20
10	Cherry/plum	8	11
11	Beech	20	15
12	Birch	12–14	10
13	White beam/rowan	8–12	11
14	Cupressus macrocarpa	18–25	20

summarised findings of several authors that the transverse settlement trough immediately following tunnel construction is well described by:

$$S_v = S_{\max} \cdot e^{\frac{-y^2}{2 \cdot i^2}} \tag{4.15}$$

S_v is settlement
S_{max} is the maximum settlement on the tunnel centre-line
y is the horizontal distance from the tunnel centre-line
i is the horizontal distance from the tunnel centre-line to the point of inflection of the settlement trough

The volume of the surface settlement trough (per metre length of tunnel) V_s is

$$V_s = \sqrt{2 \cdot \pi} \cdot i \cdot S_{\max} \tag{4.16}$$

Various investigators found that for tunnels in clay

$$i = (0.5 \pm 0.1) \cdot z_o \tag{4.17}$$

z_o is depth of tunnel axis below ground surface

For tunnels in sand

$$i = (0.35 \pm 0.1) \cdot z_o \tag{4.18}$$

For a two layer case

$$i = K_1 \cdot z_1 + K_2 \cdot z_2 \tag{4.19}$$

K_1 is the trough width factor (0.5 ± 0.1) or (0.35 ± 0.1) for the soil type in layer 1 of thickness z_1

K_2 is the trough width factor (0.5 ± 0.1) or (0.35 ± 0.1) for the soil type in layer 2 of thickness z_2

At a depth z below the ground surface, above a tunnel axis at depth z_o, the trough width parameter i is according to Mair et al. (1993) for tunnels in clay

$$i = \frac{0.175 + 0.325 \cdot (1 - z/z_o)}{1 - z/z_o} \cdot (z_o - z) \tag{4.20}$$

Similar relationships are observed for tunnels in sand. Besides i, V_s is necessary for determination of S_{max} from Eq. (4.16). Mair and Taylor (1997) summarised the results by many authors for V_s in percentage of the cross sectional area of tunnel opening:

1. V_s = 1–2 % in stiff clay
2. V_s = 0.5–1.5 % in London Clay for construction with sprayed concrete linings (NATM)
3. V_s = 0.5 % in sand for closed face tunnelling, using earth pressure balance or slurry shield machine. V_s = 1–2 % even in soft clay excluding consolidation settlement.
4. For two or more tunnels, common assumption is that the ground movements are superimposed for each individual tunnel acting independently. However, for tunnels in close proximity, the second tunnel exhibited approximately double the volume loss of the first tunnel.

Attewell (1978) and O'Reilly and New (1982) proposed that, for tunnels in clay, ground displacement vectors are directed towards the tunnel axis so that the horizontal displacements S_h are calculated as

$$S_h = \frac{y}{z_o} \cdot S_v \tag{4.21}$$

y is the horizontal distance from the tunnel centre-line

z_o is depth of tunnel axis from the ground surface

S_v is settlement (Eq. 4.15)

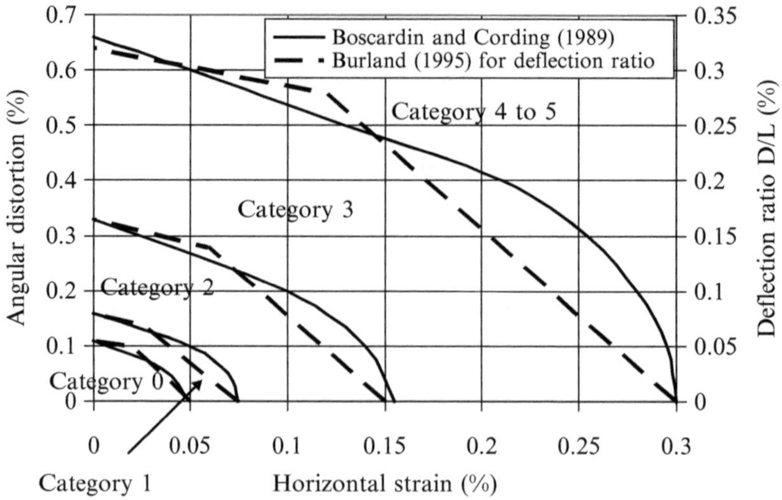

Fig. 4.18 Building damage categories related to horizontal strain, angular distortion and deflection ratio in hogging mode

This assumption leads to the distribution of surface horizontal ground movement as

$$S_h = 0.61 \cdot K \cdot S_{\max} \cdot 1.65 \cdot \frac{y}{i} \cdot e^{\frac{-y^2}{2 \cdot i^2}} \tag{4.22}$$

K is the trough width factor (Eqs. 4.17 and 4.18)
S_{max} is the maximum settlement as described above
y is the horizontal distance from the tunnel centre-line
i is the horizontal distance from the tunnel centre-line to the point of inflection of the settlement trough as described above for the settlement

From Eq. (4.22) it follows that the distribution of horizontal strain (as the first derivative of the horizontal displacement) is

$$\varepsilon_h = S_h \cdot \left(\frac{1}{y} - \frac{y}{i^2} \right) \tag{4.23}$$

S_h *is* surface horizontal ground movement
y is the horizontal distance from the tunnel centre-line
i is the horizontal distance from the tunnel centre-line to the point of inflection of the settlement trough as described above for the settlement

The values of the horizontal strain are use by Boscardin and Cording (1989) as well as Burland (1995) in combination with other parameters to define category of damage experienced by a building due to occurrence of the surface subsidence as shown in Fig. 4.18. Relationship between category of damage and tensile strain in buildings is given in Table 4.3. Angular distortion and deflection ratio Δ/L are indicated in Fig. 4.19.

Table 4.3 Relationship between category of damage and limiting tensile strain in buildings

Category of damage	Degree of severity	Limiting tensile strain (%)
0	Negligible	0–0.05
1	Very slight	0.05–0.075
2	Slight	0.075–0.15
3	Moderate	0.15–0.3
4–5	Severe to very severe	>0.3

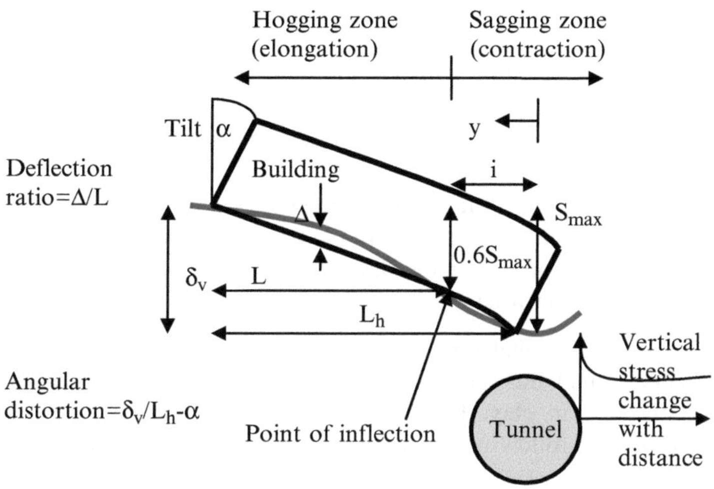

Fig. 4.19 Parameters used for subsidence and building damage definition

Similar approach is used for deep excavations as for tunnels. According to Gaba et al. (2003), S_{max} in stiff clay is 0.35 % of the maximum excavation depth for low stiffness walls and 0.1 % of the maximum excavation depth for high stiffness walls. In sand, $S_{max} = 0.3$ % of the maximum excavation depth. The maximum horizontal movement recorded is 0.4 % of the maximum excavation depth for low stiffness walls and 0.15 % of the maximum excavation depth for high stiffness wall. The maximum distance of excavation influence on ground surface subsidence is 4 times the maximum excavation depth in stiff clay and 2 times the maximum excavation depth in sand. Clough et al. (1989) provided a chart for the maximum lateral wall movement dependent on system stiffness and factor of safety against basal heave.

4.6.2.4 Collapse of the Openings That Remained After Mining Including Extraction of Gas and Oil

Ground subsidence at the beginning of mining and gas/oil extraction can be considered using the same method used for tunnelling. As mining last longer than construction of a tunnel, consolidation settlement will occur. Mair and Taylor (1997) summarised the following cases concerning long-term post construction

Table 4.4 Coefficients to calculate axial stresses at the edge of an opening

Shape							
A	5	4	3.2	3.1	3	2	1.9
B	2	1.5	2.3	2.7	3	5	1.9

settlements caused by tunnelling in clay, which are also applicable to mining and oil/gas extraction.

1. In soft clay when tunnelling with earth pressure balance machine shields when over-pressurization at the tunnel face happens or when tail void grouting pressures are high. These positive excess pore water pressures are generated within about one tunnel diameter as sketch in Fig. 4.19.
2. If a tunnel lining is permeable relative to the permeability of clay, the tunnel acts as a drain and the resulting consolidation settlements lead to a significantly wider surface subsidence trough than the short term trough associated with the volume loss during tunnel construction.

It should be mentioned that construction of an opening in ground induces the arching effects around the opening and increase in radial stresses with excess pore water pressure in clay. For elastic isotropic continua, Hoek and Brown (1980) suggested the following equations for the axial stress σ at the edge of an opening

$$\begin{aligned}\frac{\sigma_{side}}{\sigma_{vertical}} &= B - k\\ \frac{\sigma_{roof}}{\sigma_{vertical}} &= A \cdot k - 1\end{aligned} \tag{4.24}$$

σ_{side} is the vertical axial stress on the side of an opening
$\sigma_{vertical}$ is the vertical stress at the level of a tunnel axis before opening
B, A are constant given in Table 4.4
κ is the ratio between the horizontal and vertical stress before opening
σ_{roof} is the horizontal axial stress at the top of an opening

The biggest problem connected with collapse of openings underground is great speed of the events and creation of sharp edges along the rim. The propagation of collapsed ground towards the surface can be from nearly vertical to a conical shape with approximate inclination of the side at about two vertical to one horizontal.

4.6.2.5 Tectonic Fault Movement and Ground Slumping

- Normal (and oblique downward) moving tectonic faults, which exhibit lowering of ground level on one side of the fault surface, in extensional tectonic regions can create depressions of the order of centimetres to metres. Wells and Coppersmith (1994) provided the following equation for the maximum surface displacement from normal tectonic faults

$$D = 10^{0.89 \cdot M_w - 5.9 \pm 0.38 \cdot SD} \tag{4.25}$$

D is the maximum surface displacement due to normal tectonic fault break
M_w is the moment magnitude
SD is the number of standard deviations

$$\begin{aligned} M_w &= \frac{2}{3} \cdot \log_{10}(M_o) - 10.7 \\ M_o &= L_f \cdot W_f \cdot S_f \cdot \mu \end{aligned} \tag{4.26}$$

M_o is the seismic moment in Nm
L_f is the length of a tectonic fault in m
W_f is the width of a tectonic fault in m
S_f is the average slip (m) on a fault during an earthquake (which is typically about $5*10^{-5} * L_f$ for intraplate earthquakes, Scholz et al. 1986)
μ is shear modulus of the Earth's crust (which is usually taken as $3.3*10^{10}$ N/m^2)

- Slumping occurs in nearly level ground as a result of accelerated creep or gas and fluid escape (offshore called pockmarks). Bowl shaped depressions can be from a few centimetres to metres. Examples of the magnitudes and rates of accelerated creep, which causes also slope instability, are shown in Fig. 4.20 from Sing and Mitchell (1968).

4.6.2.6 Biodegradation of More Than 4 % of Organic Matter

This problem occurs when peat, solid organic waste (municipal) and rotting vegetation exist in significant amount.

4.6.3 *Identification*

- Small scale depressions of ground surface are identified by precise levelling up to 0.1 mm accuracy. Large scale depressions can be identified by both

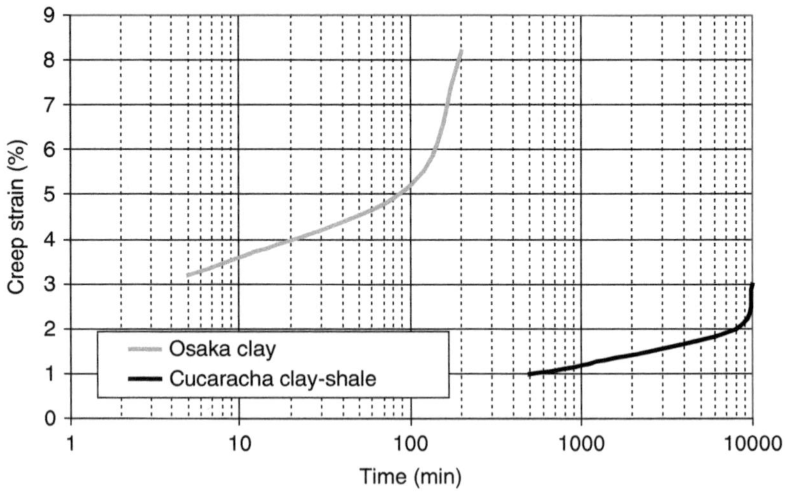

Fig. 4.20 Examples of accelerated creep strain magnitudes and rates

levelling and remote sensing method mentioned in Sect. 1.2. Locations of underground cavities can be identified using the methods mentioned in Sect. 4.4.3.

- Lowering of ground water level for deep excavations or water supply is usually subject of a permit by state environmental agency, which keeps records on ground water level change in time.
- Thawing of frozen ground occurs when it is covered by structures with internal heating, embankments, roads and other area covers.
- Shrinkage of clay on drying is accompanied by formation of cracks and the existence of trees at a site is evident from the site reconnaissance visit or from recent survey maps.
- Deep excavations and tunnels are marked on local topographical and transportation maps.
- The information on old mining conduits is held by local mining boards and industry.
- Normal tectonic fault scarps vary from mountain fronts thousands of metres high cut in bedrock to decimetre scale scarplets that displace Quaternary alluvium and colluvium (Stewart and Hancock 1990). Overlaps, step-overs and gaps are common in normal fault surface rupture. Range-front morphology can be controlled by factors other than tectonic, such as climate, lithology and tectonic structure (McCalpin 1996).
- Identification of amount of organic content in soil indicates how much biodegradation will affect ground subsidence.

4.6.4 Remediation

Avoidance of the causes of ground surface subsidence is the best method if possible. Other measures include:

- Use of water recharge wells behind ground water pumping wells
- Prevention of ice thawing by isolation of natural ground from heat sources
- Covering of clay surface to prevent clay drying and removal of trees although this could cause ground heave
- Use of controlled compensation grouting above tunnels and besides deep excavations
- Infill of deep openings against their collapse after cease of mining and injection of water into oil/gas fields
- Bridging of known tectonic faults by structures not very sensitive to subsidence (e.g. simply supported single spans)
- Use of deep foundations (piles) over known ground slumps and biodegradable organics in soil

More information on geo-structures and works are provided in Chaps. 5 and 6.

4.7 Heave of Soil and Rock

4.7.1 Description

Increase of the level of existing ground surface can be caused by:

- Formation of ice lenses during ground freezing
- Unloading or increase in moisture content of clay leading to softening and decrease of clay strength and stiffness
- Chemical reactions such as grow of gypsum crystals by oxidation of pyrite in shale rock and sulphate reaction in lime and cement stabilised soil
- Tectonic processes such as faulting and valley bulging

4.7.2 Extent

- Formation of ice lenses causes about 9 % increase in the volume occupied by previously unfrozen water, from which ice was formed
- Amount of clay heave depends on unloading amount or on clay plasticity and the initial moisture content before wetting. Known locations of expansive clays on wetting are shown in Fig. 4.21. Heave value because of unloading can be calculated from the expression used for calculation of plate settlement (EN 1997–2:2007(E))

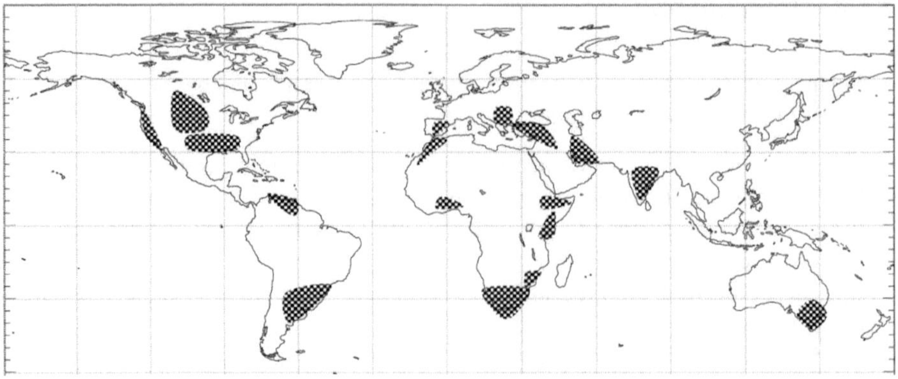

Fig. 4.21 Known locations where heaving occurs (Adopted from Donaldson 1969)

$$h = \frac{\Delta p}{E} \cdot \frac{\pi \cdot b}{4} \cdot \left(1 - \nu^2\right) \cdot C \tag{4.27}$$

h is the amount of heave
Δp is the decrease in vertical effective overburden pressure at a depth of excavation
E is soil modulus in recompression (can be taken equal to the maximum axial stiffness modulus from Eqs. (2.5), (2.11), (2.13), (2.14), (2.25), (2.40), (2.42) depending on ground type present at a location and on undrained or drained condition considered)
b is an equivalent diameter for the excavation area
ν is Poisson's ratio in undrained =0.49 or drained =0.25 condition considered
C is depth correction factor according to Burland (1969)

$$\begin{aligned}
&For\ \nu = 0.49\\
&C \approx 0.95 - 0.075 \cdot \sin\left(\frac{depth}{5 \cdot b} \cdot \frac{\pi}{2}\right)\ for\ \frac{depth}{b} \le 5\\
&C \approx 0.87\ for\ \frac{depth}{b} > 5\\
&For\ \nu = 0.25\\
&C = 0.95 - 0.125 \cdot \sin\left(\frac{depth}{5 \cdot b} \cdot \frac{\pi}{2}\right)\ for\ \frac{depth}{b} \le 5\\
&C \approx 0.81\ for\ \frac{depth}{b} > 5
\end{aligned} \tag{4.28}$$

- Heave induced by the chemical reactions involving sulphate is not frequent but is problematical when occurs

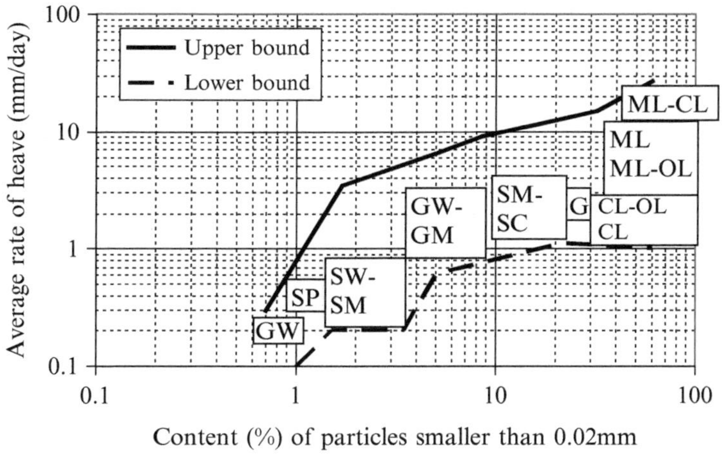

Fig. 4.22 Results of laboratory tests of heave on freezing of various soil types represented by their symbols (Adopted from Kaplar 1974)

- Reverse (thrust) and oblique upward moving tectonic faults, which exhibit rising of ground level on one side of the fault surface, in compressional tectonic regions can create grabens with their heights ranging from centimetres to metres.
- Valley bulging effects occur in soft rock such as shale and mudstones (claystone, siltstone, marl)

4.7.3 Identification

4.7.3.1 Ice Lenses Formation

Casagrande (1932) proposed that soil subjected to unfavourable frost effects (and growth of ice lenses) is:

1. Uniformly graded if containing more than 10 % of particles with diameters smaller than 0.02 mm
2. Well graded if containing more than 3 % of particles with diameters smaller than 0.02 mm.

The existence of high ground water level facilitates the formation of ice lenses by enabling greater capillary rise, which increases soil saturation with water.

A particular problem can represent high rates of soil heave on freezing as indicated in Fig. 4.22.

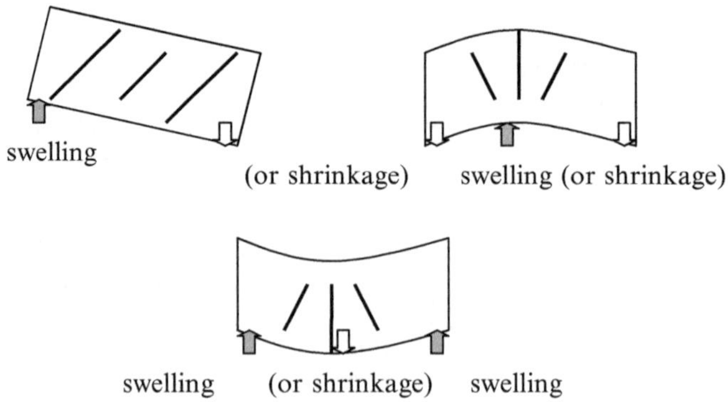

Fig. 4.23 Patterns of cracking induced by different locations of soil swelling (*grey arrows*) or shrinkage (*white arrows*) in walls of buildings

4.7.3.2 Clay Heave

Patterns of cracks induced by clay swelling (or shrinkage) in walls of building are shown in Fig. 4.23. Holtz and Kovacs (1981) provided a boundary between the susceptibility to collapse or expansion of soil:

$$\begin{aligned} \gamma_d &> 3,760 - 663 \cdot \ln_e(LL) \rightarrow expansion \\ \gamma_d &< 3,760 - 663 \cdot \ln_e(LL) \rightarrow collapse \end{aligned} \tag{4.29}$$

γ_d is in situ dry density (kg/m^3)
LL is liquid limit (%)

Seed et al. (1962) provided a chart for estimation of swelling potential of clay on wetting as shown in Fig. 4.24. Van Der Merwe (1964) provided an activity chart of soil shown in Fig. 4.24. Swelling potential and swelling pressure when swelling is prevented is best determined in laboratory using oedometer tests. When potential to swelling or swelling pressure are measured in oedometer then it should be recognized that partially saturated soil can suck up water from saturated porous stones placed at the end of specimen and swell quickly so that a part of its swellability or swelling pressure would not be recorded if the specimen thickness change has not been monitored from the beginning.

Soil moisture content changes throughout year and the largest swelling and swelling pressure on wetting occur when natural soil moisture content reaches its minimum during the driest season at a location.

4.7.3.3 Chemical Reactions Involving Sulphates

- Rollings et al. (1999) described the case of formation of ettringite (calcium sulfoaluminate). The calcium and alumina came from cement and the stabilized soil's clay minerals. The source of the sulphur was well water that was mixed

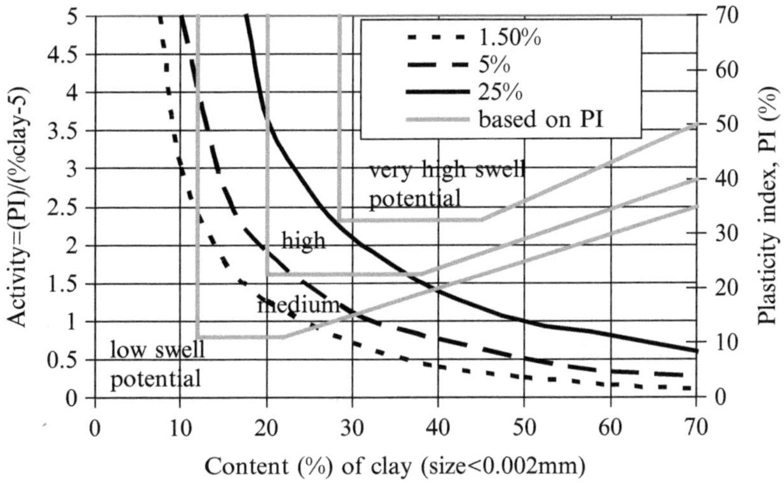

Fig. 4.24 Classification chart for swelling potential in per cent (After Seed et al. 1962 based on clay activity, and Van Der Merwe 1964 boundaries of swell ability based on plasticity index – PI)

with the cement stabilized base. Unexpected transverse bumps in the road formed within 6 months after the construction. Puppala et al. (2004) studied sulphate resistant cement stabilization methods to address sulphate-induced heave of pavements.

- Little et al. (2010) addressed sulphate induced heave in lime treated soil due to formation of ettringite/thaumasite minerals. The ettringe precipitation is expressed by the chemical formula

$$C_3A + 3CSH_2 + 26H_2O \rightarrow C_3A \cdot 3CSH_2 \cdot H_{26} \tag{4.30}$$

C_3A is tricalcium aluminate
CSH_2 is gypsum
H_2O is water

- Grattan-Bellew and Eden (1975) described the case of concrete deterioration and floor heave due to biogeochemical weathering of underlying shale. Oxidation of pyrite on weathering of underlying shale rock caused formation of gypsum, which crystal growth caused the heave of a floor. Jarosite mineral ($HFe_3(SO_4)_2(OH)_6$) was also found as the product of oxidation reactions from pyrite. Some oxidation reactions occur only in the presence of sulphur bacteria. The cementation portion of the concrete floor was leached out by acid. The chemical formula for oxidation of pyrite to sulphuric acid is (Lungren and Silver 1980) is

$$2FeS_2 + 2H_2O + 7O_2 \rightarrow 2FeSo_4 + 2H_2SO_4 \tag{4.31}$$

Sulphuric acid reaction with calcite produces hydrated gypsum (Grattan-Bellew and Eden 1975)

$$CaCO_3 + H_2SO_4 + H_2O \rightarrow CaSO_4 \cdot 2H_2O + CO_2 \quad (4.32)$$

$CaCO_3$ is calcium carbonate
H_2SO_4 is sulphuric acid
H_2O is water
$CaSO_4.2H_2O$ is hydrated gypsum with bounded two molecules of water
CO_2 is carbon dioxide

- Pye and Miller (1990) listed reported instances of heave of buildings on pyritic shale or compacted shale fill in Canada, France, Norway, Sweden and the United States.
- Steel slag and older blast furnace slugs can exhibit volume expansion of 10 % or more on exposure to moisture, caused by hydration of free calcium and magnesium oxides in steel slag or because of formation of ettringate and thaumasite in older blast furnace slugs (Charles 1993)

4.7.3.4 Tectonic Faults and Valley Bulging

- Reverse (thrust) and oblique upward moving tectonic faults, which exhibit rising of ground level on one side of the fault surface, in compressional tectonic regions can create depressions of the order of centimetres to metres. Wells and Coppersmith (1994) provided the following equation for the maximum surface displacement from reverse tectonic faults

$$D = 10^{0.29 \cdot M_w - 1.84 \pm 0.42 \cdot SD} \quad (4.33)$$

D is the maximum surface displacement due to reverse (thrust) and oblique upward moving tectonic fault break
M_w is the moment magnitude (Eq. 4.26)
SD is the number of standard deviations

- Valley bulging is popular term for grow of anticlines in valleys during present time. The induced movement is sketched in Fig. 4.25. Hutchinson and Coope (2002) describe a case of valley bulging in the U.K.

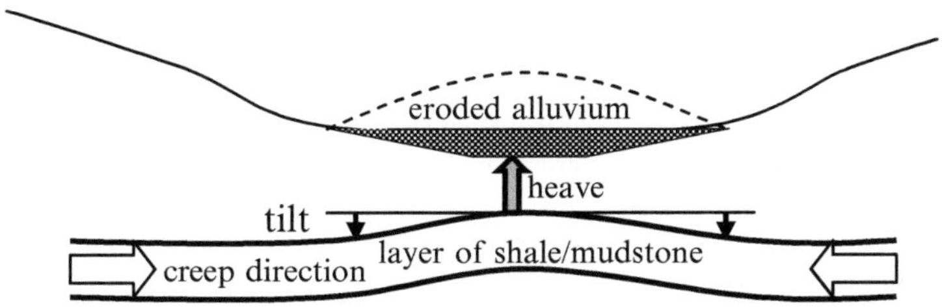

Fig. 4.25 Sketch of ground movements and tilt because of valley bulging in underlying shale or mudstone masked by overlying alluvium erosion process

4.7.4 Remediation

- Prevention of ground freezing and formation of ice lenses by covering of ground surfaces or detachment of structures from freezing ground using suspended floors above ground and on piles or on very rigid foundation beams
- Use of tensioned piles (permanent) or anchors (temporary) against heave induced by unloading of clay
- Prevention of increase of moisture content in swellable clay or detachment of structures from swelling ground using suspended floors above ground and on piles or on very rigid foundation beams
- Avoidance of sulphate reactions
- Using of structures not very sensitive to tectonic movements (e.g. simply supported single spans).

More information on geo-structures and works are provided in Chaps. 5 and 6.

4.8 Slope Instability

4.8.1 Definition

According to type and speed of movement, the following slope instabilities exist (most of them mentioned by Varnes 1978, 1984):

- Toppling (mostly rock)
- Falls (of soil or rock blocks)
- Avalanches (of soil or rock masses)
- Flows or spreads (of liquefied soil or block debris)
- Slides (in soil or rock) of planar, circular, polygonal or wedge shape
- Slumps in soil
- Turbidity currents offshore

Falls, avalanches, flows or spreads and turbidity currents are chaotic in nature and difficult to analyse. They are also very fast.

Table 4.5 Types, frequency and minimum triggering magnitudes of earthquakes to cause slope failures

Type of slope failure	Frequency of occurrence during earthquakes	Minimum triggering earthquake magnitude M_L
Rock falls, disrupted soil slides	Very frequent	4.0
Rock slides		4.0
Soil lateral spreads	Frequent	5.0
Soil slumps, soil block slides		4.5
Soil avalanches		6.5
Soil falls	Moderately	4.0
Rapid soil flows, rock slumps	Frequent	5.0
Sub aqueous landslides	Rare	5.0
Slow earth flows		5.0
Rock block slides		5.0
Rock avalanches		6.0

4.8.2 Extent

The sizes vary from a few metres via mountain sides (e.g. Ambraseys and Bilham 2012, the Usoy slide with the volume of 2.4 km^3) and small regions offshore (e.g. Bugge et al. 1988, for the Storegga slide with the volume of 3,880 km^3). Slope failures can be caused by:

- Water saturation from rainfall or other sources
- Erosion at the toe
- Loading at the crest
- Weathering of rock and soil
- Tectonic movements
- Ground water level increase including water pressure rise (artesian or during liquefaction)
- Creep, freezing, swelling, collapse
- Ground vibration from machinery or blasting
- Earthquakes
- Volcanoes

Keefer (1984) studied the effects of 40 historical earthquakes on slope failures and the results of his finding are given in Table 4.5. Table 4.5 shows that earthquake magnitudes less than 4.0 are of no engineering significant and that slope failures occur more frequently in non-cohesive materials.

Keefer (1984) and Rodriguez et al. (1999) plotted the graph of the combinations of earthquake magnitudes and fault distances at which different types of landslides occurred. A combined graph is shown in Fig. 4.26 from which it follows that no slope failures are expected beyond 300 km epicentral distance even from earthquakes with magnitudes up to 8 and that disrupted landslides are more prone to earthquake effects than coherent landslides, spreads and flows. These types of slope movements are sketched in Fig. 4.27.

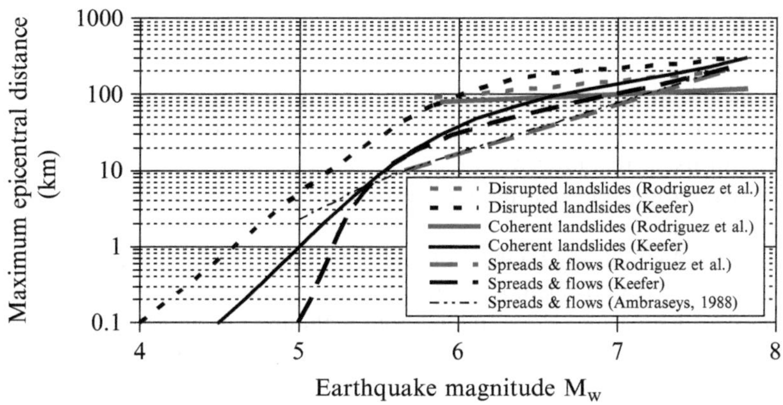

Fig. 4.26 The maximum epicentral distances to various slope failures dependent on earthquake moment magnitudes M_w

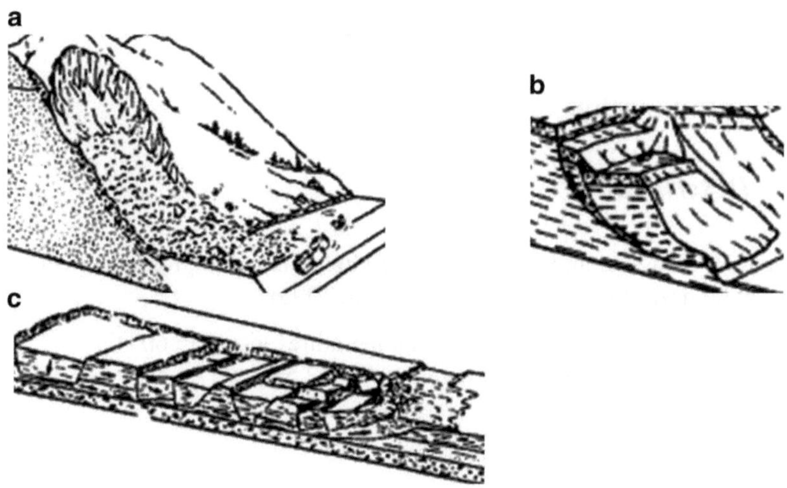

Fig. 4.27 Some types of slope failures (**a**) disrupted landslide, (**b**) coherent landslide, (**c**) lateral spread (Adapted from Varnes 1978)

4.8.3 Identification

Slope failures are identified by:

- Anomalous topography including arcuate or linear scarps, backward-rotated masses, benched or hummocky topography, bulging toes and ponded or deranged drainage
- Unusual vegetation type, age or position
- Discontinuous stratigraphy unless it has been caused by tectonic faulting

Calculations of factors of safety of slope stability and considerations of the effects of slope instabilities are provided in Chap. 5.

4.8.4 Remediation

- Toppling by eliminating driving forces (such as water pressure in open cracks, freezing of it, growth of tree roots), increasing resistance forces by anchoring, bolting or using retaining walls
- Falls by removing or securing in place of loose pieces or providing arrest fences to catch them
- Avalanches by building bunkers or retaining walls including dams
- Flow or spreads by channelling within conduits, building bunkers or retaining walls
- Slides by decreasing slope inclination including addition of toe berms or gabions, introducing drainage of ground water, anchoring or bolting, retaining walls including reinforced soil, soil replacement or soil mixing
- Slumps by soil mixing or bridging them with viaducts
- Turbidity currents offshore by deep buried cables or pipelines or covering them

More information on geo-structures and works are provided in Chaps. 5 and 6.

4.9 Contamination

4.9.1 Description

Commercial, industrial and agricultural activities can cause ground contamination, which can be hazardous (Tchobanoglous et al. 1993):

1. Flammable (e.g. liquids ignitable at temperatures less than 60 °C)
2. Corrosive (liquids having the range 3<pH>12.5)
3. Reactive (explosive)
4. Toxic

   - Non-metals: selenium – Se,
   - Metals: barium – Ba, cadmium – Cd, chromium – Cr, lead – Pb, mercury – Hg, silver – Ag,
   - Organic compounds: benzene (benzol) – C_6H_6, ethylbenzene (phenylethane) – $C_6H_5C_2H_5$, toluene (methylbenzene) – $C_6H_5CH_3$,
   - Halogenated compounds: chlorobenzene (phenylchloride) – C_6H_5Cl, chloroethene (vinyl chloride) – CH_2CHCl, dichloromethane (methylene chloride) – CH_2CL_2, tetrachloroethane (tetracholoroethylene, perchloroethylene)- CCl_2CCl_2,
   - Pesticides, herbicides, insecticides: endrin – $C_{12}H_8OCl_6$, lindane – $C_6H_6Cl_6$, methoxychlor – $Cl_3CCH(C_6H_4OCH_3)_2$, toxaphene – $C_{10}H_{10}Cl_8$, silvex – $Cl_3C_6H_2OCH(CH_3)COOH$

5. Carcinogenic

   - Non-metals: arsenic – As,

- Metals: cadmium –Cd, chromium- Cr, organic compounds: benzene (benzol) – C_6H_6,
- Halogenated compounds: dichloromethane (methylene chloride)-CH_2CL_2,
- Pesticides, herbicides, insecticides: endrin – $C_{12}H_8OCl_6$
- Radioactive elements

4.9.2 Extent

Maximum recommended/permissible concentrations/activity of contaminants in water (Rowe et al. 1997) and soil with occupational exposure limits (Barry 1991; Smith 1991) are given in Table 4.6. Ground gases are considered separately in Sect. 4.15.

4.9.3 Identification

Typical standard tests include (Cairney 1995):

- "Total elements": Arsenic, Boron, Cadmium, Chromium, Copper, Lead, Nickel, Zinc
- Anions: Chloride, Sulphate, Sulphide
- Other: pH, Phenols, Toluene extractables
- Possible additions: "Coal tars", Cyanides, "Mineral oils", Polyaromatic Hydrocarbons (PAHs)

BS 10175 (2011) provides more information on investigation of potentially contaminated sites. Individual tests are standardised in ASTM C1255-11; D4646-03; D5730-04; D5831-09; D7203-11; D7352-07; D7458-08; E1527; E2600-10.

4.9.4 Remediation

Types of available treatment with their advantages and disadvantages given in Table 4.7 based on Grasso (1993) include:

- Soil vapour extraction
- Chemical extraction and soil washing
- Solidification and stabilisation
- Chemical destruction
- Bioremediation
- Thermal process

Privett et al. (1996) described barriers, liners and cover systems for containment.

Excavation and redeposition of contaminated soil can be used for small quantities of not very toxic materials.

Table 4.6 Recommended/maximum concentration/activity of contaminants

Type	Recommended/maximum concentration/activity (according to drinking water standards)	ICRCL (1987) for soil	Occupational exposure limit
Inorganic (recommended)			
Total dissolved solids	500 mg/l	–	–
Chloride	250 mg/l	–	–
Copper	1 mg/l	–	–
Hydrogen sulphide	0.05 mg/l	–	–
Iron	0.3 mg/l	–	–
Manganese	0.05 mg/l	–	–
Sodium 200 mg/l	200 mg/l	–	–
Sulphate	250 mg/l	–	–
Zinc	5 mg/l	–	–
Organic (recommended)			
Acetone	1 mg/l	–	–
Phenols	0.002 mg/l	–	–
Toluene	0.024 mg/l	–	–
Inorganic (maximum permissible)			
Zinc	5 mg/l	300 mg/kg	$ZnCl_2$ 1 mg/m^3 (8 h), 2 mg/m^3 (10 min) ZnO 5 mg/m^3 (8 h), 10 mg/m^3 (10 min)
Antimony	10 μg/l	–	–
Arsenic	50 μg/l	10 mg/kg gardens, 40 mg/kg (parks, playing fields)	0.2 mg/m^3 (8 h)
Asbestos	–	Crocidolite and amosite 0.2 fibre/ml, chrysotile 0.5 fibre/ml	Blue>brown>white for mesothelioma development in 20–30 years
Barium	1 mg/l	–	–
Boron	5 mg/l	3 mg/kg (water-soluble)	Tribromide (BBr_3) 1 ppm (8 h), 3 ppm (10 min) Oxide (B_2O_3) 10 mg/m^3 (8 h) 20 mg/m^3 (10 min)

Cadmium	5 µg/l	3 mg/kg (gardens), 15 mg/kg (parks, playing fields, open spaces)	Compounds 0.05 mg/m^3(8 h), oxide fume 0.05 mg/m^3 (8 h and 10 min)
Chromium (Cr^{VI})	50 µg/l	600 mg/kg (gardens, allotments), 100 mg/kg (parks, playing fields, open spaces) except type VI 25 mg/kg (all uses)	0.5 mg/m^3 (8 h) except type VI 0.05 mg/kg (8 h)
Copper	–	130 mg/kg where plants are to be grown	As fume 0.2 mg/m^3 (8 h), as dusts and mists 1 mg/m^3 (8 h), 2 mg/m^3 (10 min)
Selenium	10 µg/l	3 mg/kg (gardens, allotments), 6 mg/kg (parks, playing fields, open spaces)	0.2 mg/m^3 (8 h)
Sulphur	–	Sulphate 2,000 mg/kg, Sulphide 250 mg/kg, Sulphur 5,000 mg/kg	–
Lead	10 µg/l	500 mg/kg (gardens, allotments), 2,000 mg/kg (parks, playing areas, open spaces)	0.15 mg/m^3 (8 h) except tetraethyl 0.1 mg/m^3 (8 h)
Mercury	1 µg/l	1 mg/kg (gardens, allotments), 20 mg/kg (parks, playing fields, open spaces)	0.05 mg/kg^3 (8 h), 0.15 mg/m^3 (10 min), as alkyl 0.01 mg/m^3 (8 h), 0.03 mg/m^3 (10 min),
Nickel	–	70 mg/kg	Insoluble compounds 1 mg/m^3 (8 h), 0.3 mg/m^3 (10 min), soluble compounds 0.1 mg/m^3 (8 h), 0.3 mg/m^3 (10 min)
Nitrate as nitrogen	10 mg/l	–	–
Silver	50 mg/l	–	–
Zinc	–	300 mg/kg	$ZnCl_2$ 1 mg/m^3 (8 h), 2 mg/m^3 (10 min) ZnO 5 mg/m^3 (8 h), 10 mg/m^3 (10 min)
Organic (maximum permissible)			
Coal tar	–	50 mg/kg (gardens, allotments, play areas), 1,000 mg/kg (landscaped areas, buildings, hard cover)	–
Cyanide	100 µg/l	Free CN 25 mg/kg (gardens, allotments, landscaped areas), 100 mg/kg (buildings, hard cover) Complex CN 250 mg/m^3, Thiocyanite 50 mg/kg	(except HCN, cyanogens, cyanogens chloride) 5 mg/m^3 (8 h)

(continued)

Table 4.6 (continued)

Type	Recommended/maximum concentration/activity (according to drinking water standards)	ICRCL (1987) for soil	Occupational exposure limit
DDT	30 μg/l	–	–
Endrine	0.2 μg/l	–	–
Lindane	4 μg/l	–	–
Methoxychlor	100 μg/l	–	–
Toxapene	2 μg/l	–	–
2,4-D	100 μg/l	–	–
2,4,5-TP silvex	10 μg/l	–	–
Polyaromatic hydrocarbons (PAHs)	–	50 mg/kg (gardens, allotments, play areas), 1,000 mg/kg (landscaped areas, buildings, hard cover)	–
Phenols	–	5 mg/kg (gardens, allotments, landscaped areas, buildings, hard cover)	5 ppm (8 h), 1 ppm (10 min)
Polychlorinated biphenyls (PCBs)	–	–	$C_{12}H_7Cl_3$ (42 % Cl) 1 mg/m^3 (8 h), 2 mg/m^3 (10 min), $C_6H_2Cl_3$ (54 % Cl) 0.5 mg/m^3 (8 h), 1 mg/m^3 (10 min)
Synthetic detergents	500 μg/l	–	–
1,2-Dichloroethane	5 μg/l	–	–
1,4-Dichlorobenzine	5 μg/l	–	–
1,4-Dioxane	412 μg/l	–	–
Dichloromethane (methylene chloride)	50 μg/l	–	–
Tetrachloroethylene	500 μg/l	–	–

Trichloroethylene	50 µg/l	–	–
Trichloromethane (chloroform)	100 µg/l	–	–
Xylenes	300 µg/l	–	–
Vinyl chloride	2 µg/l	–	–
Radionuclides and radioactivity			
Radium 226	1pCi/l	–	ICRP (1977)
Strontium 90	8pCi/l	–	50 millisievert/year (workers), 1 millisievert/year (public, mean annual), 5 millisievert/year (public, short periods),
Gross beta activity	1000pCi/l	–	–
Gross alpha activity	15pCi/l	–	–

Note: *ppm* parts per million

Table 4.7 Contaminated land remedial measures

Use	Advantages	Disadvantages
Soil vapour extraction		
Contaminants vapour pressure >0.5 mm Hg	Requires few mechanical parts	Contaminants with low vapour pressure not removed efficiently
	Low operating and maintenance cost	
Soil excavation is impossible	Employs conventional equipment, labour and materials	Additional treatment may be needed
		Vapour are impeded b fine grained soil
	Can be installed rapidly	Resulting vapour pollute air and usually require treatment
Large volume of contaminated soil need treatment	Shows immediate results	
	Does not require reagents	Soil must be permeable and fairly homogeneous for the treatment to be effective
	Provides permanent remediation	
		Toxicity of contaminant is not altered
Soil air permeability 10^{-2} to 10^{-5} cm/s	Not limited to depth to ground water	Cleanup times may be unpredictable due to complexity of soil strata
	No excavation required	
Biological degradation of the contaminant may need enhancement	Not disruptive to business or residents	
	May be used beneath buildings	
	May enhance biodegradation or bioventing	
Chemical extraction/soil washing		
A low soil/liquid partitioning coefficient can be induced through the use of additives and when diffusive or mass transport considerations are not limiting	Clean-up level of <1 ppm achieve for some contaminants over a wide range of soil types and influent contaminant concentrations	Contaminant not destroyed
		Fines and washing fluids require secondary treatmentor must be disposed of as a hazardous waste
	Acceptable to public	
A significant volume of reduction can be achieved through soil gradation	Relatively low cost	Sludges from waste water treatment must be treated
	Can utilize a closed treatment system	Chemical specific
	Can effectively pretreat soil for bioremediation	Large organic fraction may require use of additional pre-treatment steps
Solidification/stabilisation		
Soil is contaminated with heavy metals or other inorganics	Can immobilize certain compounds, thus allowing delisting as contaminated ground	Compounds not destroyed
		Can deteriorate over time
	Can utilize process by-products (e.g. fly ash)	Typically requires excavation and material handling

		Requires curing period
Sites are easily accessible	Results in a table structurally sound end product	Contaminants will diffuse slowly (leaching)
Soil contains low organic contents		May increase volume
Chemical destruction		
Hydrolysis	For hydrolysis	For hydrolysis:
Soil is contaminated with nitrides	Primary reagent is water	Mass transfer limitations
Soil is contaminated with epoxides and esters	Alkaline conditions can be produced inexpensively	PH conditions that may leach minerals out of soil
	For dechlorination	For dechlorination
	Treats soil previously considered untreatable	Mass transfer limited
	Lower cost for removal of PCBs	Low permeability of treated soil
		Limited field data
		Moisture content of soil
		For chemical oxidation
Dechlorination is used when:	For chemical oxidation	Dependent on oxidant used
Soil contaminated with activated halogenated aromatic compounds	Dependent on oxidant used	
Soil contaminated with halogenated aliphates		
Volumes greater than 1,000 m^3		
Chemical Oxidation is used when present:		
Unsaturated aliphatic hydrocarbons		
Halogenated phenolic compounds		
Halogenated aromatic compounds		
Sulphides and other organic contaminants		
Bioremediation		
Biopiles/land farming/composting	Land farming:	Bioremediation
Slurry reactors	Simple to implement	pH between 5.5 and 8.5
Bioreactors	Less expensive	Moisture 25–85 %
	Applicable for slowly degrading contaminants	Oxygen $\geq$0.5 ppm
		Temperature 10–45 °C

(continued)

Table 4.7 (continued)

Use	Advantages	Disadvantages
		Ratio o carbon : nitrogen : phosphorus 120:10:1
Soil treatment	Usually used for moderate to low contaminate concentrations	Land farming:
		Volatile contaminants must be pre-treated
	Soil slurry	Particulate matter may cause a dust generation problem
	Proper operating conditions can be achieved for effective treatment	Presence of toxic metal ions requires pre-treatment to avoid toxicity to microbes
	Cost are over 50 % less than incineration	
	Effective control of volatile emissions are provided	Requires proper control of pH, water content and nutrients
		Environmental temperature must be favourable
	Bioreactors	Only for aerobic treatment of organic waste
	Lower capital cost for implementation	Requires large area for treatment for significant volume of contaminated soil
	No request for excavation and transportation of soil	
	Not causing cross media contamination	May require bottom lining of large area
	Can be used for large scale contaminations	Less rapid method than other bioremediation techniques
		Soil slurry
		Excavation and transportation of soil
	Bioventing	Requires handling equipment
	Allows direct extraction of organic compounds and enhances biodegradation	Decanted water must be treated
	Relatively inexpensive when no air vented treatment is required	Requires that the chemical can be easily desorbed from soil matrix and stabilised in the water phase
	Minimal exposure of the public	Requires energy consumption
		Bioreactors
		Difficult to monitor and quantify the extent
		Can produce biodegradation products
		Can take a very long time
		Requires favourable environmental conditions
		My be required to be combined with containment

		Bioventing
		Minimum air permeability of 10^{-5} cm/s
		Prone to aging
		Contaminant materials are mobilized by the movement of air and water around the contaminated zone
		Treatment time cannot be guaranteed
		Public acceptance may require effort
Thermal processes		
Heat content of waste >2500BTU/lb	Hazardous compounds destroyed	Some materials are not incinerable
Waste contains <1,000 ppm of halogens	Volume and weight reduction	Some organics generate toxic PIC's
	Short residence times	High capital and operating costs
	Control of air discharge	Skilled operators are required
	Heat recovery	Supplemental fuel is often required
	Possibility of on-site operation	Public opposition
	Applicable to numerous waste types	
	General regulatory acceptance	
Barriers, linear and cover systems		
Control leachate migration	Cut and fill barriers	Cut and fill barriers
Control water ingress	Relatively well developed techniques	Excavation of contaminated soil
Control gas/vapour migration	Very flexible, many systems exist	Displacement barriers
Control migration of contaminated solids e.g. dust	Constructible in most soil including rock	Boulders, rock, bulky waste may impede installation
Provision of growing medium for vegetation	Displacement barriers	Penetration depth dependent on soil type
Prevention of physical contact	No excavation of contaminated soil	Incapable of forming a deep key into rock
	Well known installation techniques	Not yet viable long term solution
	Mix in place barriers	Mix in place barriers
Provision of support to structures	No excavation of contaminated soil	Performance dependent on soil type
	Constructible in most types of ground	Not well established

(continued)

Table 4.7 (continued)

Use	Advantages	Disadvantages
Protection of services	Very quick construction	Contaminant migration cannot be reduced to very low levels
	Injected barriers	Not necessarily a long term solution
	No excavation of contaminated ground	Injected barriers
	Well developed technique	Very dependent on soil type
	Suitable for short term	Unlikely to achieve low permeability
	Liners	Liners
	Can be adjacent to site specific requirement	Requires rigorous health and safety protection measures for workers and public since contaminants are disturbed
	Good construction control is possible	Earthmoving is required
	Low permeability cove system provides total encapsulation of contaminated ground	Cannot be used under existing structures
	More cost effective when used for shallow contamination	Likely to be regarded as landfill site by the regulatory bodies and subjected to their rules, licences, monitoring and care requirements
	Covers	
	Cost effective	
	Contaminants not disturbed	Likely to be uneconomic for deep contamination over a small plan area and high ground water level
	Uses established construction methods and equipments	
	Can be adjusted to site specific type of contamination and assessment of risk to potential targets	Covers
		Only covering sites not cleaning them
		Lateral migration of contamination may occur
	Good construction control is possible	May be damaged by later construction works

4.10 Vibration

4.10.1 Description

- Fast change of ground displacement in time having amplitude, time period of repetition and duration (number of cycles) can be continuous, intermittent, transient, pseudo steady state.
- Affects humans, equipment, structures and soil.
- Caused most frequently by construction/demolition and mining industry, traffic, machinery, earthquakes and volcanoes.
- Causes strain (ratio between peak particle velocity and velocity of wave propagation) and stress (a product of unit density of material, peak particle velocity and velocity of wave propagation).
- Can be amplified by resonance (inertial interaction in structures) and attenuated by material and radiation damping (kinematic interaction in structures). The amplification factor for amplitudes of the harmonic motion of a single degree of freedom oscillator SDOFO (e.g. Clough and Penzien 1993) is:

$$\begin{aligned} \frac{A_o}{A_i} &= \sqrt{\frac{1+(2\beta_t\xi)^2}{\left(1-\beta_t^{\,2}\right)^2+(2\beta_t\xi)^2}} \\ \xi &= \frac{c}{2\cdot\sqrt{k\cdot m}} \\ \beta_t &= \frac{\omega_d}{\omega_o} \end{aligned} \tag{4.34}$$

A_o is amplitude of output motion
Ai is amplitude of input motion
β_t is the tuning ratio,
ω_d is the circular frequency of an input motion,
ω_o is the circular frequency of the output motion,
ξ is the damping ratio as a portion of the critical damping, which prevents oscillations
c is viscous coefficient
k is stiffness of SDOFO
m is mass of SDOFO

The amplification factor for different damping ratios and tuning factor are shown in Fig. 4.28.

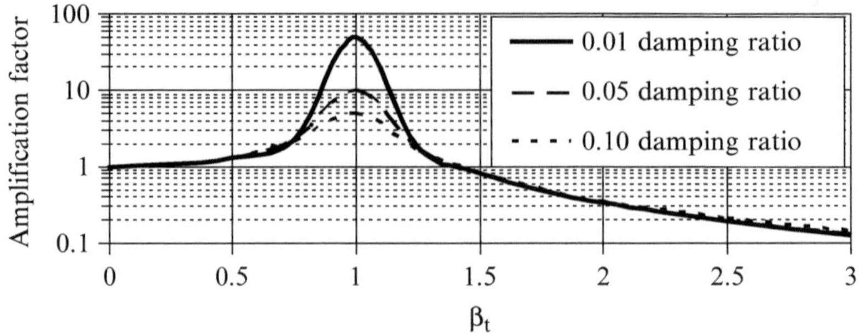

Fig. 4.28 Amplification factor between the output and input peak accelerations of a single degree of freedom oscillator

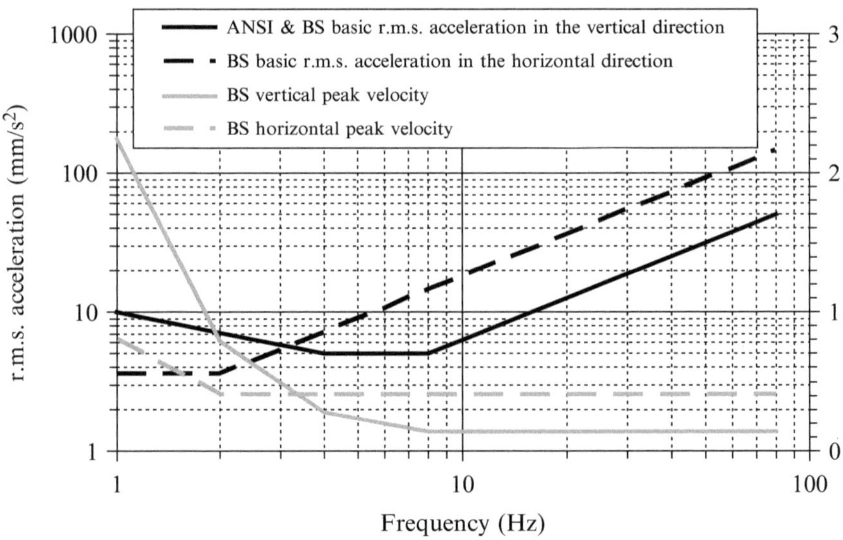

Fig. 4.29 Limiting for humans basic root mean square accelerations and componential peak velocities for the vertical and horizontal directions versus frequency of vibration in buildings according to ANSI S3.29 (1983) and BS 6471 (1992)

4.10.2 *Extent*

4.10.2.1 Effect on Humans

ANSI S3.29 (1983) and BS 6472 (1992) recommend the same basic root mean square (r.m.s.) accelerations in the vertical direction for critical working areas such as hospital operating theatres and precision laboratories shown in Fig. 4.29. The r.m.s. acceleration is the square root of the average of sum of squares of componential accelerations. Both codes recommend the multiplication factor of 4

Table 4.8 Multiplication factors of the basic r.m.s. acceleration in residential buildings

	Continuous vibration		Impulsive vibration (duration <2 s) with up to three occurrences	
Time	ANSI S3.29 (1983)	BS 6472 (1992)	ANSI S3.29 (1983)	BS 6472 (1992)
Day (7–22 h)	1.4–4	2–4	90	60–90
Night (22–7 h)	1–1.4	1.4	1.4	20

of the basic r.m.s. acceleration for offices, and 8 for workshops for continuous and intermittent vibrations and repeated impulsive shock according to ANSI S3.29 (1983) and 128 for both offices and workshops for impulsive vibration excitation (with duration less than 2 s) with up to three occurrences a day. These two codes differ only concerning the multiplication factors of the basic r.m.s. accelerations for residential buildings as shown in Table 4.8. In addition, BS 6472 (1992) recommends the use of the same multiplication factors for the peak velocity.

4.10.2.2 Effect on Equipment

Excessive vibration can cause malfunction and damage of sensitive equipment. Manufacturers of equipment specify tolerable levels of vibrations for their equipment. In order to compare some of these levels with the acceptable levels of vibration for humans and structures, the following list is provided for example from Dowding (2000).

- IBM 3380 hard disk drive: 18 mm/s between frequencies from 1 to 200 Hz (0.3 g in the vertical direction, 0.1 g for 5 Hz, 0.3 g for 16 Hz, and 0.4 g above 20 Hz in the horizontal direction),
- operating theatre (ISO): 0.13 mm/s between frequencies from 60 to 1,000 Hz,
- analytical balance: 0.076 mm/s between frequencies from 45 to 1,000 Hz,
- electronic microscope (Phillips): 0.025 mm/s between frequencies from 50 to 1,000 Hz

Amick (1997) and BS 5228-2 (2009) provide the following limits:

- optical microscope with magnification 400 times, microbalances, optical balances, proximity and projection aligners etc.: 0.050 mm/s at above 8 Hz
- optical microscope with magnification 1,000 times, inspection and lithography equipment (including steppers) to 3 μm line width: 0.025 mm/s at above 8 Hz
- most lithography and inspection equipment (including electron microscopes) to 1 μm detail size: 0.0125 mm/s at above 8 Hz
- electron microscopes (TEM's and SEMs) and E-Beam systems: 0.006 mm/s at above 8 Hz
- long path laser based small target systems 0.003 mm/s at above 8 Hz

It is worth mentioning that footfall induced floor vibration velocity is in the range from 1.1 to 3.8 mm/s between frequencies from 5 to 10 Hz according to

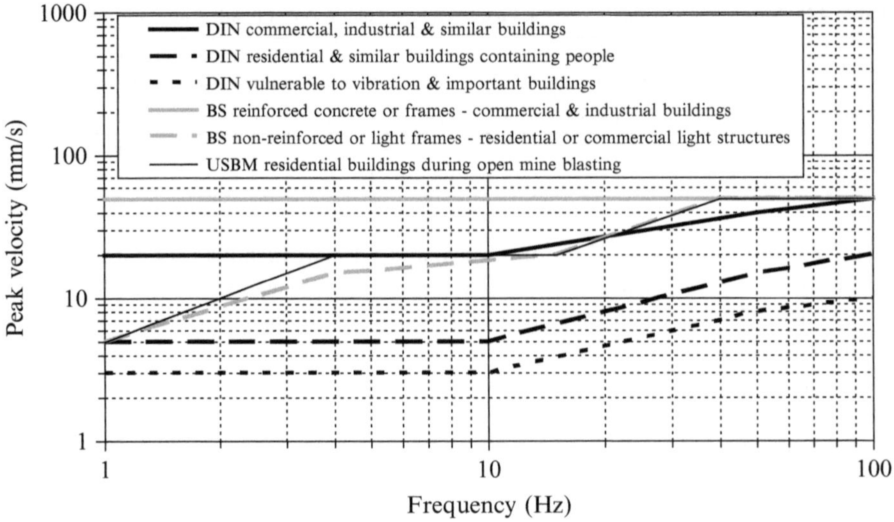

Fig. 4.30 Peak velocity of foundations/basements for appearance of cosmetic cracking in buildings due to transient vibration

Dowding (2000). New (1986), instead, reports peak particle velocities between 0.02 and 0.5 mm/s from footfalls, 0.15–3.0 mm/s from foot stamping, 3–17 mm/s from door slamming and 5–20 mm/s from percussive drilling in buildings. Not only precise equipment but also other industrial machine manufacturers specify tolerable levels of vibration. For example,

- large compressor (MAN) foundation velocity <2.8 mm/s in operational condition and <6 mm/s in accidental case between frequencies from 25 to 190 Hz
- gas turbine (EGT) foundation velocity <2 mm/s and that a peak to peak amplitude of any part of the foundation is less than 50 μm at the operating frequency of 250 Hz.

4.10.2.3 Effect on Structures

- German DIN 4150-3 (1999) specifies peak velocities of foundations by transient vibrations causing so called cosmetic damage (opening of cracks in plaster on walls, increase of existing cracks, and detachment of non-structural partitions from structural walls and columns) as shown in Fig. 4.30.
- British BS 7385-2 (1993) specifies peak velocities of building bases arising from transient vibrations causing cosmetic damage to buildings as shown in Fig. 4.30. For non-reinforced or light frames, at frequencies below 4 Hz, a maximum displacement of 0.6 mm (zero to peak) should not be exceeded.

Table 4.9 Threshold peak particle velocities in mm/s for minor or cosmetic damage according to BS 5228-2 (2009)

Vibration type	Reinforced or framed structures. Industrial and heavy commercial buildings	Not reinforced or light framed structures. Residential or light commercial buildings	Slender and potentially sensitive masonry walls	Propped or tied walls or mass gravity walls	Underground services (for elderly and dilapidated brickwork sewers to use 20–50 % reduction)
Intermittent vibration	50	15 @ 4 Hz	10 @ the toe	[a]	30
		20 @ 15 Hz			
		50 @ 40 Hz	40 @ the crest		
Continuous vibration	50 % lower than the intermittent vibration limits		Reduced 1.5–2.5 the intermittent vibration limits		15

[a]50–100 % greater than for slender and potentially sensitive masonry walls

- USBM (U.S. Bureau of Mines) RI 8507 (1980) specifies peak velocities causing visible damage to residential houses as a result of open mine blasting as shown in Fig. 4.30.
- British standard BS 5228-2 (2009) recommends the threshold peak particle velocities for minor or cosmetic (i.e. non-structural) damage shown in Table 4.9.

4.10.2.4 Effect on Soil

Vibration can cause:

- Collapse of soil structure
- Liquefaction
- Thixotropy is defined as an isothermal, reversible, time-dependent process which occurs under constant volume when a material softens instantly, as a result of disturbance including shaking, and then gradually returns to its original strength when allowed to rest. It should be noted that thixotropy occurs under constant soil volume unlike liquefaction, which requires decrease in soil volume. Clay with natural water content close to the water content corresponding to its liquid state is known to be subjected to almost complete shear strength loss when disturbed. Long distance flow type failures in so called quick clay are well known (e.g. Ter-Stepanian 2000). Seed and Chan (1959) demonstrated that thixotropic strength regain is also possible for soil with water content at or near the limit of its plasticity. More information on formation of quick clay in Sweden, for example, is provided by Rankka et al. (2004).

4.10.3 Identification

Two types of instruments are used:

- Geophones measure ground velocity and consist of a permanent magnet, coils, top and bottom springs, steel casing and cable connector.
- Accelerometers have a range of ±50 times the gravitational acceleration and a near linear response from about 1 Hz to 10 kHz, but are not suitable for low-frequency measurements when the outputs are integrated to obtain velocity. Further disadvantages of the accelerometers are that they require a power supply and are more susceptible to background noise than geophones (Hiller and Crabb 2000). Bormann (2002) states "*However, the latest generation accelerometers are nearly as sensitive as standard short-period (SP) seismometers and also have a large dynamic range. Consequently, for most traditional short period networks, accelerometers would work just as well as 1 Hz SP seismometers although the latter are cheaper. In terms of signal processing, there is no difference in using a seismometer or an accelerometer*".

Table 4.10 contains formulas for calculation of ground vibration velocities from different sources.

4.10.4 Remediation

1. Minimization at source
   - Base isolation by elastomeric/lead rubber bearings or sliding friction pendulum
   - Energy dissipation by dampers
2. Ground wave propagation barriers
   - Stiff barriers such as piles, diaphragms
   - Soft barriers such as cut-off trenches without infill or with a soft infill
3. Recipient isolators and energy dampers
   - Passive such as base isolators
   - Active such as bracing systems, active mass dampers, variable stiffness or damping systems and smart materials installed within structures

   Srbulov (2010a) provides more information on ground vibration engineering.

Table 4.10 Formulas for calculation of peak ground velocity

Vibration source	Formula	Source
Vibratory rollers	$V_{res} = k_s \cdot \sqrt{n_d} \cdot \left(\frac{A_r}{x_r + w_d}\right)^{1.5}$ V_{res} is the resultant particle velocity (mm/s) that is obtained from the three componential velocities k_s =75 for an average value i.e. a 50 % probability of the vibration level being exceeded, =143 for a 33 % probability of the vibration level being exceeded and =276 for a 5 % probability of the vibration level being exceeded, n_d <2 is the number of vibration drums, A_r is the nominal amplitude of the vibrating roller (mm) in the range from 0.4 to 1.7 mm, x_r is the distance along the ground surface from the roller (m) in the range from 2 to 110 m, w_d is the width of the vibrating drum (m) in the range from 0.75 to 2.2 m. Equation is applicable for a travel speed of approximately 2 km/h. For significantly different operating speeds of rollers, v_{res} in Equation could be scaled by the ratio between $2^{1/2}$ and (roller speed in km/h)$^{1/2}$ according to Hiller and Crabb (2000)	Hiller and Crabb (2000), BS 5228-2 (2009)
Dropping heavy weights	$V_{res} \leq 92 \cdot \left(\frac{\sqrt{M_d \cdot H_d}}{x_i}\right)^{1.7}$ V_{res} is the resultant particle velocity (mm/s) M_d is the tamper mass (tonnes), H_d is the drop height (m), x_i is the distance from impact (m). BS 5228-2 (2009) recommends to use 0.037 multiplying coefficient instead of 92 but for the product M_d H_d expressed in J instead of tm like Mayne (1985). Also BS 5228-2 (2009) limits the values of x_i in the range from 5 to 100 m.	Mayne (1985), BS 5228-2 (2009)
Vibrating stone columns	$V_{res} = \frac{k_c}{x^{1.4}}$ V_{res} is the resultant particle velocity (mm/s) k_c = 33 (for 50 % probability of exceedance), 44 (for 33.3 % probability of exceedance), 95 (for 5 % probability of exceedance), x is the horizontal distance range from 8 to 100 m	BS 5228-2, (2009)
Tunnelling	$V_{res} \leq \frac{180}{r^{1.3}}$ V_{res} is the resultant particle velocity (mm/s) r is the slant distance range from 10 to 100 m	BS 5228-2 (2009)
Point vibration source at ground surface	$V_p = \frac{f \cdot (1-\nu)}{G} \cdot P_f$ V_p is the peak particle velocity ν is Poisson's ratio, G is ground shear modulus, which is dependent on the maximum shear modulus and shear stain f is the (predominant) vibration frequency P_f the maximum force amplitude at the ground surface	Based on Wolf (1994)

(continued)

Table 4.10 (continued)

Vibration source	Formula	Source
Vibratory pile hammers	$V_{res} = \frac{k_v}{x^{1.3}}$ V_{res} is the resultant particle velocity (mm/s) $k_v = 60$ (for 50 % probability of exceedance), 136 (for 33.3 % probability of exceedance) and 266 (for 5 % probability of exceedance) x is the horizontal distance along the ground surface in the range from 1 to 100 m	BS 5228-2 (2009)
Impact pile hammers	$V_{res} \leq k_p \frac{\sqrt{W}}{r_s^{1.3}}$ V_{res} is the resultant particle velocity (mm/s) W is the nominal energy (J) of an impact hammer in the range from 1.5 to 85 kJ, r_s is the radial (slant) distance (m) between source and receiver (for the pile depth range from 1 to 27 m and the horizontal distance along the ground surface range from 1 to 111 m) $k_p = 5$ at pile refusal, otherwise in the range from 1 to 3 for loose to very stiff/dense soil	BS 5228-2 (2009)
Blasting	$V_p \approx \sqrt{\frac{2}{\rho} \cdot \frac{E_o}{4/3 \cdot r_s^3 \cdot \pi \cdot e^{\kappa \cdot r_s}}}$ V_p is the peak ground velocity E_o is energy released at a point source i.e. pile tip, ρ is ground unit density $4/3 r_s^3 \pi$ is the volume of ground between the source and the site k is material damping coefficient (e.g. Fig. 4.31)	Ambraseys and Hendron (1968)
	$V_p = 714 \cdot \left(\frac{\sqrt{W_e}}{D_b}\right)^{1.6}$ V_p is the peak particle velocity (mm/s) D_b is distance (m) to blast location W_e mass (kg) of explosive used.	U.S. Bureau of Mines (1971)

4.11 Earthquakes

4.11.1 Description

The effects are listed in decreased frequency of occurrence:

- Ground acceleration, velocity and displacement causing inertial forces, mechanical energy input into structures, deformation and failures of structures and utilities leading to fires, luck of water, electricity and gas supply
- Ground deformations and slope failures interrupting transport and lifelines
- Soil liquefaction leading to ground flows and spreads, pile failures, shallow footing sinking and tilting, shafts and pipelines uplifting
- Tectonic fault break of ground surface
- Tsunamis

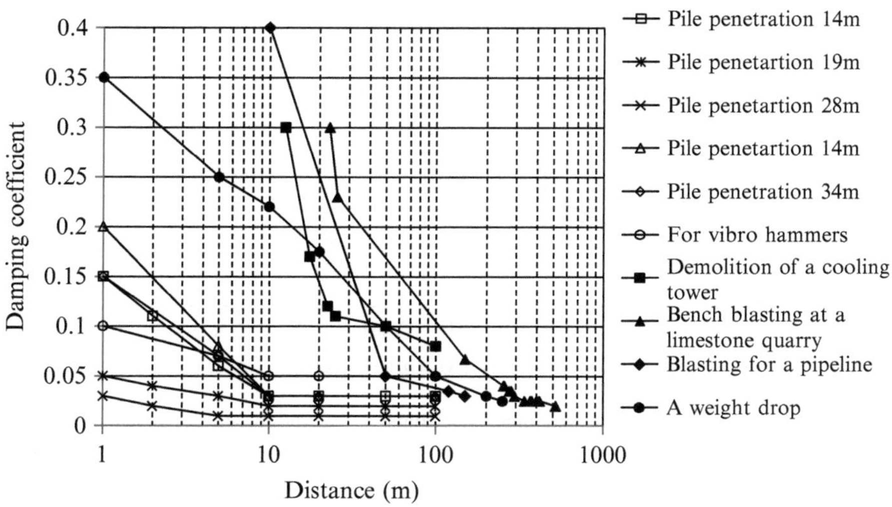

Fig. 4.31 Examples of damping coefficients inferred from back analyses of measured peak particle velocities (Srbulov 2010a)

4.11.2 Extent

4.11.2.1 Ground Acceleration, Velocity and Displacement

Site specific probabilistic and deterministic seismic hazard assessments provide information on the peak ground motion parameters, response spectral values, influential earthquake magnitudes and their epicentral distances.

Quick tentative information about the seismicity at a location can be gained from:

- local seismic codes (some are available at http://www.iaee.or.jp/worldlist.html)
- Global seismic hazard map produced by the global seismic hazard assessment program (GSHAP) of U.N. (1999)
- European-Mediterranean seismic hazard map produced by SESAME project (2003)
- NEHRP recommended provisions for seismic regulations for new buildings and other structures produced by the Building Seismic Safety Council U.S. (1997)

Older publications and seismic codes contain information on earthquake intensity rather than acceleration. The peak horizontal ground acceleration can be estimated based on earthquake intensity as shown in Fig. 4.32. The peak horizontal ground accelerations are different from the peak bedrock (at depth) accelerations specified in codes and seismic maps. Srbulov (2003) plotted the ratios between the surface and at depth accelerations reported in literature, Fig. 4.33. From Fig. 4.33 it follows that the ground surface accelerations are greater than at depth accelerations for the at depth accelerations of up to 3–4 m/s^2 and vice versa. The reason for this is soil

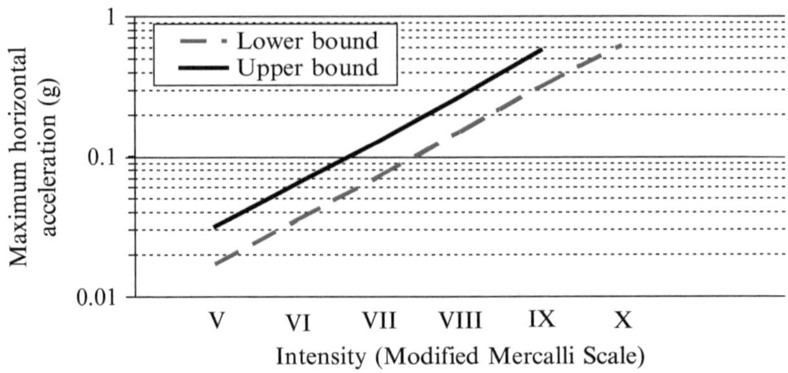

Fig. 4.32 Correlation between maximum horizontal acceleration and earthquake intensity (MIL-HDBK, 1997, Soil dynamics and special design aspects. U.S. Department of Defence Handbook 1007/3)

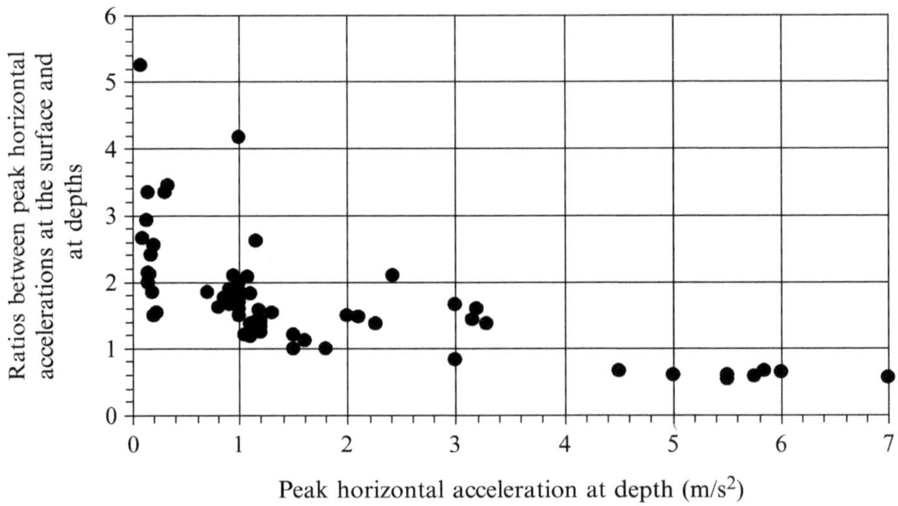

Fig. 4.33 Ratios between reported in literature peak horizontal accelerations at the ground surface and at depth

yield at greater accelerations. EN 1998-1 (2004) specifies the acceleration amplification factors in the range from 1 to 1.4, which corresponds only to the at depth acceleration range from 3 to 4 m/s^2.

EN 1998-1:2004(E) provides the following formula for design ground displacement d_g corresponding to the design ground acceleration a_g when site specific studies are absent.

$$d_g = 0.025 \cdot a_g \cdot S \cdot T_c \cdot T_d \tag{4.35}$$

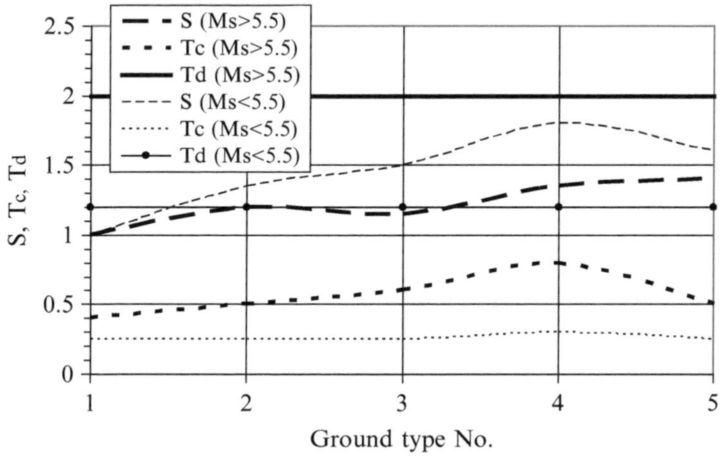

Fig. 4.34 Values of ground parameters depending on ground type from Fig. 2.7 (type 5 is like types 3 or 4 with thickness varying between about 5 and 20 m, underlain by type 1) and surface wave earthquake magnitude M_s

a_g is design ground acceleration at a location
S is the soil factor shown in Fig. 4.34 based on the ground types defined in Fig. 2.7
T_c is the upper limit of the period of the constant spectral acceleration branch (Fig. 4.34)
T_d is the value defining the beginning of the constant displacement response range of the spectrum (Fig. 4.34)

4.11.2.2 Ground Deformation and Slope Failures

Ishihara and Yoshimine (1992) provided a chart of post liquefaction volumetric strain and shear strain as the functions of factor of safety against liquefaction and relative density (i.e. SPT N_1 or CPT q_{c1}). However, strain variations are large for the factor of safety range between 1 and 0.9 for relatively small change of the relative density.

Section 4.8.2 refers to slope failures. Chapter 5 contains descriptions of simplified methods that can be used for considerations of kinematics of moving slopes.

4.11.2.3 Soil Liquefaction and Flow Failures

Section 4.3.2 refers to soil liquefaction and 4.1.3 to global hydraulic failure. Chapter 5 contains description of simplified methods that can be used for an estimation of the extent of a flow failure.

4.11.2.4 Tectonic Fault Break of Ground Surface

Equations (4.25) and (4.33) contain expressions for calculation of maximum surface displacement caused by normal, reverse and oblique tectonic fault types according to Wells and Coppersmith (1994). The maximum surface displacement for strike-slip tectonic fault types is:

$$D = 10^{1.03 \cdot M_w - 7.03 \pm 0.34 \cdot SD} \tag{4.36}$$

D is the maximum surface displacement due to strike-slip tectonic fault break
M_w is the moment magnitude (Eq. 4.26)
SD is the number of standard deviations

4.11.2.5 Tsunamis

Most frequently occur in Pacific ocean and north-eastern Indian ocean; less frequently in eastern Atlantic ocean and eastern Mediterranean sea as a result of uplift of huge water mass by reverse tectonic faults offshore.

The higher known tsunami wave height was about 525 m and occurred as a result of rock fall of about $30.6 \times 10^6 m^3$ from about 914 m height into Lituya Bay in Alaska in 1958 (Tocher 1960). The fall was triggered by a magnitude 8 (intensity XI) earthquake. Ground displacements of 1.05 m upward and 6.3 m in the horizontal plane were measured on the surface breaks along the Fairweather fault 10–17 km southeast of Lituya Bay's Crillon Inlet (Tocher and Miller 1959).

4.11.3 Identification

4.11.3.1 Ground Acceleration, Velocity and Displacement

One (old versions) and three (modern versions) componential accelerometers, geophones and seismographs are used for measurements of amplitudes of ground motion (acceleration, velocity and displacement respectively). Bormann (2002) provides more information on the instruments.

Douglas (2011) provides an overview of ground motion prediction equations in the period from 1964 to 2010. Examples of attenuation relationships are given in Table 4.11 and shown in Figs. 4.35, 4.36, and 4.37.

Amplification of ground acceleration by structural vibration is described by acceleration response spectra, defined in seismic codes, as a function of structural

Table 4.11 Examples of attenuation relationships for the peak horizontal ground acceleration, velocity and displacement caused by earthquakes

Attenuation relationship	Reference
The peak horizontal ground acceleration a_{ph} in m/s^2 dependent on the earthquake moment magnitude M_w, and the minimal distance d from the location of interest to the surface projection of a fault (or epicentral distance where the location of the causative fault has not been reported): $\log_{10}(a_{p,h}) = 2.522 - 0.142 \cdot M_w +$ $(-3.184 + 0.314 \cdot M_w) \cdot \log_{10}\sqrt{d^2 + 7.6^2} + S + F$ $S = 0.137$ for soft soil sites (with the transversal wave velocity range between 180 and 360 m/s to a depth of 30 m) $S = 0.05$ for stiff soil sites (with the transversal wave velocity range between 360 and 750 m/s to a depth of 30 m) S = 0 otherwise $F = -0.084$ for normal and strike-slip faulting earthquakes, $F = 0.062$ for reverse (thrust) faulting earthquakes, $F = -0.044$ for unspecified faulting earthquakes F = 0 otherwise. The standard deviation is 0.222–0.022 M_w. One standard deviation is added when it is expected that the effect of one of the following factors may increase the peak acceleration above the average value. Rupture directivity and fling step near tectonic faults Confining and focusing by sediment basin edges Topographic amplification by ridges and canyons Wave bouncing from Moho surface (the Earth's crust and mantle boundary)	Ambraseys et al. (2005)
The peak horizontal ground velocity $v_{p,h}$ in cm/s depending on the earthquake moment magnitude M_W, and the minimal distance d from the location of interest to the surface projection of the fault (or the epicentral distance where the location of the causative fault has not been reported) $\log_{10}(v_{p,h}) = -1.26 + 1.103 \cdot M_w - 0.085 \cdot {M_W}^2 +$ $(-3.103 + 0.327 \cdot M_w) \cdot \log_{10}\sqrt{d^2 + 5.5^2} + S + F$ $S = 0.266$ for soft soil sites (with the transversal wave velocity range between 180 and 360 m/s to a depth of 30 m), $S = 0.079$ for stiff soil sites (with the transversal wave velocity range between 360 and 750 m/s to a depth of 30 m) S = 0 otherwise, $F = -0.083$ for normal and strike-slip faulting earthquakes, $F = 0.0116$ for reverse (thrust) faulting earthquakes F = 0 otherwise. The standard deviation is 0.344–0.04 M_w. One standard deviation is added when it is expected that the effect of one of the following factors may increase the peak acceleration above the average value. Rupture directivity and fling step near tectonic faults Confining and focusing by sediment basin edges Topographic amplification by ridges and canyons Wave bouncing from Moho surface (the Earth's crust and mantle boundary)	Akkar and Bommer (2007)

(continued)

Table 4.11 (continued)

Attenuation relationship	Reference
The peak horizontal ground displacement $d_{p,h}$ in cm dependent on the surface wave magnitude M_s, and the minimal distance d to the surface projection of the fault (or epicentral distance where the location of the causative fault has not been reported):. $\log_{10}(d_{p,h}) = -1.757 + 0.526 \cdot M_s -$ $1.135 \cdot \log_{10}\sqrt{d^2 + 3.5^2} + S + 0.32 \cdot P$ $S = 0.114$ for soft soil sites (with the transversal wave velocity range between 180 and 360 m/s to a depth of 30 m), $S = 0.217$ for stiff soil sites (with the transversal wave velocity range between 360 and 750 m/s to a depth of 30 m) S = 0 otherwise, P is a variable that takes a value of 0 for mean peak displacement and 1 for 84-percentile values of exceedance of the mean peak displacement. One standard deviation (P = 1) is added when it is expected that the effect of one of the following factors may increase the peak acceleration above the average value. Rupture directivity and fling step near tectonic faults Confining and focusing by sediment basin edges Topographic amplification by ridges and canyons Wave bouncing from Moho surface (the Earth's crust and mantle boundary)	Bommer and Elnashai (1999)

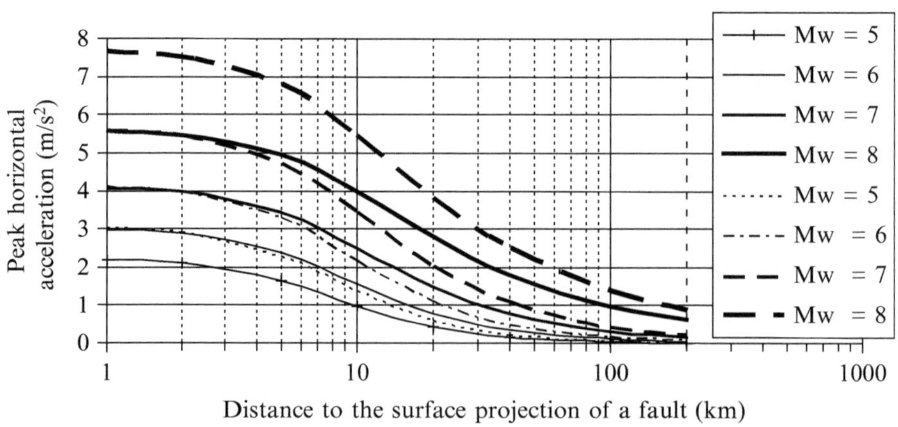

Fig. 4.35 The average peak horizontal accelerations of rock (*continuous lines*) and of soft soil (*dashed lines*) from unspecified faulting according to Ambraseys et al. (2005)

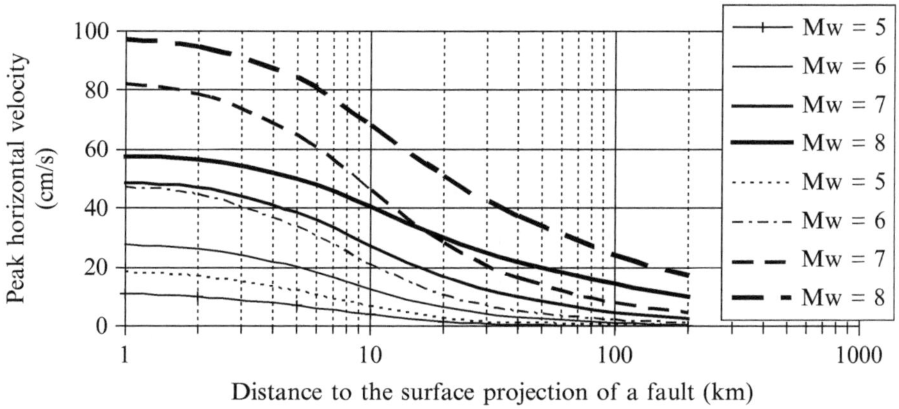

Fig. 4.36 The average peak horizontal velocities of rock (*continuous lines*) and of soft soil (*dashed lines*) from unspecified faulting according to Akkar and Bommer (2007)

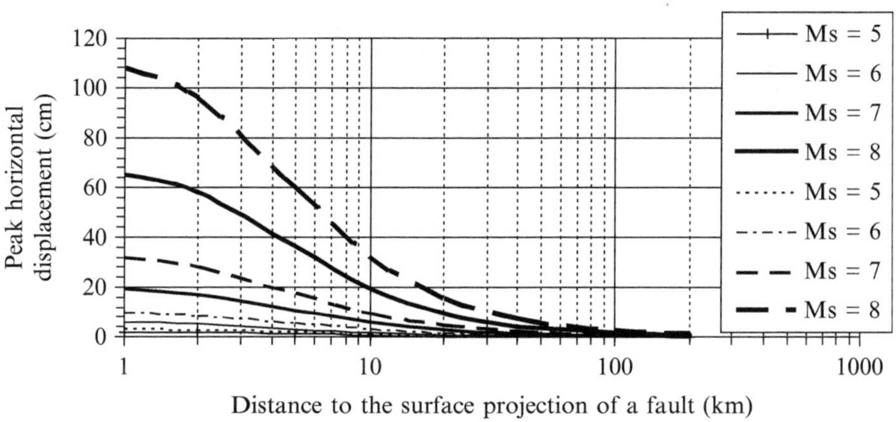

Fig. 4.37 The average peak horizontal displacements of rock (*continuous lines*) and of soft soil (*dashed lines*) from unspecified faulting according to Bommer and Elnashai (1999)

vibration period. EN 1998-1 (2004) provides for building with heights up to 40 m the value of the period of the first vibration mode as:

$$T_1 = C_t \cdot H^{3/4} \tag{4.37}$$

T_1 is the period of the first vibration mode

$C_t = 0.085$ for moment resistant space steel frames, $=0.075$ for moment resistant space concrete frames and for eccentrically braced steel frames, $=0.05$ for all other structures

H is the height of a building (m) from the foundation or from the top of a rigid basement

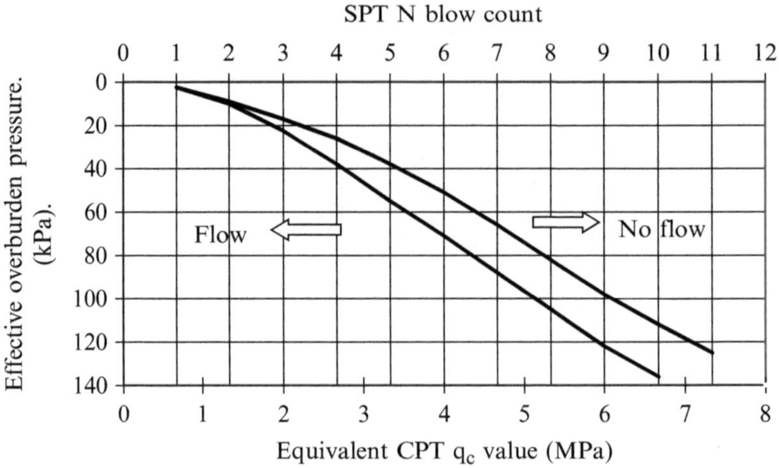

Fig. 4.38 Boundary range between flow and no flow condition based on Ishihara (1993) for fines contents less than 30 %

4.11.3.2 Ground Deformation and Slope Failure

Past events can be inferred from geological maps based on topographic features (geomorphology) and from disturbed stratigraphy in trenches and galleries as mentioned in Sect. 4.8.3. New slope movements can be monitored using geodetic survey, inclinometers and extensometers, described in Chap. 6.

4.11.3.3 Soil Liquefaction and Flow Failure

Simplified method for identification of potential of soil liquefaction due to an earthquake is described in Sect. 4.3.3. Ishihara (1993) proposed the boundary between flow and no flow conditions as shown in Fig. 4.38, but factors other than effective overburden pressure and SPT blow count N influence occurrence or absence of flow type failures as described in Chap. 5.

4.11.3.4 Tectonic Fault Break of Ground Surface

The movements tend to reoccur so the past events can be identified from regional geological maps, satellite survey (SPOT/LANDSAT), aerial photographs and digital elevation models (DEM, SRTM) as well as by geological mapping. Trifonov et al. (1999) provided a map of active tectonic faults of the Arabian-Eurasian and Indian-Eurasian collision regions. Skobelev et al. (2004) plotted active tectonic faults in Africa.

4.11.3.5 Tsunamis

Sensitive ocean bottom pressure sensors are used for detection of tsunamis (e.g. Ambraseys and Synolakis 2010).

4.11.4 Remediation

- Ground acceleration velocity and displacement effect by designing structures for their effect or providing base isolators and dampers as mentioned for the vibration effects in Sect. 4.10.4
- Deformation and slope failure effect as mentioned in Sect. 4.8.4
- Soil liquefaction and soil flow effect as mentioned in Sect. 4.3.4
- Tectonic fault movement effect by avoiding placing structures across the faults or by making them flexible enough to accommodate estimated fault movements
- Tsunami effect by using an efficient monitoring and alarm system

4.12 Volcanoes

4.12.1 Description

Accompanying hazards are (e.g. Blong 1984):

- Volcanic ash fall
- Volcanic gases: mainly water vapour, carbon dioxide (CO_2), sulphur dioxide (SO_2), smaller amounts of hydrogen sulphide (H_2S), hydrogen (H_2), carbon monoxide (CO), hydrogen chloride (HCl), hydrogen fluoride (HF) and helium (He)
- Lahars (a hot or cold mixture of water and rock fragments flow several tens of metres per second)
- Volcanic slides with their sizes exceeding 100 km^3 and the movement velocity exceeding 100 km/h can generate lahars, tsunamis in lakes and ocean as well as bury river valleys to form lakes
- Lava flow can move as fast as 30 km/h when confined within a channel or lava tube on a steep slope but typically 1 km/h causing burying structures, causing fire, trigger pyroclastic flows
- Pyroclastic flows are high density mixtures of hot, dry rock fragments and hot gases erupted from a vent at high speed to destroy by direct impact, bury sites, melt snow and ice to form lahars, burn forests and buildings
- Tephra are fragments of volcanic rock and lava blasted into the air by explosions or carried upward by hot gases after an eruption

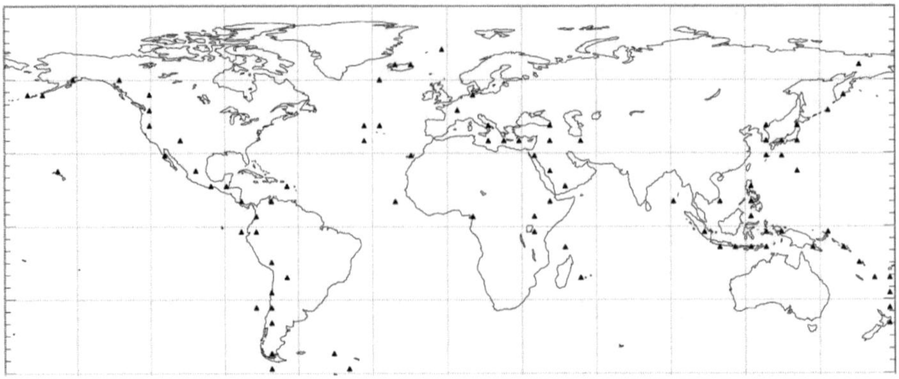

Fig. 4.39 Known locations of volcanoes as triangles (From various sources)

4.12.2 *Extent*

Figure 4.39 shows approximate locations of known volcanoes containing molten rock as lava on the surface or magma at depth.

4.12.3 *Identification*

Molten rock volcanoes are formed along mid ocean ridges undergoing spreading and at the subduction zones of continental tectonic plates. Typical conical shape is formed by repeated depositions of lava flows in time.

Mud volcanoes containing methane gas are connected with the existence of hydrocarbons in Azerbaijan (including Caspian sea), Turkmenistan, Georgia, on the Kerch and Taman peninsulas, on Sakhalin Island, in West Kuban, Italy, Romania, Iran, Pakistan, India, Burma, China, Japan, Indonesia, Malaysia, New Zealand, Mexico, Colombia, Trinidad and Tobago, Venezuela and Ecuador. Mud volcanoes can erupt with explosion of methane gas and fire onshore.

Sand volcanoes of relatively small size are formed by ejection to the ground surface of liquefied sand during earthquakes.

4.12.4 *Remediation*

Due to the extent and energy of this natural hazard, avoidance is the best measure if it is possible. Some limited remedial measures for various volcanic outflows are possible as mentioned in Sect. 4.8.4 for avalanches and flows.

4.13 Frozen Ground

4.13.1 Description

The following hazards are listed by Andersland and Ladanyi (1994)

- Ground heave on freezing
- Ground subsidence on thawing
- Formation of ice wedges close to the permafrost surface occur down to 10 m depth and up to 3 m wide
- Pingos – conical mounds or hills occurring in continuous and discontinuous permafrost zones
- Formation of thermo-karst resulting from differential melting of ground ice in permafrost (mounds, caverns, disappearing stream, funnel shaped pits, elongated troughs, large flat floored valleys with steep sides)
- Patterned ground in periglacial environments forming circles, polygons, nets, steps and strips

Permanently frozen ground with ice lenses is a high strength and stiffness and low permeability material and for this reason artificial ground freezing is undertaken sometimes although it is a slow and very expensive process. When used, it causes subsidence on thawing of ice lenses formed during ground freezing.

4.13.2 Extent

Figure 4.15 shows regions affected by low temperatures seasonally or permanently.

The volume expansion of saturated soil on freezing is equal to 0.09n, where n is the porosity of the soil. Ice lens when formed increases its volume about 9 % with respect to the volume of water before freezing. The volume of ground contraction on thawing is also equal to 0.09n. Nixon and Ladanyi (1978) proposed the following empirical relationship for the vertical strain

$$\varepsilon_f = 0.9 - 0.868 \cdot \left(\frac{\rho_f}{\rho_w} - 1.15\right)^{0.5} \pm 0.05 \qquad (4.38)$$

ε_f is the vertical thaw deformation
ρ_f is the unit density of frozen ground
ρ_w is the unit density of water

Uneven (one sided) formation of ice lenses and their thawing causes differential settlement and distortion (tilt) to which structures are more sensitive than to uniform settlement. Skempton and MacDonald (1956) suggested the following limits for maximum settlement:

Isolated foundations on clay	65 mm
Isolated foundations on sand	40 mm
Rafts on clay	65–100 mm
Rafts on sand	40–65 mm

Bjerrum (1963) suggested the following limits of angular distortions:

Sensitive machinery	1:750
Structural frames with diagonals	1:600
When cracking is not permissible	1:500
When first cracking occurs or difficulties with overhead cranes	1:300
Tilting of high buildings is evident	1:250
Considerable cracking in panel and brick walls with height: length ratio	1:4
Also structural damage occurs	1:150

Observations based on the weight of buildings lifted by frost heaving of soil indicate heave pressure of 760 kPa. Penner (1970) measured pressure in excess of 1,800 kPa on a 300 mm diameter anchor plate.

Drouin and Michel (1971) reported an average thermal expansion strain for ice of 52×10^{-6} per temperature change of 1 °C.

4.13.3 *Identification*

Besides low temperatures, soil type is the second important factor. Casagrande (1932) proposed that soil subjected to unfavourable frost effects are:

1. Uniformly graded if contain more than 10 % of particles with diameters smaller than 0.02 mm
2. Well graded if contain more than 3 % of particles with diameters smaller than 0.02 mm.

The third factor contributing to formation of ice lenses in ground is supply of ground water. High ground water level is not the only water supply source but also capillary rise. The maximum capillary rise h_c (Holtz and Kovacs 1981) is approximated by the relationship

$$h_c = \frac{0.03}{d} \tag{4.39}$$

h_c is the maximum capillary rise (m)
d is the effective pore diameter (mm), which is equal to about 20 % of the effective grain size – D_{10} (the grain diameter in mm corresponding to 10 % of the total weight passing a sieve with the opening size D)

Laboratory freezing and thawing test results are the most reliable representatives of local ground conditions although tentative recommendations exist (example Fig. 4.22). Frost susceptibility is considered low for the average rate of heave below 2 mm/day, high for more than 4 mm/day and very high for more than 8 mm/day.

4.13.4 Remediation

- Lowering of foundation depth below the depth of frost penetration is most common for shallow foundation
- For roads, other traffic areas and retaining walls, use well compacted coarse grained and well drained sub-base to minimize capillary rise and prevent formation of ice lenses
- Prevention of permafrost thawing or frost penetration from then ground surface by use of thermal isolation around and under building
- Use of suspended floors above ground with provision of cavity between ground and structure
- Use of piles with sleeves to isolate their shafts from uplifting and downdrag forces during freezing and thawing

4.14 Unexploded Ordnance (UXO)

4.14.1 Description

Stone et al. (2009) divide military ordnance into the following categories:

- Rockets
- Projectiles
- Grenades
- Mortars
- Mines
- Bombs

Explosives used in construction and mining for blasting rocks may have failed to detonate.

4.14.2 Extent

Main locations are:

- War sites
- Damping ground of ammunition including offshore sites
- Military bases and exercise locations (past and present)
- Weapons manufacture and storage areas

4.14.3 Identification

Sources are:

- Local knowledge
- Past records and news
- Specialist's survey (mostly military)

Limited information exists in public domain such as by Taylor et al. (1991).

4.14.4 Remediation

Mostly removal if safe otherwise detonation in place.

4.15 Ground Gases

4.15.1 Description

The potential hazards from ground gases (types) are:

- Toxicity – acute and chronic (carbon dioxide, carbon monoxide, hydrogen sulphide, hydrogen cyanide, phosphine, sulphur dioxide)
- Eco-toxicity (hydrogen chloride, hydrogen fluoride, oxides of nitrogen and sulphur)
- Fire and explosion (methane, hydrogen, hydrogen sulphide)
- Asphyxiation (carbon dioxide)
- Odour (hydrogen sulphide, methanethiol, dimethyl sulphide, ethanol, carbon disulphide)
- Carcinogenic (radon from granite rocks)
- Corrosiveness in solution (carbon dioxide, sulphur dioxide)

4.15.2 Extent

Ground gases originate in marshes, mines, landfills, sewers, fresh water and marine sediments. Table 4.12 contains details of hazards due to particular gases from Barry (1991).

Methane hydrates exist in ocean floor sediment at depths greater than 300 m. Their presence can be detected by geophysical surveys (Dean 2010). Drilling through gas hydrates can cause them to sublimate because of heating by warm drilling fluids and hot recovered hydrocarbons leading to severe damage of foundations and slope instability even for slope inclination range from 2° to 5°.

4.15.3 Identification

Crowhurst (1987) and IWM (1998) described advantages and disadvantages of gas detectors as given in Table 4.13.

4.15.4 Remediation

Wearing protective equipment (gas masks and breathing apparatus) during the site investigation stage and good ventilation during the construction stage.

The following permanent measures are used:

- Containment (by lining of the sides, base and capping as well as by gas extraction to reduce pressure)
- Collection (by wells, layers, pipes, extraction plant, condensate system)
- Treatment (utilisation and flaring)

The main types of containment materials are (LFTGN 03 2004):

- Compacted clay
- Bentonite –soil mixtures
- Geosynthetic – clay liners (GCL)
- Geomembranes (high density polyethylene HDPE membranes, medium density polyethylene MDPE membranes, linear low density polyethylene LLDPE membranes)
- Dense asphaltic concrete (DAC)
- Composites of above

Common types of capping materials used include (LFTGN 03 2004):

- Compacted clay
- Bentonite enhanced sand
- Geomembranes HDPE, LLDPE, GCL

Table 4.12 Ground gases properties and hazards

Type	Properties	Effect on humans	Occupational exposure limits	Other effects
Carbon dioxide CO_2	Colourless Odourless Denser than air, exists in air at 0.03 % (300 ppm) Dissolves in water Non-combustible	>3 % difficult breathing and headaches >5–6 % as above but more severe 12–25 % unconsciousness >25 % death	5,000 ppm (8 h), 15,000 (10 min)	Strong solution corrosive to metals and concrete
Carbon monoxide CO	Colourless, almost odourless Highly inflammable at 12–75 % concentration Slightly soluble in water	>200 ppm headache after 50 min >500 ppm headache after 20 min 1,000 to 10000 ppm headache, dizziness and nausea in 3 to 15 min, death in 10 to 45 min 10,000 to death within a few minutes	50 ppm (8 h), 300 ppm (10 min)	Phytotoxic to plants
Hydrogen cyanide HCN	Colourless Smell of bitter almonds Highly inflammable with lower explosive limit 6 % in air, soluble in water (white liquid below 26.5 °C)	Highly toxic <18 ppm poisoning 18–36 ppm for several hours causes slight weakness, headache, confusion and nausea >100 ppm for several minutes causes collapse, respiratory failure and possible death >300 ppm immediately fatal	10 ppm (10 min)	–
Hydrogen sulphide H_2S	Colourless Odour of rotten egg (threshold 0.5 parts per billion) Sweetish taste Soluble in water Highly inflammable (lower explosive limit 4.5 % in air)	Highly toxic >20 ppm causes loss of smell, 20–150 ppm causes irritation of the eyes and respiratory tract >400 ppm toxic >700 ppm life threatening	10 ppm (8 h), 15 ppm (10 min)	Phytotoxic to plants

(continued)

Table 4.12 (continued)

Type	Properties	Effect on humans	Occupational exposure limits	Other effects
Methane CH_4	Colourless Odourless Lighter than air Inflammable (lower explosive limit in air 5 %, upper explosive limit in air 15 %)	Asphyxiant as it replaces air but non-toxic	–	Causes plant root death by replacing oxygen
Phosphine PH_3	Colourless Garlic odour Denser than air Spontaneously inflammable at room temperature	Highly toxic causing headache, fatigue, nausea, vomiting, jaundice and ataxia, Irritant, odour threshold 2 ppm	0.3 ppm (8 h), 1 ppm (10 min)	–
Sulphur dioxide SO_2	Colourless Sharp pungent odour Denser than air Soluble in water Non-combustible	Toxic 0.3 to 1 ppm detectable by most, 6–12 ppm irritates eyes and mucous membranes,	2 ppm (8 h), 5 ppm (10 min)	Phytotoxic to plants, corrosive to concrete and metals when sulphurous acid is formed

Note: *ppm* parts per million

Table 4.13 Main properties of gas detectors

Type	Gas	Advantages	Disadvantages
Infra-red	Methane Carbon dioxide Hydrocarbons	Fast response Simple to use Wide detection range Less prone to cross interference No change to gas sample	Prone to zero drift Pressure, temperature, moisture sensitive Not specifically sensitive to methane but hydrocarbon bond only Sensitive to contamination
Flame ionisation	Methane Flammable gases and vapours	Highly sensitive Fast response	Needs oxygen Accuracy affected by presence of other gases Limited detection range Gas sample destroyed Inappropriate for explosive gases
Electrochemical	Oxygen Hydrogen Sulphide Carbon monoxide	Low cost	Limited life Requires frequent calibration Can lose sensitivity due to moisture, corrosion and saturation Poor response when contaminated with other gases

(continued)

Table 4.13 (continued)

Type	Gas	Advantages	Disadvantages
Paramagnetic	Oxygen	Accurate Robust Not subjected to interference	Prone to drift and gas contaminants Expensive Responds to pressure not to concentration
Catalytic oxidation (pellistor)	Methane Flammable gases and vapour	Fast response Low detection range Responds to any flammable gas Not expensive Easily operated	Accuracy affected by other flammable gases Inaccurate if oxygen is deficient Prone to ageing, deterioration and moisture without detection Gas sample destroyed
Thermal conductivity	Methane Flammable gases and vapour	Fast response Full detection range Independent of oxygen amount Can be combined with other detectors	Accuracy affected by the presence of other gases Poor sensitivity Error prone at low concentrations
Semiconductor	Toxic gases	Good differentiation of some toxic gases Less sensitive to saturation High sensitivity to low concentrations Long-term stability	Not sensitive to flammable gases Not differencing among gases Humidity affects accuracy and response
Chemical	Carbon dioxide Carbon monoxide Hydrogen sulphide Water vapour Other gases	Simple to use Not expensive	Crude (for indication only) Prone to interference
Photo-ionisation	Organic gases	Very sensitive	Susceptible to cross contamination High cost

Potential deterioration in gas collection well and pipelines performance includes (LFTGN 03 2004):

- Increase in density and decrease in conductivity of soil or waste due to gas extraction
- Blockage by soil particles around wells and bacterial growth
- Overdrawing of gas well or pipes by inducing excessively high flow rates

Gas collection layers are made of (LFTGN 03 2004):

- Aggregate (sand, gravel, crushed stone)
- Geocomposites (e.g. geo-nets)

More details about soil liners, wells and pipelines are given in Chap. 5.

4.16 Rock Burst

4.16.1 Description

Sudden failure of continuous and competent rock mass causing ejection of rock pieces up to a metre in size as in the case of blasting by explosives is termed rock burst. The phenomenon is caused by rock ability to decrease its strength abruptly on reaching its ultimate value, rock strength dependence on confining stress, which is lost during excavations in rock, and by built up of elastic energy in rock mass due to excavation.

4.16.2 Extent

Tang et al. (2010) reviewed several known cases of rock burst as summarised in Table 4.14. Rock bursts occurred in many types of rock typically in stronger rock at greater depths. At shallow depths, large horizontal stresses (stress anisotropy) could be the main contributory factor.

4.16.3 Identification

Tang et al. (2010) listed many rock burst theories and described the use of micro seismic monitoring technique because it has been found that precursory micro cracking existed prior to most rock bursts. The problem with the monitoring could be the ambient noise created by rock excavation. Palmstrom (1995) suggested the use of rock mass index for characterisation of rock burst potential. Strength anisotropy in rock may cause that the criteria proposed may not always be representative.

4.16.4 Remediation

Kaiser and Cai (2012) described ground control measures and burst resistant rock support, which selection process is iterative, requiring design verification and modification based on field observations. The problem is that the observation method is not applicable to very fast ground movements, which leave no time for intervention. Palmstrom (1995) mentioned that the experience in Scandinavia

Table 4.14 Summary of rock burst cases

Project	Rock type	Depth	Location
Gold mines in South Africa	(Massive quartzite)	–	–
Simplon hydraulic tunnel in the Alps	–	2,200 m	–
Shimizu tunnel in Japan	–	1,000–1,300 m	–
Kanestu tunnel in Japan	Quartz diorite	730–1,050 m	Tunnel face after blasting
Ruhr mining area in Germany	Coal field	–	–
Tastgol iron mine in the former USSR	–	–	–
Rubin copper mine in Poland	–	–	–
Couer d'Alene lead, zinc and silver mines in Idaho US	–	–	–
Sudbury copper and nickel mines in Canada	–	–	–
Makassar gold mine in the Kirkland Lake area, Ontario	–	–	–
El Teniente copper mine in Chile	–	–	–
A road tunnel in Norway	–	–	–
A headrace tunnel in Sweden	–	–	–
Hydraulic tunnels of Forsmark nuclear plant in Sweden	Granite gneiss	5–15 m	–
Ritsem traffic tunnel in Sweden	Mylonite	130 m	–
Shengli coal mine at Fushun in China	–	–	–
The headrace tunnel of YuzixiI hydro power station on Minjiang river in China	Granodiorite and diorite	250–260 m	Working face
The headrace tunnel with 10 m diameter of Tianshengqiao II hydro power station on Nanpanjiang River in China	Massive limestone and dolomite	120–160 m	Side wall 4–10 m away from the working face
Erlangshan tunnel in Sichuan – Tibet	Sandy mudstone, marl and quartzite	270–570 m	Sidewalls, spandrel, vault
Dongguashan copper mine in China	–	790–850 m	Sidewalls, roof
Quinling railway tunnel in China	Granite and gneiss	900 m	–
Canling tunnel for the Taizhou – Jiyun highway Zhejiang province	Tuff	768 m	Sidewall, near the vault
Pubugou hydro power station under ground power house	Granite	–	Upper corner of sidewall
Jinping II hydro power station tunnel	Marble	2,000 m	Working face of the tunnel vault and haunch
Lujialing tunnel along Chongqing – Yichang highway	Tuff	120–600 m	The upper corners an sidewalls

indicates that rock burst is less developed in blasted tunnels than tunnel boring machined tunnels. Increased development of joints and cracks from additional blasting in the periphery of the tunnel is sometimes used in Scandinavia to reduce rock burst problem according to Palmstrom (1995).

The importance of the shape and size of an excavation controlling the magnitude of the additional stresses caused by excavation and on the stability of excavated shape has been shown by several authors (e.g. Hoek and Brown 1980). Selmer-Olsen (1988) mentions that in high anisotropic stress regimes with rock burst, reducing the radius of an excavation in the roof where the largest in situ tangential stress occurs can reduce the extent of rock support.

4.17 Summary

Description, extent, identification and remediation of ground hazards considered are given in Table 4.15.

Table 4.15 Summary of geo-hazards considered

Description	Extent	Identification	Remediation
Uplift (buoyancy) of fine grained soil Uplift (boiling) of coarse grained soil	Several situations shown in Fig. 4.1	Uplift (buoyancy) based on water pressure and total pressure at the bottom of soil layer or structure Uplift (boiling) when the critical hydraulic gradient in upward direction is exceeded (~1)	Uplift (buoyancy) by decreasing the water pressure, increasing the weight or anchoring a structure Uplift (boiling) by increasing the surcharge on soil, decreasing hydraulic gradient
Internal erosion (suffusion) of coarse grained soil Concentrated erosion (piping) Surface erosion by rainfall and waves	Internal erosion (suffusion) in loose to medium dense coarse grained soil particularly with gap or wide gradation Concentrated erosion (piping) in coarse grained soil by retrogressive washing out of particles, in fine grained soil by change of chemistry of water flow	Internal erosion (suffusion) Eq. (4.2) Concentrated erosion (piping) in coarse grained soil from Fig. 4.3, in fine grained soil from Eqs. (4.3) and (4.4)	For internal erosion (suffusion) Decreasing the hydraulic gradients Placing granular filters at the surface where outflow exists, Placing perforated drain pipes within soil Use geotextile barriers For concentrated erosion (piping) In coarse grained soil: increase its density, mixing it,

(continued)

Table 4.15 (continued)

Description	Extent	Identification	Remediation
			decreasing hydraulic gradient, use filter In fine grained soil: proper compaction or mixing Surface erosion Vegetating slopes Various covers, from geomembranes to rip-rap
Liquefaction	Caused by: Rapid sedimentation, loading and moving landslides Artesian pressure Slumping Chemical weathering Periglacial environment Earthquakes Pile driving Soil vibratory compaction Blasting and compaction by explosives Geophysical survey Leaching of salty water in clay Loss of natural cementation in clay	From gradation Figs. 4.6 and 4.7 for coarse grained soil Saturated clay with the sensitivity greater than 16 Coarse-grained soil during earthquake Figs. 4.8 and 4.9, Eqs. (4.10), (4.11), and (4.12) for level ground. Equation (4.13) in place of Eq. (4.11) for sloping ground	Vibratory compaction Dynamic compaction Blasting at greater depths Compaction grouting Stone columns Mixing with binder Jet grouting Dewatering Gravel drains
Dissolving and caves in Halite (rock salt) Gypsum Chalk Limestone (& dolomite)	Known locations shown in Figs. 4.10, 4.11, and 4.12	Local geological maps Geophysical cross-hole, electromagnetic and resistivity profiling, ground penetration radar (standards in Sect. 2.1.8)	Large cavities unfilled by grout and concrete Smaller random cavities can be spanned by geogrid reinforced fills, raft foundations, capping slabs
Collapse of soil structure of loess, residual soil, compacted soil, after thawing	Known locations of major loess deposits are shown in Fig. 4.13, of residual tropical	Oedometer tests in laboratory Shaking and vibratory table tests in laboratory	Excavation and replacement (re-compaction when possible) if shallow

(continued)

Table 4.15 (continued)

Description	Extent	Identification	Remediation
	soil in Fig. 4.14, of frozen soil in Fig. 4.15	Freezing and thawing tests in laboratory Field inundation test Field test with vibratory equipment Digging trenches in situ	Wetting in situ if thick Densification from ground surface if shallow Deep densification if thick Piles or caissons Compaction or chemical grouting New rafts
Subsidence of ground surface	Caused by: Lowering or rising of ground water level and thawing of frozen ground Shrinkage of clay on drying or extraction of moisture by tree roots Deep excavation and tunnelling Collapse of the openings after mining and extraction of gas and oil Tectonic fault movement and ground slumping Biodegradation of more than 4 % of organic matter	Precise levelling, remote sensing, geophysical methods State environmental and mining agencies Structural cracking and deformation Site reconnaissance and survey maps Geological maps InSAR	Water recharge wells besides pumping wells Ground isolation from heat Covering of clay surface and removing of trees Compensation grouting Infill of deep openings Use of structures less sensitive to differential settlement Piles
Heave of soil an rock	Caused by: Formation of ice lenses on ground freezing Unloading or increase of moisture content of clay Chemical reactions involving sulphates Tectonic faulting and valley bulging	Frost susceptibility based on gradation Clay heave based on Eq. (4.27), Fig. 4.24 Chemical reactions described in Sect. 4.7.3.3 Tectonic fault movement by InSAR Valley bulging in shale/mudstone	Use isolators against ground wetting and freezing, detachment of structure from heaving or frozen Tension piles or anchors against clay heave Avoid sulphate reactions Structures not very sensitive to differential movements

(continued)

Table 4.15 (continued)

Description	Extent	Identification	Remediation
Slope instability (toppling, falls, avalanches, flow or spread, slides, slumps, turbidity currents offshore)	Caused by: Water saturation Erosion at the toe Loading at the crest Weathering Tectonic movement Ground water level rise and pressure increase Creep, freezing, swelling, collapse Vibration from machinery or blasting Earthquakes Volcanoes	Anomalous topography Unusual vegetation type, age, position Discontinuous stratigraphy	Eliminating driving forces and/or increasing resisting forces Supporting in place or removing loose peaces For avalanches, flows and spreads bunkers, retaining walls or dams For slides, decreasing slope inclination, ground water lowering, anchoring or bolting, retaining walls including reinforced soil, replacement or soil mixing Slumps by soil mixing or viaducts Turbidity currents offshore by deep burial or covering of cables and pipelines ground, piles or very rigid foundation beams
Contamination	Maximum recommended/permissible concentrations/activity with occupational exposure limits given in Table 4.6	Typical standard tests types in Sect. 4.9.3	Soil vapour extraction Chemical extraction and soil washing Solidification and stabilisation Chemical destruction Bioremediation Thermal process

(continued)

Table 4.15 (continued)

Description	Extent	Identification	Remediation
Vibration and effects	On humans Fig. 4.29 and Table 4.8 On equipment in Sect. 4.10.2.2. On structures Fig. 4.30, Table 4.9 On soil (collapse, liquefaction, thixotropy)	Geophones Accelerometers Attenuation relationships, Table 4.10	Minimization at source by isolation and use of dampers Wave propagation barriers Recipient isolation and use of dampers
Earthquakes and their effects	Insight from: Local seismic codes Global seismic hazard maps International recommendations	Accelerometers Probabilistic and deterministic seismic hazard assessment	Earthquake resistant structures Measures according to relevant effect
Volcanoes and their effects	Fig. 4.39	By conical shape Mud volcanoes by the presence of methane gas in hydrocarbons	Avoidance or limited measures as described for relevant effects
Frozen ground and ice effects	Fig. 4.15	Low temperatures Soil gradation Ground water level and capillary rise Laboratory freezing and thawing tests Tentative recommendations Fig. 4.22	Foundation depth below frost penetration Soil replacement by well compacted coarse grained and well drained material for roads, traffic areas and retaining walls Thermal isolation Suspended floors Piles
Unexploded ordnance (UXO)	Located at: War sites Ammunition damps Military bases and exercise locations Weapons manufacture and storage areas	Local knowledge Past records ad news Specialist's survey	Removal if safe Detonation in place if possible

(continued)

Table 4.15 (continued)

Description	Extent	Identification	Remediation
Ground gases	Originated at: Marshes Mines Landfills Sewers Fresh water and marine sediments Table 4.12 describes hazards	Gas detectors in Table 4.13	Protective equipment (gas masks and breathing apparatus) during site investigation Good ventilation during construction. Containment (by lining of the sides, base and capping as well as by gas extraction to reduce pressure) Collection (by wells, layers, pipes, extraction plant, condensate system) Treatment (utilisation and flaring)
Rock burst	Table 4.14 for examples	Continuous rock Brittle strength Loss of confining stress	Use rock support (sprayed concrete, anchors, bolts) Optimisation of shape and size of excavation Sometimes additional blasting around excavation to create discontinuities

References

Aitcison GD, Wood CC (1965) Some interactions of compaction, permeability and post-construction deflocculation affecting the probability of piping failures in small dams. In: Proceedings of the 6th international conference on soil mechanics and foundation engineering, Montreal, vol 2, pp 442–446

Akkar S, Bommer JJ (2007) Empirical prediction for peak ground velocity derived from strong-motion records from Europe and the Middle East. Bull Seismol Soc Am 97:511–530

Ambraseys NN (1988) Engineering seismology. Earthq Eng Struct Dyn 17:1–105

Ambraseys N, Bilham R (2012) The Sarez-Pamir earthquake and landslide of 18 February 1911. Seismol Res Lett 83(2):294–314

Ambraseys NN, Hendron AJ (1968) Dynamic behaviour of rock masses. In: Stagg KG, Zienkiewicz OC (eds) Rock mechanics in engineering practice. Wiley, London

Ambraseys NN, Menu JM (1988) Earthquake induced ground displacements. Earthq Eng Struct Dyn 16:985–1006

Ambraseys NN, Synolakis C (2010) Tsunami catalogues for the Eastern Mediterranean, revisited. J Earthq Eng 1(3):309–330

Ambraseys NN, Douglas J, Sarma SK, Smit PM (2005) Equations for the estimation of strong ground motions from shallow crustal earthquakes using data from Europe and the Middle East: horizontal peak ground acceleration and spectral accelerations. Bull Earthq Eng 3:1–53

Amick H (1997) On generic vibration criteria for advanced technology facilities: with a tutorial on vibration data representation. J Inst Environ Sci XL(5):35–44

Andersland OB, Ladanyi B (1994) An introduction to frozen ground engineering. Chapman & Hall, New York

ANSI S3.29 (1983) Guide to the evaluation of human exposure to vibration in buildings. American national Standards Institute, Acoustical Society of America, New York, Secretariat for Committees 51, 52 and 53

ASTM C1255-11 Standard test method for analysis of uranium and thorium in soils by energy dispersive x-ray fluorescence spectroscopy. American Society for Testing and Materials, Philadelphia, PA

ASTM D4221-11 Standard test methods for dispersive characteristics of clay soil by double hydrometer. American Society for Testing and Materials, Philadelphia, PA

ASTM D4253-00(2006) Standard test methods for maximum index density and unit weight of soils using a vibratory table. American Society for Testing and Materials, Philadelphia, PA

ASTM D4646-03(2008) Standard test method for 24-h batch-type measurement of contaminant sorption by soils and sediments. American Society for Testing and Materials, Philadelphia, PA

ASTM D4647-06e1 Standard test method for identification and classification of dispersive clay soils by the pinhole test. American Society for Testing and Materials, Philadelphia, PA

ASTM D5730-04 Standard guide for site characterization for environmental purposes with emphasis on soil, rock, the vadose zone and groundwater. American Society for Testing and Materials, Philadelphia, PA

ASTM D5831-09 Standard test method for screening fuels in soils. American Society for Testing and Materials, Philadelphia, PA

ASTM D7203-11 Standard practice for screening trichloroethylene (tce)-contaminated media using a heated diode sensor. American Society for Testing and Materials, Philadelphia, PA

ASTM D7352-07 Standard practice for direct push technology for volatile contaminant logging with the membrane interface probe (MIP). American Society for Testing and Materials, Philadelphia, PA

ASTM D7458-08 Standard test method for determination of beryllium in soil, rock, sediment, and fly ash using ammonium bifluoride extraction and fluorescence detection. American Society for Testing and Materials, Philadelphia, PA

ASTM E1527 Standard practice for environmental site assessments: phase I environmental site assessment process. American Society for Testing and Materials, Philadelphia, PA

ASTM E2600-10 Standard guide for vapour encroachment screening on property involved in real estate transactions. American Society for Testing and Materials, Philadelphia, PA

Attewell PB (1978) Ground movement caused by tunnelling in soil. In: Gedddes JD (ed) Proceedings of the international conference on large movements and struct, Pentech Press, London, pp 812–948

Barry D (1991) Hazards in land recycling. In: Fleming G (ed) Recycling derelict land. Thomas Telford, London, pp 28–63

Bell FG, Culshaw MG (1998) Some geohazards caused by soil mineralogy, chemistry and micro fabric: a review. In: Maumd JG, Eddlestone M (eds) Geohazards in engineering geology, The Geological Society engineering geology special publication no. 15. Geological Society, London, pp 427–442

Bertram GE (1940) An experimental investigation of protective filters. Harvard University Graduate School, Publication No 267

Bjerrum L (1963) Discussion on Section 6. In: Proceedings – European conference for soil mechanics and foundation engineering, Wiesbaden, vol 2, pp 135–137 (also Norwegian Geotechnical Institute Publication No 98)

Blong RJ (1984) Volcanic hazards – a sourcebook on the effects of eruptions. Academic, Sydney

Bommer JJ, Elnashai AS (1999) Displacement spectra for seismic design. J Earthq Eng 3:1–32
Bormann P (ed) (2002) New manual of seismological observatory practice. GeoForschungsZentrum, Potsdam
Boscardin MD, Cording EJ (1989) Building response to excavation-induced settlement. J Geotech Eng ASCE 115(1):1–21
Bray JD, Sancio RB, Durgunoglu T, Onalp A, Youd TL, Stewart JP, Seed RB, Cetin OK, Bol E, Baturay MB, Christensen C, Karadayilar T (2004) Subsurface characterization at ground failure sites in Adapazari, Turkey. J Geotech Geoenviron Eng ASCE 130(7):673–685. http://dx.doi.org/10.1061/(ASCE)1090-0241(2004)130:7(673)
BS 1377-4 (1990) Soils for civil engineering purposes. Compaction-related tests (AMD 8259) (AMD 13925). British Standards Institution, London
BS 1377-5 (1990) Methods of test for soils for engineering purposes. Compressibility, permeability and durability tests (AMD 8260). British Standards Institution, London
BS 10175 (2011) Investigation of potentially contaminated sites – code of practice. British Standards Institution, London
BS 5228-2 (2009) Code of practice for noise and vibration control on construction and open sites – Part 2: Vibration. British Standard Institution, London
BS 5930:1999+A2:2010 Code of practice for site investigations. British Standards Institution, London
BS 6471 (1992) Guide to evaluation of human exposure to vibration in buildings (1hz to 80hz). British Standards Institution, London
BS 6472 (1992) Guide to evaluation of human exposure to vibration in buildings (1Hz to 80Hz). British Standard Institution, London
BS 7385–2 (1993) Evaluation and measurement for vibration in buildings, Part 2: Guide to damage levels from ground borne vibration. British Standard Institution, London
Bugge T, Beldderson RH, Kenyon NH (1988) The Storegga slide. Phil Trans R Soc A 325:357–388
Burdick JL, Rice EF, Arvind P (1978) Chapter 1: Cold regions: descriptive and geotechnical aspects. In: Geotechnical engineering for cold region. McGraw-Hill, New York
Burland JB (1969) Reply to discussion. In: Proceedings of the conference on in Situ investigations of soil and rock. Institute of Civil Engineering, London, UK 62
Burland JB (1995) Assessment of risk of damage to buildings due to tunnelling and excavations. Invited Special Lecture to IS-Tokyo'95: 1st international conference on earthquake geotechnical engineering
Cadergen HR (1997) Seepage, drainage and flow nets, 3rd edn. Wiley, New York
Cairney T (1995) The re-use of contaminated land. A handbook of risk assessment. Wiley, New York
Casagrande A (1932) A new theory on frost heaving. Discussion. Highw Res Board HRB Proc 11:168–182
Charles JA (1993) Building on fill: geotechnical aspects. Building Research Establishment Report, UK
CIRIA C683 (2007) Chapter 5: Physical properties and design tools. In: Rock manual, the use of rock in hydraulic engineering, 2nd edn. Construction Industry Research and Information Association, London
Clough RW, Penzien J (1993) Dynamics of structures, 2nd edn. McGraw Hill, New York
Clough GW, Smith EM, Sweeney BP (1989) Movement control of excavation support systems by iterative design procedure. ASCE Found Eng Curr Princ Pract 1:869–884
Crawford CB (1968) Quick clays of eastern Canada. Eng Geol 2(4):239–265
Crowhurst D (1987) Measurement of gas emissions from contaminated land. Building Research Establishment Report, UK
Dean ETR (2009) Offshore geotechnical engineering – principles and practice. Thomas Telford, London

Dean ETR (2010) Offshore geotechnical engineering: principles and practice. Thomas Telford, London
DIN 4150-3 (1999) Erschütterungen im Bauwesen – Teil 3: Einwirkungen auf bauliche Anlage Norm Ausgabe, Deutsch, Bestellen beim Beuth Verlag
Donaldson GW (1969) The occurrence of problems of heave and factors affecting its nature. In: Proceedings of the 2nd international research and engineering conference on expansive clay soil, Texas
Douglas J (2011) Ground motion prediction equations, 1964–2010. Pacific Earthquake Engineering Research Report PEER 2011/102. http://peer.berkeley.edu/publications/peer_reports/reports_2011/webPEER-2011-102-Douglas.pdf
Dowding CH (2000) Construction vibration. Reprinted 1996 version. Prentice Hall
Driscoll R (1984) The influence of vegetation on the swelling and shrinking of clay soils in Britain. In: Wakeling TRM (ed) The influence of vegetation on clays. Thomas Telford, London
Drouin M, Michel B (1971) Les poussees d'origine thermique exercees par les couverts de glace sur les structures hydrauliques. Rapport S-23. Faculte des Sciences, Genie Civil, Universite Laval, Quebec, Canada
Earth Manual (1980) A guide to the use of soils as foundations and as construction materials for hydraulic structures, 2nd edn. U.S. Department of the Interior, Water and Power Resources Service, Engineering and Research Centre, Denver
Ehgashira K, Ohtsubo M (1982) Smectite in marine quick-clays of Japan. Clay Clay Miner 30(4):275–280
Ekstrom A, Olofsson T (1985) Water and frost – stability risks for embankments of fine grained soils. In: Symposium on failures in earthworks. Institution of Civil Engineers, London, pp 155–166
EN 1997-1 (2004) Eurocode 7: Geotechnical design – Part 1: General rules. European Committee for Standardization, Brussels
EN 1998-1:2004(E) Eurocode 8: design of structures for earthquake resistance – Part 1: General rules, seismic actions and rules for buildings. European Committee for Standardization, Brussels
EN 1998-5:2004(E) Eurocode 8: design of structures for earthquake resistance – Part 5: Foundations, retaining structures and geotechnical aspects. European Committee for Standardization, Brussels
Ford D, Williams PW (1989) Karst geomorphology and hydrology. Unwin Hyman, London
Fookes PG (ed) (1997) Tropical residual soils. The Geological Society, London
Gaba AR, Simpson B, Powrie W, Beadman DR (2003) Embedded retaining walls – guidance for economic design. Construction Industry Research and Information Association Report C580, London
Geertsema M, Torrance JK (2005) Quick clay from the Mink Creek landslide near Terrace, British Columbia: geotechnical properties, mineralogy, and geochemistry. Can Geotech J 42(3):907–918
Grasso D (1993) Hazardous waste site remediation, source control. Lewis Publishers, Boca Raton
Grattan-Bellew PE, Eden WJ (1975) Concrete deterioration and floor heave due to biological weathering of underlying shale. Can Geotech J 12:372–378
Hamada M, O'Rourke T (eds) (1992) Case studies of liquefaction and lifeline performance during past earthquakes. National Centre for Earthquake Engineering Research, State University of New York at Buffalo, Technical Report NCEER-92-0001&2
Harder LFJ, Boulanger RW (1997) Application of K3 and K2 correction coefficients. In: Proceeding of the NCEER workshop on evaluation of liquefaction resistance of soils. National Center for Earthquake Engineering Research State Union of New York at Buffalo, pp 167–190
Harr ME (2003) Groundwater and seepage (3rd print). McGraw-Hill International
Hiller DM, Crabb GI (2000) Ground borne vibration caused by mechanised construction works. Transportation Research Laboratory Report 429, UK

Hoek E, Brown ET (1980) Underground excavation in rock. The Institution of Mining and Metallurgy, London
Hoffmans GJCM, Verheij HJ (1997) Scour manual. Balkema, Rotterdam
Holtz RD, Kovacs WD (1981) An introduction to geotechnical engineering. Prentice-Hall, Englewood Cliffs
Hryciw RD, Vitton S, Thomann TG (1990) Liquefaction and flow failure during seismic exploration. J Geotech Eng Am Soc Civil End 116:1881–1899
Hutchinson JN, Coope GR (2002) Cambering and valley bulging, periglacial solifluction and late glacial coleopteran at Dowdeswell, near Cheltenham. In: Proceedings of the Geologists' Association, UK, vol 113, pp 291–300
ICRCL (1987) Guidance on the assessment and redevelopment of contaminated land, 2nd edn. Interdepartmental Committee on the Redevelopment of Contaminated Land, Department of the Environment, London
ICRP (1977) Sowby FD (rd) Recommendations of the International Commission on Radiological Protection. Publications No. 26, Annals of the ICRP. Pergamon Press, Oxford, 1:3
Idriss IM (1990) Response of soft soil sites during earthquakes. In: Proceedings of the H. Bolton seed memorial symposium. BiTech Publishers, Vancouver, vol 2, pp 273–290
Ishihara K (1993) Liquefaction and flow failure during earthquakes. Geotechnique 43(3):351–415
Ishihara K, Yoshimine M (1992) Evaluation of settlements in sand deposits following liquefaction during earthquakes. Soils Found 32(1):173–188
IWM (1998) The monitoring of landfill gas, 2nd edn. Institute of Waste Management, Northampton
Kaiser PK, Cai M (2012) Design of rock support system under rock burst condition. J Rock Mech Geotech Eng 4(3):215–227
Kaplar CW (1974) U.S. Army Cold Regions Research Laboratory, CRREL Report 250
Keefer DK (1984) Landslides caused by earthquakes. Bull Geol Soc Am 95:406–421
Koerner RM (1985) Construction and geotechnical methods in foundation engineering. McGraw-Hill, New York
Koerner RM (1998) Designing with geosynthetics, 4th edn. Prentice Hall, Englewood Cliffs
Kokusho T (2007) Liquefaction strengths of poorly-graded and well-graded granular soils investigated by lab tests. In: Pitilakis KD (ed) Earthquake geotechnical engineering, proceedings of the 4th international conference on earthquake geotechnical engineering-invited lectures. Geotechnical, geological and earthquake engineering series, vol 6. Springer
Kramer SL (1996) Geotechnical earthquake engineering. Prentice Hall, Englewood Cliffs
Lane EW (1935) Security from under-seepage: masonry dams on earth foundations. Trans ASCE 100:1235–1272
LFTGN 03 (2004) Guidance on the management of landfill gas. Environmental Agency, UK
Liao S, Whitman RV (1986a) Overburden correction factor for SPT in sand. J Geotech Eng ASCE 112(3):373–377
Liao S, Whitman RV (1986b) Catalogue of liquefaction and non-liquefaction occurrences during earthquakes. Research Report, Department of Civil Engineering, Massachusetts Institute of Technology, Cambridge, MA
Little DN, Nair S, Herbert B (2010) Addressing sulphate-induced heave in lime treated soil. J Geotech Geoenviron Eng ASCE 136(1):110–118
Liu L, Dobry R (1997) Seismic response of shallow foundation on liquefiable sand. J Geotech Geoenviron Eng ASCE 123(6):557–567
Luettich SM, Giroud JP, Bachus RC (1992) Geotextile filter design guide. J Geotext Geomembr 11(4–6):19–34
Lungren DG, Silver D (1980) Ore leaching by bacteria. Annu Rev Microbiol 34:263–283
Mair RJ, Taylor RN (1997) Bored tunnelling in the urban environment. In: Proceedings of the XIV international conference on soil mechanics and foundation engineering. Balkema, Hamburg, vol 4, pp 2353–2385

Mair RJ, Taylor RN, Bracegirdle A (1993) Subsurface settlement profiles above tunnels in clay. Geotechnique 43(2):315–320

May RW, Ackers JC, Kirby AM (2002) Manual of scour at bridges and other hydraulic structures. Construction Industry Research and Information Association (CIRIA) Report C551

Mayne PW (1985) Ground vibrations during dynamic compaction. In: Gazetas G, Selig ET (eds) Vibrations problems in geotechnical engineering. Proceedings of ASCE convention in Detroit, Michigan, pp 247–265

McCalpin J (1996) Paleoseismology. Academic, New York

Mitchell JK, Houston WN (1969) Causes of clay sensitivity. J Soil Mech Found Div ASCE 95(3):845–869

Mortimore RN (1990) Chalk. Thomas Telford, London

New BM (1986) Ground vibration caused by civil engineering works. U.K. Transport and Road Research Laboratory Research Report 53

New-mark NM (1965) Effect of earthquakes on dams and embankments. Geotechnique 15:139–160

Nixon JF, Ladanyi B (1978) Thaw consolidation. In: Andersland OB, Anderson DM (eds) Geotechnical engineering in cold regions. McGraw-Hill, New York, pp 164–215

O'Reilly MP, New BM (1982) Settlements above tunnels in the United Kingdom – their magnitude and prediction. Tunnelling 82, London, IMM 173–181

Obermeier SF (1996) Using liquefaction-induced features for paleoseismic analysis. In: McCalpin J (ed) Paleoseismology. Academic Press, San Diego

Olson SM (2001) Liquefaction analyses of level and sloping ground using field case histories and penetration resistance. PhD thesis, The Graduate College, University of Illinois at Urbana-Champaign (http://www.geoengineer.org/onlinelibrary/paper_details.php?paperid=1530)

Palmstrom A (1995) Characterising rock burst and squeezing by the rock mass index. In: Proceedings of the conference on design and construction of underground structures. New Delhi, vol 2, pp 1–9

Penner E (1970) Frost heaving forces in Leda clay. Can Geotech J 7(1):8–16

PHRI (1997) Handbook on liquefaction remediation of reclaimed land. Port and Harbour Research Institute, Ministry of Transport, Japan. Translated by Waterways Experiment Station, US Army Corps of Engineers, USA. Balkema, Rotterdam/Brookfield

Preene M, Roberts TOL, Powrie W, Dyer MR (2000) Groundwater control – design and practice. Construction Industry Research and Information Association (CIRIA) Report C515

Privett KD, Matthews SC, Hodges RA (1996) Barriers, liners and cover systems for containment and control of land contamination. Construction Industry Research and Information Association Special Publication 124

Puppala AJ, Griffin JA, Hoyos LR, Chomtid S (2004) Studies of sulphate-resistant cement stabilization methods to address sulphate-induced soil heave. J Geotech Geoenviron Eng ASCE 130(4):391–402

Pye K, Miller JA (1990) Chemical and biochemical weathering of pyritic mudrocks and shale embankments. Q J Eng Geol 25:365–381

Rankka K, Andersson-Skold I, Hulten K, Larson R, Leroux V, Dahlin T (2004) Quick clay in Sweden. Swedish Geotechnical Institute Report 65

Rodriguez CE, Bommer JJ, Chandler RJ (1999) Earthquake-induced landslides: 1980–1997. Soil Dyn Earthq Eng 18:325–346

Rollings RS, Burkes JP, Rollings MP (1999) Sulphate attack on cement-stabilised sand. J Geotech Geoenviron Eng ASCE 125(5):364–372

Rowe RK, Quigley RM, Booker JR (1997) Clayey barrier systems for waste disposal facilities. E & FN SPON, London

Scholz CH, Aviles C, Wesnousky S (1986) Scaling differences between large intraplate and interplate earthquakes. Bull Seismol Soc Am 76:65–70

Seed HB, Chan CK (1959) Thixotropic characteristics of compacted clay. Trans Am Soc Civ Eng 124:894–916

Seed HB, Idriss IM (1971) Simplified procedure for evaluating soil liquefaction potential. J Geotech Eng Div ASCE 97(9):1249–1273
Seed HB, Woodward RJ, Lundgran R (1962) Prediction of swelling potential of compacted clays. J Soil Mech Found Eng ASCE 55:107–131
Seed HB, Tokimatsu K, Harder LF, Chung RM (1985) The influence of SPT procedures in soil liquefaction evaluations. J Geotech Eng ASCE 111(2):1425–1445
Selmer-Olsen R (1988) General engineering design procedures. Norwegian Tunnelling Today, Tapir, pp 53–58
Sherard JL, Dunnigan LP, Decker RS (1977) Some engineering problems with dispersive clays. In: Sherard JL, Decker RS (eds) Proceedings of a symposium on dispersive clays, related piping and erosion in geotechnical projects, ASTM Special Publication, vol 623, pp 3–12
Sing A, Mitchell JK (1968) General stress–strain-time functions for soils. J Soil Mech Found Div ASCE 94(1):21–46
Skempton AW, MacDonald DH (1956) The allowable settlement of buildings. In: Proceedings of the Institution of Civil Engineers UK 5(3), pp 737–784
Skobelev SF, Hanon M, Klerkx J, Govorova NN, Lukina NV, Kazmin VG (2004) Active faults in Africa: a review. Tectonophysics 380:131–137
Smith M (1991) Data analysis and interpretation. In: Fleming G (ed) Recycling derelict land. Thomas Telford, London, pp 88–144
Srbulov M (2003) An estimation of the ratio between peak horizontal accelerations at the ground surface e and at depth. Eur Earthq Eng XVII(1):59–67
Srbulov M (2010a) Ground vibration engineering: simplified analyses with case studies and examples. Springer, London
Srbulov M (2010b) Simple considerations of liquefaction due to earthquakes and flow failures of slopes in sandy soil. Ingegneria Sismica XXVII(3):7–16
Stewart IS, Hancock PL (1990) What is fault scarp? Episodes 13:256–263
Stone K, Murray A, Cooke S, Foran J, Gooderhan L (2009) Unexploded ordnance (UXO) – a guide for the construction industry. Construction Industry Research and Information Association Report C681
Sundborg A (1956) The river Klaralven. A study of fluvial processer. Geografiska Ann 38(3):238–316
Tang C, Wang J, Zhang J (2010) Preliminary engineering application of micro seismic monitoring technique to rock burst prediction in tunnelling of Jinping II project. J Rock Mech Geotech Eng 2(3):193–208
Task Force 25 (1991) Joint Committee Report of AASHTO-AGC-ARTBA, American Association of State, Highway and Transportation Officials, Washington, DC
Taylor T, Fragaszy RJ, Ho CL (1991) Projectile penetration in granular soil. J Geotech Eng ASCE 117(4):658–672
Tchobanoglous G, Theisen H, Vigil SA (1993) Integrated solid waste management: engineering principles and management issues. McGraw-Hill, New York
Ter-Stepanian G (2000) Quick clay landslides: their enigmatic features and mechanism. Bull Eng Geol Environ 59:47–57
Tocher D (1960) The Alaska earthquake of July 10, 1958: movement on the Fairweather fault and field investigation of southern epicentral region. Bull Seismol Soc Am 50(2):267–292
Tocher D, Miller DJ (1959) Field observations on effects of Alaska earthquake of 10 July, 1958. Science 129(3346):394–395
Trifonov VG, Vostrikov GA, Trifonov RV, Karakhanian AS, Soboleva OV (1999) Recent geodynamic characteristics in the Arabian-Eurasian and Indian-Eurasian collision region by active fault data. Tectonophysics 308:119–131
U.S. Bureau of Mines (1971) Blasting vibrations and their effects on structures. Bulleting 656 by Nicholls HR, Johnson CF, Duvall WI
U.S. Navy (1971) Design manual – soil mechanics, foundations and earth structures, NAVFAC DM-7, Washington, DC

USBM RI 8507 (1980) by Siskind DE, Stagg MS, Kopp JW, Dowding CH, Structure response and damage produced by ground vibrations from surface blasting. U.S. Bureau of Mines, Washington, DC

Van Der Merwe DH (1964) The prediction of heave from the plasticity index and percentage clay fraction of soil. S Afr Inst Civ Eng 6:103–107

Varnes DJ (1978) Slope movement types and processes. In: Schuster RL, Krizek RJ (eds) Landslides – analysis and control, Special report (National Research Council (U.S.). Transportation Research Board), 176. National Academy of Sciences, Washington, DC, pp 11–33

Varnes DJ (1984) Landslide hazard zonation: a review of principles and practice. UNESCO, Paris

Wells DL, Coppersmith KL (1994) New empirical relationships among magnitude, rupture length, rupture width, rupture area, and surface displacement. Bull Seismol Soc Am 84:974–1002

Wolf JP (1994) Foundation vibration analysis using simple physical models. PTR Prentice Hall, Englewood Cliffs

Youd TL, Idriss IM, Andrus RD, Arango I, Castro G, Christian JT, Dobry R, Finn WDL, Harder LF Jr, Hynes ME, Ishihara K, Koester JP, Liao SSC, Marcuson WFIII, Martin GR, Mitchell JK, Moriwaki Y, Power MS, Robertson PK, Seed RB, Stokoe KHII (2001) Liquefaction resistance of soils: summary report from the 1996 NCEER and 1998 NCEER/NSF workshop on evaluation of liquefaction resistance of soils. J Geotech Geoenviron Eng ASCE 127(10):817–833

Chapter 5
Geo-structures

Abstract This chapter contains descriptions of types, stabilities/capacities, movements and executions of the following geo-structures:

- Ground slopes
- Foundations
- Retaining walls
- Anchors, bolts and nails
- Reinforced soil
- Tunnels and shafts
- Pipes
- Landfills
- Fill and tailing dams
- Road and railway subgrade
- Offshore foundations

5.1 Ground Slopes

5.1.1 Types

The following slope instabilities are considered:

- Rock falls
- Avalanches and fast debris spread
- Soil flows
- Slides

Figure 5.1 shows sketches of the above types of instabilities.

M. Srbulov, *Practical Guide to Geo-Engineering: With Equations, Tables, Graphs and Check Lists*, Geotechnical, Geological and Earthquake Engineering 29,
DOI 10.1007/978-94-017-8638-6_5,

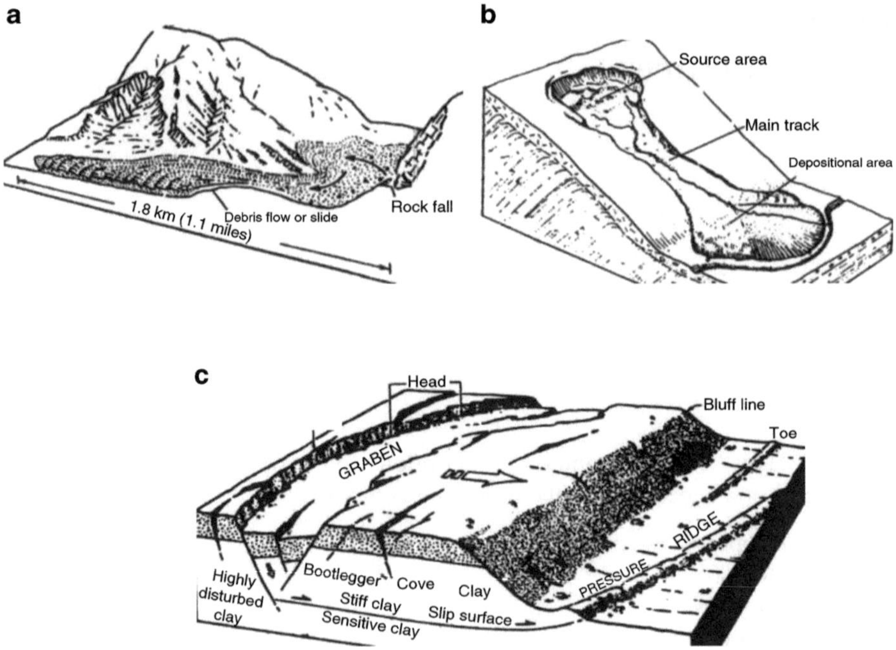

Fig. 5.1 Sketches of (**a**) rock fall and fast debris flow – avalanche, (**b**) soil flow, (**c**) soil block type slide (Adapted from Schuster RL and Krizek RJ (eds) Landslides – analysis and control (1978) Transportation Research Board of the U.S. National Academy of Sciences, Special Report 176)

5.1.2 Stability/Capacity

5.1.2.1 Rock Falls

Triggering of rock fall occurs when the overturning moments of destabilising forces (earthquake inertial force, water/ice/swelling pressure in the back crack, uplift at the base) exceed the overturning moments of stabilising forces (self weight and anchor/cable force) around the face edge of a rock block as sketched in Fig. 5.2.

5.1.2.2 Avalanches and Fast Debris Spreads

- In general case, triggering of fast debris spreads and avalanches is caused by three dimensional slope instability of a rock mass. Wedge shaped three dimensional slope instability (Fig. 5.3) is a simplest form considered.
- Equal area projection is used to visualise traces of the wedge bounding planes (Hoek and Bray 1981), Fig. 5.4.
- Only the intersection of a plane with the lower half of the reference sphere is sufficient to be projected on the horizontal plane, because the other half is symmetric. The azimuth of direction and the inclination of the line of

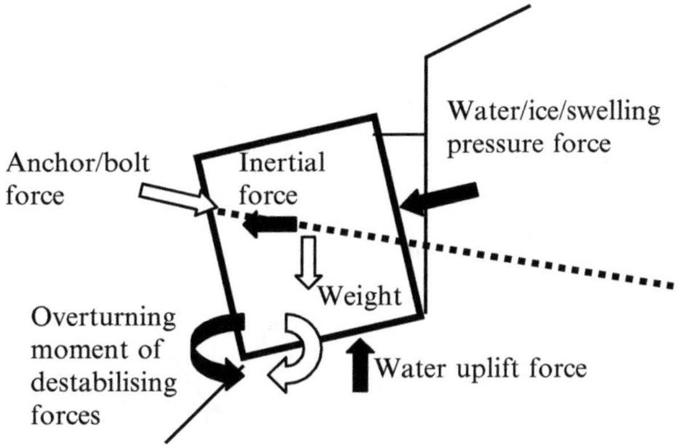

Fig. 5.2 Sketch of stabilising (*white arrows*) and destabilising (*black arrows*) forces and overturning moments around the face edge of a block

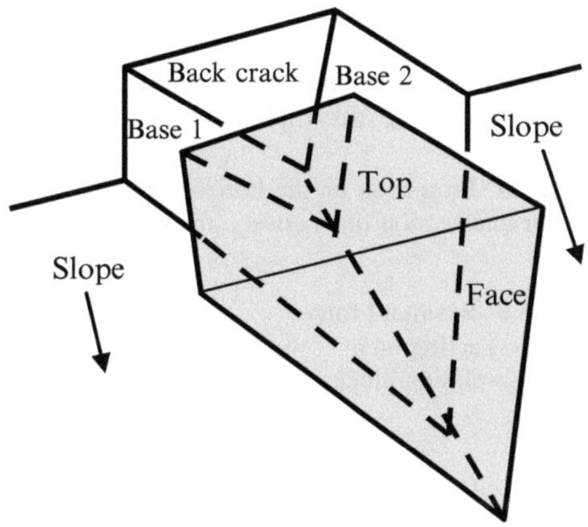

Fig. 5.3 Geometry of a tetrahedral translational wedge with the bounding planes

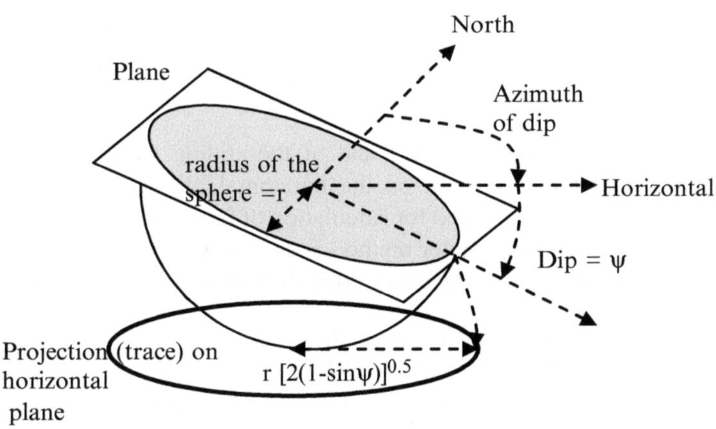

Fig. 5.4 Method of construction of an equal area projection by intersection of a plane with a reference sphere with the radius r

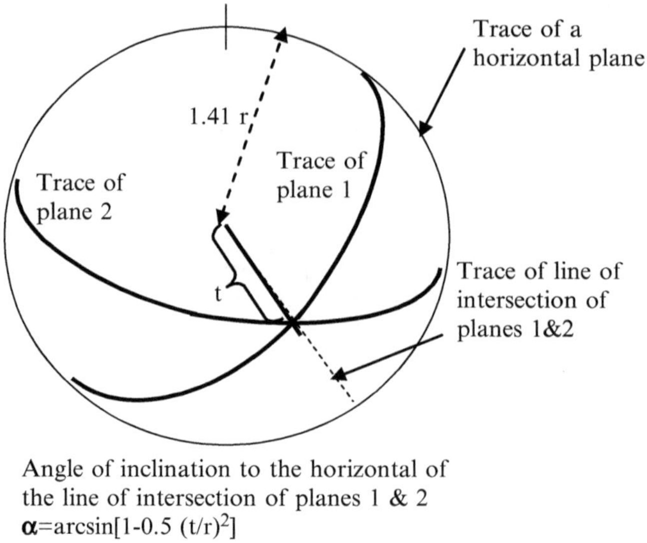

Fig. 5.5 Projection on the horizontal plane of the lower traces of the planes 1 and 2 and of the lower intersection line between them

Fig. 5.6 Resultant forces (*arrows*) acting on a wedge – sliding block

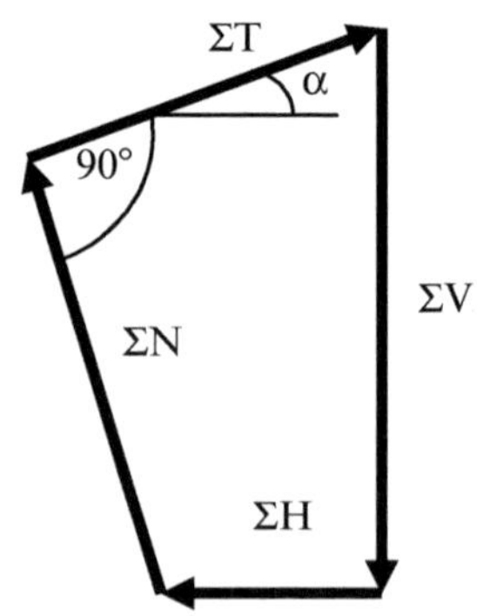

intersection of the planes 1 and 2 and of the sliding wedge (block) can be determined from the projection on the horizontal plane, Fig. 5.5. Hoek and Bray (1981) described a method for calculation of factor of safety against sliding of a wedge. Quicker and simpler method is as follows.

- From the sketch of forces acting on a wedge slide (Fig. 5.6), it is possible to write:

$$\begin{aligned}\Sigma N &= \Sigma V \cdot \cos\alpha - \Sigma H \cdot \sin\alpha \\ \Sigma T &= \Sigma V \cdot \sin\alpha + \Sigma H \cdot \cos\alpha \\ \Sigma R &= \Sigma c \cdot A + \Sigma N \cdot \tan\phi - \Sigma U \cdot \tan\phi\end{aligned} \tag{5.1}$$

ΣN is the sum of forces normal to the line of intersection of two sliding planes
ΣV is the sum of all vertical forces acting on a wedge

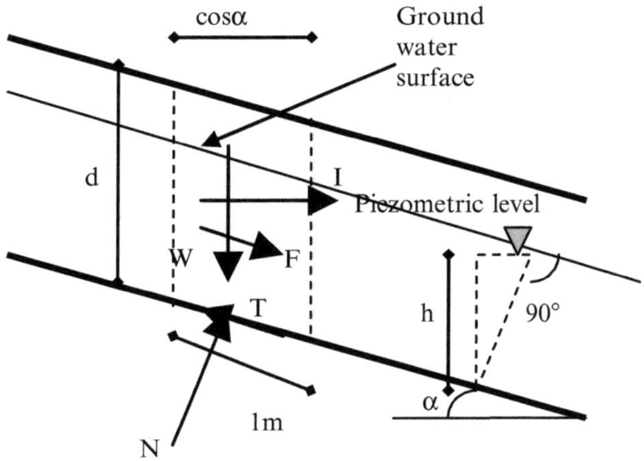

Fig. 5.7 Forces (*arrows*) acting on a planar slip surface

α is angle of inclination to the horizontal of the intersection line between sliding planes 1 and 2
ΣH is the sum of all horizontal forces acting on a wedge in the direction of a vertical plane passing through the line of intersection of the planes 1 and 2
ΣT is the sum of forces parallel to the line of intersection of two sliding planes
ΣR is the sum of all resistant forces acting along planes 1 and 2
c is the cohesion existing at the planes 1 and 2
A are the contact areas of the planes 1 and 2
ϕ is the angle of friction existing at the planes 1 and 2
U are the uplift forces (if any) acting at the planes 1 and 2
The condition for wedge instability is $\Sigma T > \Sigma R$.

More detailed approach has been used by Srbulov (2008) among others.

5.1.2.3 Soil Flow

The triggering condition is determined according to Sect. 4.3.3 using Eq. (4.13) and Figs. 4.8 and 4.9.

5.1.2.4 Slides

For slopes which soil achieve their peak shear strength at similar deformations and their peak shear strength does not decrease significantly with further increase in deformations:

- **Planar slides** (Fig. 5.7) factor of safety F_s against sliding is:

$$F_s = \frac{c + [\gamma \cdot d \cdot \cos\alpha \cdot (\cos\alpha - k_h \cdot \sin\alpha) - \gamma_w \cdot h] \cdot \tan\phi}{\gamma \cdot d \cdot \cos\alpha \cdot (\sin\alpha + k_k \cdot \cos\alpha) + \gamma_w \cdot {}^{\mathrm{h}}/\cos\alpha \cdot \tan\alpha} \tag{5.2}$$

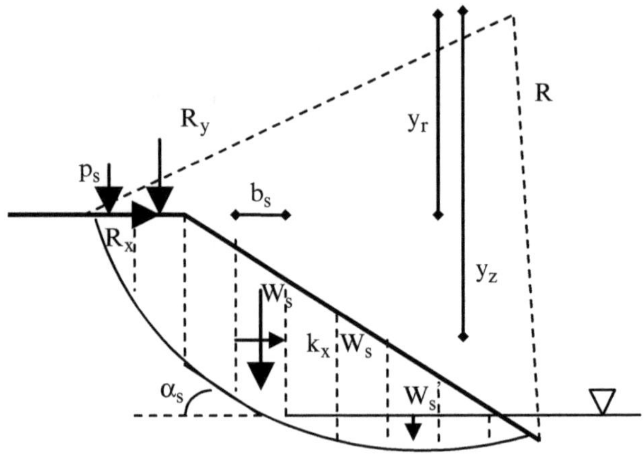

Fig. 5.8 Cross section through a circular cylindrical slip surface with marked basic parameters

γ is ground unit weight
d is the vertical depth to slip surface
α is the inclination to the horizontal of the slip surface
γ_w is water unit weight
h is the piezometric height above the slip surface
k_h is the ratio between the horizontal and the gravitational acceleration, the vertical acceleration can be accounted for by multiplying γ by $(1 \pm k_v)$, where k_v is the ratio between the vertical and the gravitational acceleration
c is cohesion
ϕ is friction angle
Other forces shown in Fig. 5.7 are:
Ground weight $W = \gamma\ d\ cos\alpha$
Water filtration force $F = \gamma_w\ h/\cos\alpha\ \tan\alpha$
Inertial force $I = k_h\ \gamma\ d\ cos\alpha$
Normal force $N = \gamma\ d\ cos\alpha\ (cos\alpha\ -\ k_h sin\alpha)$
Tangential force $T = \gamma\ d\ cos\alpha(sin\alpha\ +\ k_h\ cos\alpha)\ +\ \gamma_w\ h/\cos\alpha\ \tan\alpha$
Vertical surcharge along slope surface has not been considered

When $F_s = 1$, critical slope inclination α_c or critical horizontal acceleration ratio $k_{h,c}$ can be determined by an iterative procedure.

- **Circular cylindrical slides** (Fig. 5.8) factor of safety F_s against sliding is according to the routine method by Bishop (1955) for static condition:

$$F_s = \frac{\sum [c' \cdot b_s + (W_s - u_w \cdot b_s) \cdot \tan\phi'] \cdot \dfrac{1}{\cos\alpha_s + \sin\alpha_s \cdot \dfrac{\tan\phi'}{F_s}}}{\sum (W_s \cdot \sin\alpha_s)} \tag{5.3}$$

c' is soil cohesion in terms of effective stresses in drained conditions,
b_s is the width of a vertical slice into which a potential sliding mass is divided,
W_s is the total weight of a vertical slice,
u_w is pore water pressure at the base of a vertical slice,
ϕ' soil friction angle at the base of a vertical slice,
α_s is inclination to the horizontal of the base of a vertical slice
Σ is the sum for all vertical slices considered

Equation (5.3) contains the factor of safety F_s implicitly and is solved iteratively.

Equation (5.3) is derived for static condition and does not take into account inertial force in soil during an earthquake. This effect can be accounted for by considering apparent ground surface inclination for an angle necessary to rotate the resultant of inertial and gravitational force so that the resultant force acting on soil is vertical. Maksimovic (2008) suggests and extension of Eq. (5.3) when external loads and inertial forces act on a soil mass.

$$
\begin{aligned}
F_s &= \frac{\Sigma[c' \cdot b_s + (R_v - u_s \cdot b_s) \cdot \tan\phi'] \cdot m_\alpha}{\Sigma(R_v \cdot \sin\alpha_s + M_x)} \\
R_v &= W_s + W_s' + R_y + p_s \cdot b_s + k_y \cdot (W_s + W_z) \\
M_x &= R_x \cdot {}^{y_r}/_R + k_x \cdot y_z \cdot {}^{(W_s+W_z)}/_R \\
m_\alpha &= \frac{1}{\cos\alpha_s + \sin\alpha_s \cdot \tan\phi'/F_s}
\end{aligned}
\tag{5.4}
$$

R_x, R_y are the resultants of horizontal and vertical forces respectively (if any) acting on a vertical slice. Equation (5.4) can be used when $F_s = 1$ to calculate the horizontal force R_x acting on a rotating structure (e.g. contiguous or secant piles wall)
u_s is pore water pressure above steady state water level for a partially submerged slope,
p_s is line vertical load (if any) acting on a vertical slice,
k_y, k_x are the ratios between the vertical and horizontal inertial and the gravitational acceleration, which may vary along the slope height,
W_s is the weight of part of a vertical slice above steady state water level,
W_s' is the submerged weight of part of a vertical slice below steady state water level,
W_z is the saturated weight of part of a vertical slice below steady state water level,
y_r, y_z are liver arms of the horizontal components of external and inertial forces respectively with respect to the centre of a trial circular slip surface with the radius R
Σ is the sum for all vertical slices considered

EN 1998-5 (2004) suggests that for pseudo-static analyses of slope stability k_x in the horizontal direction is taken equal to a half of the horizontal acceleration

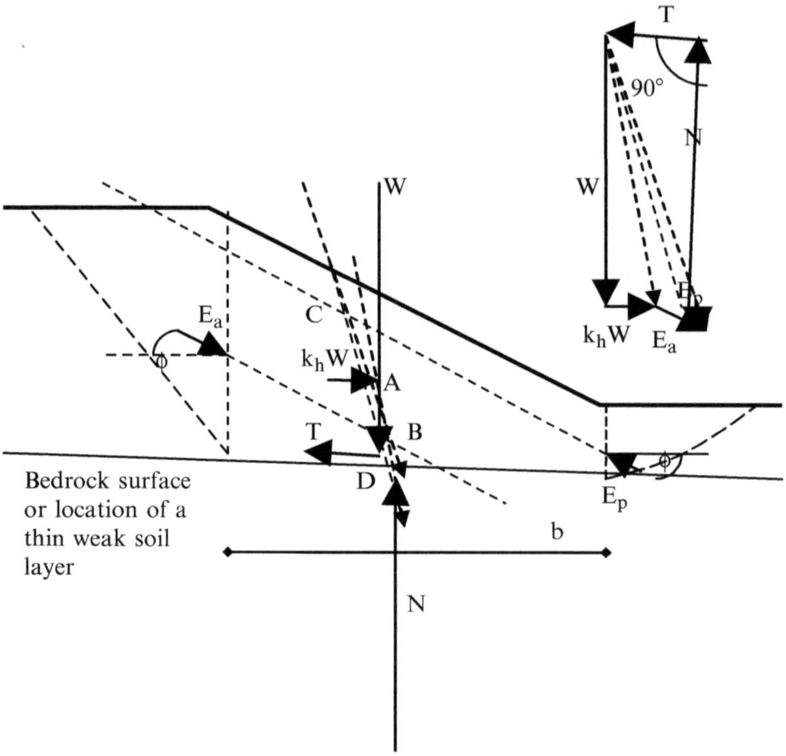

Fig. 5.9 Cross section through a prismatic slip surface with the polygon of forces (*arrows*) acting on it

divided by the gravitational acceleration and k_y in the vertical direction as a half of k_x if the vertical acceleration is greater than 0.6 of the design bedrock horizontal acceleration and 0.33 of k_x if the vertical acceleration is smaller than 0.6 of the design bedrock horizontal acceleration. EN 1998-5 (2004) in informative annex A suggests for deep seated slips where the failure surface passes near to the base that *if pseudo-static method of analysis is used, the topographic effects may be neglected.*

The minimal required factors of safety against bearing, sliding and overturning type failures are defined in local codes.

- **Prismatic slides** (Fig. 5.9) can be analysed quickly as described below for a seismic case:

1. Calculate and draw forces W, k_hW, E_a, E_p along the lines of their actions.

   W is the total self weight
   k_h is the ratio between averaged horizontal and the gravitational acceleration
   E_a is the active soil pressure force calculated from Eq. (5.64) in seismic case (from Eq. (5.62) in static case)

E_p is the passive soil pressure force calculated from Eq. (5.65) in seismic case (from Eq. (5.63) in static case)

2. Extend the direction of the resultant of forces W and k_hW passing through point A until its intersection at point B with the direction of force E_a
3. Extend the direction of the resultant of forces W, k_hW and E_a passing through point B until its intersection at point C with the direction of force E_p
4. Extend the direction of the resultant of forces W, k_hW, E_a and E_p passing through point C until its intersection at point D with the basal part of the slip surface
5. Through the ends of the polygon of forces W, k_hW, E_a and E_p (Fig. 5.9) draw lines parallel to and perpendicular at the basal part of the slip surface to find the values of the tangential T and normal N components of the resultant force acting at point D. This way, the equilibrium conditions for the acting forces and the overturning moments are satisfied.
6. The factor of safety F_s against sliding is calculated as the ratio between the resistant force $R = c\ b + (N - U)\tan\phi$ and the driving force T

c is cohesion at the basal part of the slip surface
b is the length of the basal part of the slip surface
N is the total normal force acting on the basal part of the slip surface
U is the water pressure force acting along the basal part of the slip surface over the length b
ϕ is the averaged angle of friction at the basal part of the slip surface over the length b

The static case without inertial forces ($k_h = 0$), the effect of vertical acceleration (multiplying W by factor $1 \pm k_v$) and external forces (surcharge, anchors, nails, retaining wall reaction) can be analysed by adding the forces to the polygon of forces in Fig. 5.9.

- **Wedge slides** (Fig. 5.3) factor of safety F_s against sliding is the ratio between ΣR and ΣT forces from Eq. (5.1).

For slopes composed of different soil that achieve the peak shear strength at different deformations and/or their peak shear strengths decrease significantly with further increase in deformations to much smaller residual values (heavily over consolidated and highly plastic clay), the assumption that all shear strengths are achieved simultaneously at all places in classical limit equilibrium methods is not correct. Srbulov (1987) proposed the use of local factors of safety at the bases and interfaces of wedges (or vertical slices) of a sliding body.

$$F_j = F_i \cdot \frac{\delta^{k1}{}_j}{\delta^k{}_i} \cdot \frac{\Delta^k{}_i}{\Delta^{k1}{}_j} \geq 1 \qquad (5.5)$$

F_j is the local factor of safety at place j
F_i is the local factor of safety at place i to be determined from the $3n$ equilibrium conditions of forces in two perpendicular directions and of overturning moments for the number n of wedges of a sliding body, from Eq. 5.5 at

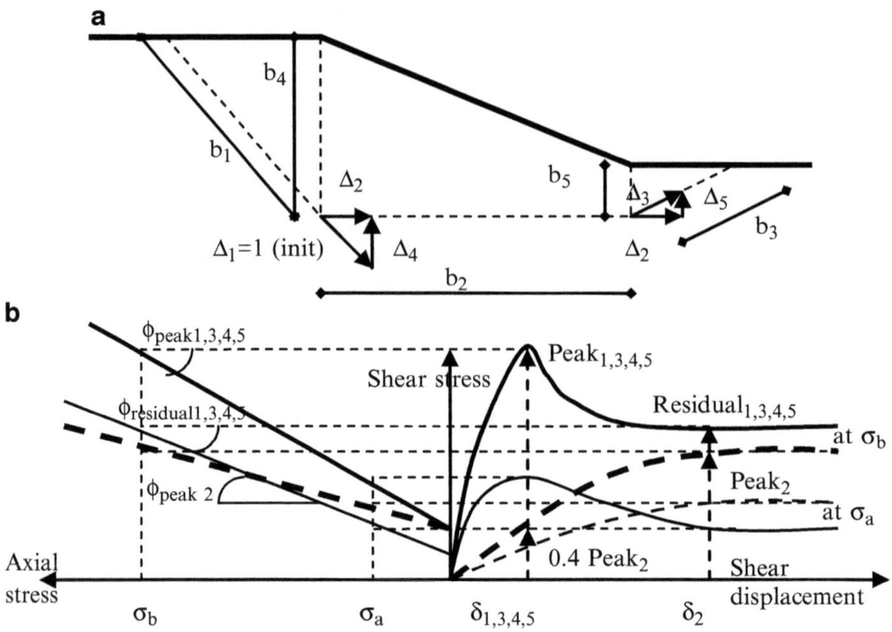

Fig. 5.10 (**a**) Kinematically possible displacements Δ_{1-5} for 3 rigid wedges of a sliding body, (**b**) Force – displacement relationships for 2 soil types along the joints 1,3,4,5 and the base 2. *Arrows* indicate different combinations of the shear strengths along the joints 1,3,4,5 and the base 2

$2n - 2$ locations and the equations $T_j = [c'_j b_j + (1 - r_{u,j})\ tan\phi j]]/F_j$ at $2n - 1$ locations, T_j is the shear force at the place j, c'_j is the cohesion at place j, b_j is the width of contact surface at place j, $r_{u,j}$ is the pore pressure coefficient at place j, ϕ_j is the friction angle at place j

$\delta^{k1}{}_j$ is the displacement at place j at which the peak shear strength is achieved (Fig. 5.10), ^{k1} is the exponent of a non-linear force-displacement relationship at place j

$\delta^{k}{}_i$ is the displacement at place i at which the peak shear strength is achieved (Fig. 5.10), ^{k} is the exponent of a non-linear force-displacement relationship at place i

$\Delta^{k}{}_i$ is the kinematically possible displacement at place i based on the kinematics of assumed rigid wedges of a sliding body (Fig. 5.10), ^{k} is the exponent of a non-linear force-displacement relationship at place i

$\Delta^{k1}{}_j$ is the kinematically possible displacement at place j based on the kinematics of assumed rigid wedges of a sliding body (Fig. 5.10), ^{k1} is the exponent of a non-linear force-displacement relationship at place j

If calculated $F_j < 1$ then $\delta^{k1}{}_j$ is increased at place j so that $F_j = 1$ with corresponding decrease in the mobilised shear strength below the peak value.

An average factor of safety F_{avr} of a sliding body

$$F_{avr} = \frac{\sum_{j=1}^{2n-1} \tau_{a,j} \cdot b_j}{\sum_{j=1}^{2n-1} \frac{\tau_{a,j} \cdot b_j}{F_j}} \tag{5.6}$$

$\tau_{a,j}$ is the peak shear strength at place j when $F_j > 1$ or the mobilised post peak shear strength for $F_j = 1$
b_j is the width of contact surface at place j
F_j is the local factor of safety at place j
n is the number of wedges of a sliding body (=3 in Fig. 5.10)

Examples of the use of the local factors of safety are provided by (Srbulov 1987, 1991, 1995, 1997, 2008).

The reasons for slope instability other than loads (self weight) and water pressures are listed in Sect. 4.8.2.

5.1.3 Movement

5.1.3.1 Rock Falls

Rock falls are highly chaotic motions. Approximate velocity of a rock fall:

$$v = \sqrt{2 \cdot g \cdot h} \tag{5.7}$$

v is the velocity of rock fall
g is the gravitational acceleration ~9.81 m/s^2
h is the height difference between the location where v_{rf} is calculated and the location where the fall originated

The product of the impact force and deflection of the barrier impacted by a rock fall must be equal to the kinematic energy of the rock fall according to the principle of conservation of energy:

$$F \cdot d = \frac{m \cdot v^2}{2} \tag{5.8}$$

F is the impact force of falling rock
d is the deflection of the barrier impacted by a rock block
m is the mass of falling rock
v is the velocity of rock block

The maximum rock jump after bouncing from a hard surface is according to the principle of conservation of energy:

$$h = \frac{v^2}{2 \cdot g} \tag{5.9}$$

h is the maximum height of rock jump above a hard surface after bouncing from it
v is the velocity of rock block just before hitting the hard surface
g is the gravitational acceleration ~9.81 m/s^2

The roll distance over a soft surface could be calculated based on the principle of energy conservation if the resistance to rolling could be assessed.

$$d = \frac{m \cdot v^2}{2 \cdot R} \tag{5.10}$$

d is the rolling distance
m is the mass of falling rock block
v is the velocity of rock block just before hitting the soft surface
R is the resistance to rolling force, which is highly dependent on the shape and size of rolling rock block and the surface stiffness

Examples of the use of a more detailed method are provided by Srbulov (2008).

Offshore flow slides may turn into turbidity currents. Kuenen (1952 and Heezen and Ewing 1952) proposed that the damages to submarine telegraphic cables following the Grand banks earthquake of 1929 were caused by turbidity current travelling at an initial velocity of 32.7 m/s near the source to 6.9 m/s at a distance of more than 880 km over the area of 390 km^2. Terzaghi (1956) argued that the stream would solidify within a relatively short distance from the source. Based on recently observed 2–5 m high and 50–100 m wavelength surface undulations in gravel at a water depth range from about 1.6 km to more than 4.5 km, Srbulov (2003) proposed that near surface ground waves propagation caused the damages to the submarine telegraphic cables.

5.1.3.2 Avalanches and Fast Debris Spreads

Avalanches and fast debris spreads are highly chaotic motions. Approximate analyses are possible using the same approach as for rock falls with consideration of contribution of internal friction on energy dissipation and the velocity decrease to around 50 % from the free rock fall. An additional complication is entrainment and deposition of material along travel path (e.g. McDougall and Hungr 2005). Davies and McSaveney (2002) considered that rock fragmentation led to higher than normal internal pressures and the longer run out. Examples of the use of a more detailed method are provided by Srbulov (2008).

5.1.3.3 Soil Flows

- **For level ground and gentle slopes,**
  Hamada et al. (1986) for uniformly graded sand of medium grain size found the flow distances as:

$$u_f = 0.75 \cdot L_l^{0.5} \cdot \beta_l^{1/3} \tag{5.11}$$

u_f is flow distance (m)
L_l is the thickness (m) of liquefied layer
β_l is the larger inclination (%) of the ground surface slope or the slope of the lower boundary of liquefied layer

Youd et al. (2002) provided revised multi-linear regression equations for prediction of lateral spread displacements. For free face conditions:

$$\begin{aligned}\log_{10} D_H = & -16.713 + 1.532 \cdot M - 1.406 \cdot \log_{10} R^* - \\ & -0.012 \cdot R + 0.592 \cdot \log_{10} W + 0.54 \cdot \log_{10} T_{15} \\ & +3.413 \cdot \log_{10}(100 - F_{15}) - \\ & 0.795 \cdot \log_{10}(D50_{15} + 0.1mm)\end{aligned} \tag{5.12}$$

For gently sloping ground conditions:

$$\begin{aligned}\log_{10} D_H = & -16.213 + 1.532 \cdot M - 1.406 \cdot \log_{10} R^* - \\ & -0.012 \cdot R + 0.338 \cdot \log_{10} S + 0.54 \cdot \log_{10} T_{15} \\ & +3.413 \cdot \log_{10}(100 - F_{15}) - \\ & 0.795 \cdot \log_{10}(D50_{15} + 0.1mm)\end{aligned} \tag{5.13}$$

D_H is the ground surface displacement in metres
M is earthquake magnitude
$R^* = R + 10^{0.89M-5.64}$
R is the horizontal distance from the location to the nearest source of seismic energy
W is the ratio of the height of the free face (slope) to the horizontal distance between the base of the free face and the location where D_H is calculated
T_{15} is the cumulative thickness (m) of saturated granular layers with $(N_1)_{60} < 15$, $(N_1)_{60}$ according to Sect. 2.1.1
F_{15} is the average fines content (%) for the granular layers comprising T_{15}
$D50_{15}$ is the average mean grain size (mm) for the granular layers comprising T_{15}
S is the ground slope inclination (%)

- **For steep slopes,** Bernoulli's equation with added flow energy loss in terms of energy heights (Fig. 5.11) is (Srbulov 2010a):

$$\frac{p}{\gamma} + z + \frac{v^2}{2 \cdot g} + \frac{\Delta H}{\Delta L} \cdot x = \frac{p_i}{\gamma} + z_i \tag{5.14}$$

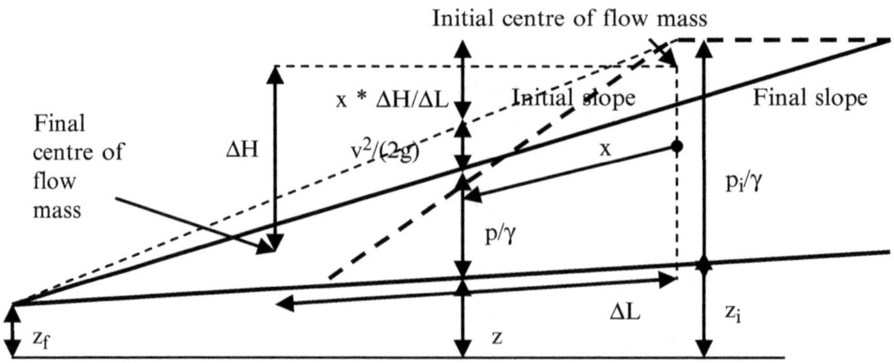

Fig. 5.11 Cross section through soil flow with markings

p/γ is so called pressure height
p is the pressure at the bottom of the flow
γ is the unit weight of flowing mass
z is elevation above a reference datum
i refers to the initial position
$v^2/(2\ g)$ is the kinetic energy height
v is the flow velocity
g is the gravitational acceleration
$\Delta H/\Delta L$ is the rate of energy height loss
x is the distance along flow path

Srbulov (2011), based on back analyses of 14 case histories, found that:

$$\begin{aligned} {}^{\Delta H}\!/_{\Delta L} &= 0.27 - 0.24 \cdot \mathit{fines\,content} \pm 0.14 \\ \mathit{Fines\,content} &\leq 0.55 \end{aligned} \tag{5.15}$$

Fines diameter <0.06 mm

The maximum flow distance x_{max} from Eq. (5.14) for $v_i = v_f = p_f = 0$

$$x_{\max} = \frac{\frac{p_i}{\gamma} + z_i - z_f}{{}^{\Delta H}\!/_{\Delta L}} \tag{5.16}$$

f refers to the final and i to the initial location of the flow

The flow duration t_f:

$$t_f = \frac{{}^{p_i}\!/_{\gamma}}{V_s} \tag{5.17}$$

p_i/γ is the initial thickness of liquefied layer (Fig. 5.11)

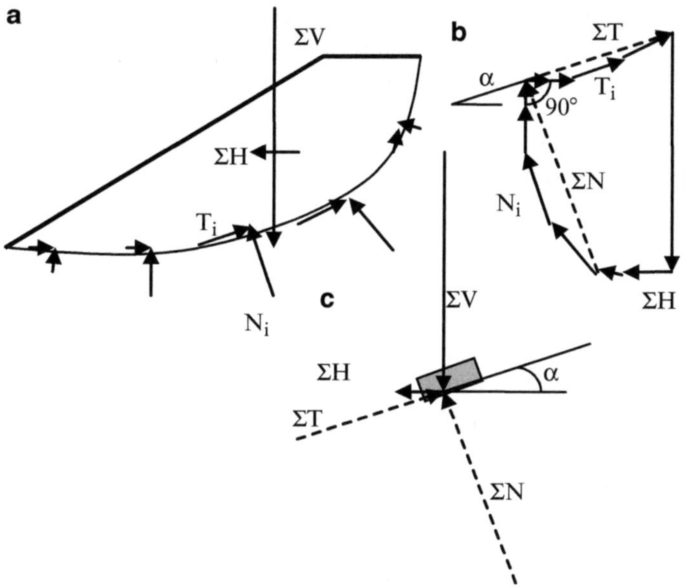

Fig. 5.12 (**a**) Forces (*arrows*) acting on a slip mass, (**b**) polygon of the forces, (**c**) equivalent sliding block of the slip mass

V_s is the velocity of particle sedimentation according to Stokes law (e.g. Das 1985):

$$V_s = \frac{\gamma_s - \gamma_w}{18 \cdot \eta_{aw}} \cdot D_{50}{}^2 \qquad (5.18)$$

γ_s and γ_w are unit weights of soil particle and water respectively (N m^{-3}).
η_{aw} is dynamic viscosity of water (N s m^{-2}). At 20 °C, $\eta_{aw} = 10^{-3}$ Nsm^{-2}, at 10 °C the viscosity is 29.8 % greater and at 30 °C 20.3 % smaller than the viscosity at 20 °C
D_{50} is an average diameter (m) of flowing soil particles

Examples of back analyses of flows of slopes are provided by Srbulov (2009, 2011).

5.1.3.4 Slides

- **Co-seismic** permanent sliding is determined based on Newmark (1965) sliding block method. The method is applicable to any slip surface shape (e.g. as demonstrated by Ambraseys and Menu 1988). The inclination of an equivalent sliding block is determined from the inclination to the horizontal of the resultant of tangential forces acting along a slip surface as sketched in Fig. 5.12.

The critical acceleration of a slope is determined using a limit equilibrium analysis of the slope stability for the factor of safety against sliding =1. Makdisi and Seed (1978) chart for estimation of permanent sliding of soil slopes is well known. Ambraseys and Srbulov (1995) provided attenuation relationships for permanent sliding for both sloping and level ground using the following assumptions and limitations:

1. Constant critical acceleration ratio is independent on the amount and rate of sliding
2. Earthquake surface wave magnitude M_s range 5–7.7.
3. Earthquake source distances to 50 km.
4. Acceleration records caused by thrust (46 %), normal (26 %) and strike slip (29 %) faults (Fig. 2.5) with the mean depth of the events 10 $\pm$ 4 km.
5. One-way horizontal component of displacement only for slopes. Down slope displacement is calculated by dividing the horizontal component with cosα, α is inclination to the horizontal of the equivalent block.
6. Horizontal ground acceleration was considered only.

One-way (down slope only) permanent horizontal component of displacements on sloping ground is according to Ambraseys and Srbulov (1995) is:

$$\log_{10}(u_1) = -2.41 + 0.47 \cdot M_s - 0.01 \cdot r_f + \log_{10}\left(\frac{\left[1 - \left({}^{k_c}/_{k_p}\right)\right]^{2.64}}{\left({}^{k_c}/_{k_p}\right)^{1.02}}\right) + 0.58 \cdot p \tag{5.19}$$

Two-way displacement on level ground is:

$$\log_{10}(u_2) = -2.07 + 0.47 \cdot M_s - 0.012 \cdot r_f + \log_{10}\left(1 - {}^{k_c}/_{k_p}\right)^{2.91} + 0.6 \cdot p \tag{5.20}$$

$u_{1,2}$ are in cm,
M_s is the surface wave magnitude of an earthquake
$r_f = (h_f^2 + d_f^2)^{0.5}$,
h_f is the hypocentral depth,
d_f is the source distance,
k_c is the ratio between the critical horizontal acceleration at which the factor of safety of slope stability is 1 and the gravitational acceleration,
k_p is the ratio between the peak horizontal ground acceleration and the gravitational acceleration, p is the number of standard deviations.
The results of these relationships are shown in Fig. 5.13 for three earthquake magnitudes and three source-to-site distances.

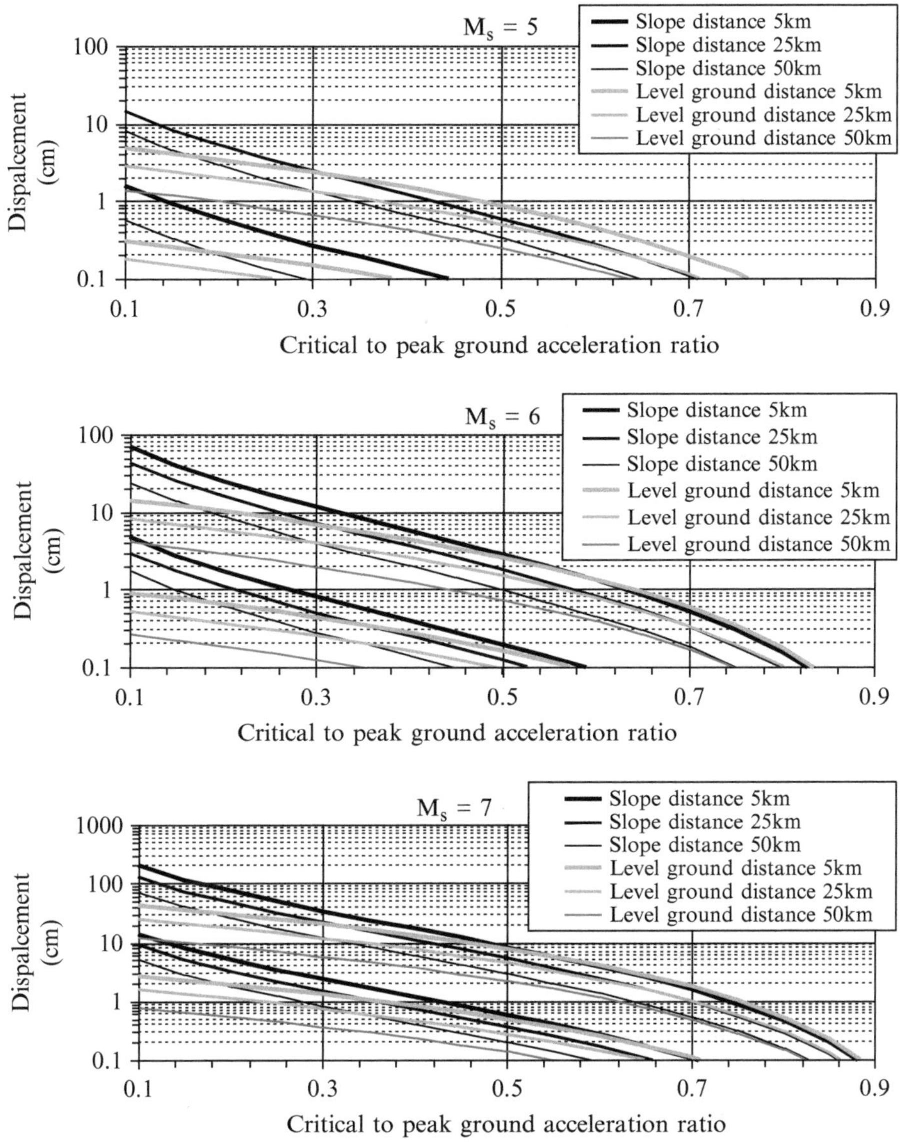

Fig. 5.13 Average $\pm$ one standard deviation of co-seismic permanent displacements of slopes and level ground

- **Post-seismic** permanent sliding of slopes when soil shear strength is reduced to the residual value or an excessive pore water pressure build-up occurred during earthquake is evaluated using two-sliding block model of planar slips according to Ambraseys and Srbulov (1995) shown in Fig. 5.14 for the case of a constant shear strength and when the angle β between the interface, the slope and the toe is the angle of symmetry between these two surfaces.

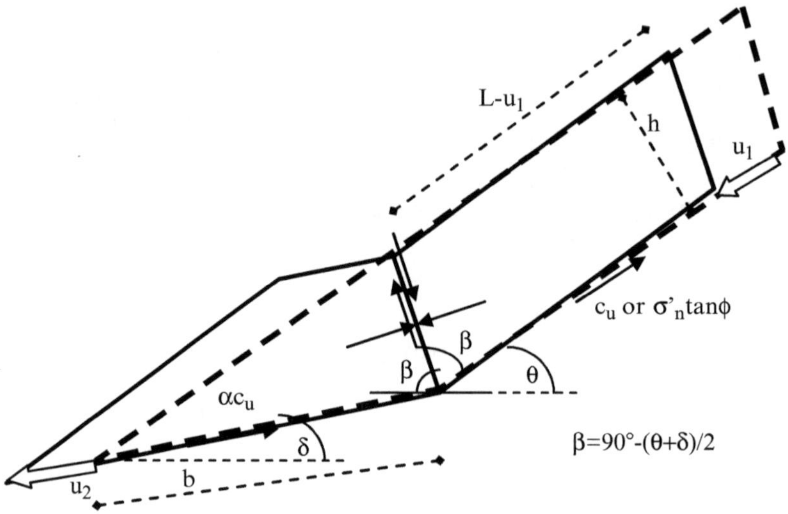

Fig. 5.14 Notions used with the two-block sliding method

The **initial factor of safety** $\boldsymbol{F_o}$ <1 at time $t = 0$ when $u_1 = 0$ is:

$$F_o = \frac{c_u \cdot \left[L + \alpha \cdot b + 2 \cdot h \cdot \tan\left(\frac{\theta - \delta}{2}\right)\right]}{\rho \cdot g \cdot h \cdot \left[L \cdot \sin\theta + \left(\frac{b}{2}\right) \cdot \sin\delta\right]} \tag{5.21}$$

Duration of sliding $\boldsymbol{T}$ in seconds until the movement stops is:

$$T = \pi \cdot \sqrt{\frac{\rho \cdot h \cdot \left(L + \frac{b}{2}\right)}{\rho \cdot g \cdot h \cdot (\sin\theta - \sin\delta) + c_u \cdot (\alpha - 1)}} \tag{5.22}$$

The maximum slip $\mathbf{u_{max}}$ at time T is:

$$u_{\max} = \frac{2 \cdot \rho \cdot h \cdot g \cdot (1 - F_o) \cdot \left(L \cdot \sin\theta + \frac{b}{2} \cdot \sin\delta\right)}{\rho \cdot g \cdot h \cdot (\sin\theta - \sin\delta) + c_u \cdot (\alpha - 1)} \tag{5.23}$$

The **maximum velocity** $\boldsymbol{V_{max}}$ **and acceleration** $\boldsymbol{A_{max}}$ of the sliding are:

$$V_{\max} = \frac{\pi \cdot u_{\max}}{2 \cdot T}$$
$$A_{\max} = \frac{\pi^2 \cdot u_{\max}}{2 \cdot T^2} \tag{5.24}$$

The **factor of safety F during and at the end of sliding** when $u_1 = u_{max}$ is the ratio between resisting and driving forces.

$$F = \frac{c_u \cdot \left[L - u_1 + \alpha \cdot (b + u_1) + 2 \cdot h \cdot \tan\left(\frac{\theta - \delta}{2}\right)\right]}{\rho \cdot g \cdot h \cdot \left[(L - u_1) \cdot \sin\theta + \left(\frac{b}{2} + u_1\right) \cdot \sin\delta\right]} \quad (5.25)$$

c_u is undrained shear strength or $\sigma'_n \tan\phi$
L is the initial length of the sliding mass along the slope
α is proportion of c_u along the toe of the slope
b is the initial toe length
h is the thickness of the sliding mass
θ is the angle of inclination to the horizontal of the slope
δ *is* the angle of inclination to the horizontal of the toe
ρ is unit density of soil
g is the gravitational acceleration

The final F > 1 because fast slides gain momentum and overshoot the place where F = 1. For back analyses of fast slides, the assumption that the final F = 1 is not correct.

Two-sliding block model is applicable to any shape of sliding body (like the single block) when kinematically possible displacements along the slip surface and its interfaces are used to find the equivalent two blocks with similar mass, length and thickness as the original sliding body, Fig. 5.15.

Examples of back analyses of post-seismic slope displacements are provided by Srbulov (2008, 2011).

5.1.4 Execution

BS 6031 (2009) in line with EN 1997-1 (2004) provide useful information. Past experience has shown that:

- Excavations at slopes must start from the crest otherwise excavations at the toe to remove material fallen into road lead to further slope instability and casualties
- Excavations in synclines formed in shell/mudstones can exhibit progressive fast failures due to release of tectonic stresses concentrated at those locations
- Progressive fast failures of slopes in heavily over consolidated and highly plastic clay can be delayed for tens of years (Skempton 1964) until pore water pressure equalisation (Skempton and Hutchinson 1969) in fact until the suction induced by clay tendency to heave on excavation is lost.
- Long slopes are provided with berms 1.5–3 m wide to minimise travel path of falling pieces as well as enable better access for construction, control of runoff water during rainfall and inspection
- Opening of the cracks at the crest and/or bulging at the toe (large deformations) are signs of the slope limited stability and possibility of failure

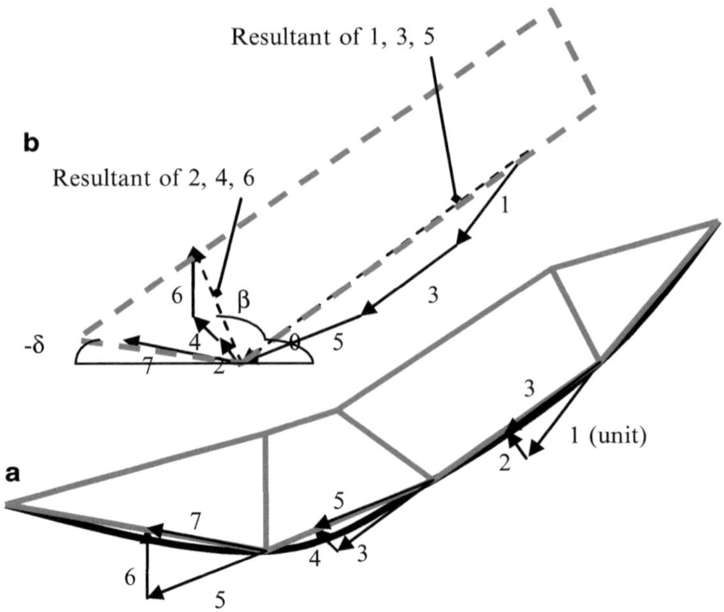

Fig. 5.15 (**a**) Multiple wedges for a concave slip surface (*grey*) approximated by a polygon with arrows of kinematically possible displacements 1–7 along the bases and interfaces of the wedges, (**b**) equivalent two blocks with similar mass, length and thickness as the multiple wedges with the inclination of angle θ based on the inclination of the resultant of kinematically possible displacements 1,3,5 and the inclination of angle β based on the inclination of the resultant of kinematically possible displacements 2,4,6

- Heavy vibratory machinery could cause liquefaction of water saturated loose to medium dense sand within a slope leading to flow failures
- Slope stabilisation involves its flattening whenever possible, provision of ballast at the toe, ground water drainage, nailing and retaining walls (from gabions, contiguous and raking piles to diaphragms with anchors).
- Drainage pipes could become clogged by siltation, grow of algae, freezing near the ground surface. Drainage trenches filled with well drained coarse grained material are preferable but have a limited depth (max. 6 m)
- Anchors are prone to corrosion (limited durability) and used mainly for temporary works.
- Slopes of concave shape in layout (mostly in mining) have increased stability due to arching effect in the horizontal direction
- Slopes are monitored using: precise levelling of the surface, extensometers at/near slope crest to monitor crack width, piezometers (stand pipe) to record ground water level and inclinometers for measurements of lateral displacements along depth. The frequency of observations depends on the speed of changes of monitored values.

5.2 Foundations

5.2.1 Types

The following types are considered:

- Shallow depth with negligible side resistance (pads, strips, rafts)
- Medium depth with significant side resistance (boxes, blocks)
- Deep with toe depths much greater than widths of piles and barrettes

5.2.2 Stability/Capacity

5.2.2.1 Shallow Depth in Soil During Static and Cyclic Load

- **In undrained conditions (short-term) in fine grained soil**

  The ultimate bearing stress q_f is:

$$\begin{aligned} & q_f = (\pi + 2) \cdot c_u \cdot s_c \cdot i_c \cdot b_c + \gamma \cdot D_f \\ & according\, to\, EN\, 1997 - 2 : 2004 \\ & q_f = (\pi + 2) \cdot c_u \cdot \left(1 + s_c + d_c - i_c - b_c - g_c\right) + \gamma \cdot D_f \\ & according\, to\, Hansen\, (1970) \end{aligned} \tag{5.26}$$

c_u is undrained shear strength
γ is total soil unit weight
D_f is foundation depth below ground surface

The values of coefficients *s, i, b, d, g* are given in Table 5.1.

Table 5.1 Shape *s*, load inclination *i*, base depth *d*, base inclination *b* and ground surface inclination *g* factors for calculation of bearing capacity of shallow foundations in terms of total stresses in undrained conditions

	EN 1997-1 (2004)	(Hansen 1970)
s_c	$1 + 0.2B'/L'$	$0.2\,(1 - 2\,i_c)\,B'/L'$
i_c	$0.5\{1 + [1 - H/(A'c_u)]^{0.5}\}$	$0.5–0.5[1 - H/(A'c')]^{0.5}$
d_c	–	$0.4D_f/B'$ for $D_f < B'$
		$0.4D_f/B'\arctan(D_f/B')$ for $D_f > B'$
b_c	$1 - 2\alpha/(\pi + 2)$	$\alpha/2.565$
g_c	–	$\beta/2.565$
m	–	–

Notes:
H is the resultant horizontal force on loaded area,
α,β are base and ground surface inclination to the horizontal in radians. Positive α upwards in direction of *H* force, positive β downwards
c_u is soil cohesion in static or cyclic condition
D_f is foundation depth
A' is the effective foundation area $= B' \times L'$. $B' = B - 2\,M/V$, $L' = L - 2\,M/V$, *B* is foundation width, *L* is foundation length, *M* is overturning moment, *V* is vertical force

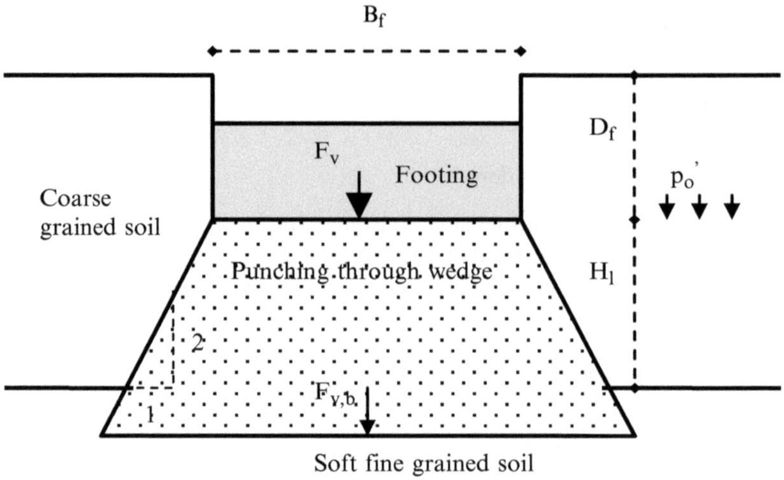

Fig. 5.16 Punching through mode of failure into a soft sub-layer

Equation (5.26) does not take into account the inertial forces in soil beneath foundation. The effect can be taken into account by considering additional angle of foundation base of ground surface inclination of arctan[0.65 a_h/g] (radians), g is the gravitational acceleration, a_h is design peak horizontal acceleration, 0.65 is factor for an effective value of acceleration. Informative Annex F of EN 1998-5 (2004) provides a general expression for checking of stability against seismic bearing capacity for a strip foundation at the surface of homogeneous soil.

When soft fine grained soil exists below a layer of coarse grained soil then punch through type failure may occur (Fig. 5.16). The vertical foundation capacity F_v in the case of punch through failure is (e.g. SNAME 2008)

$$F_v = F_{v,b} - A_f \cdot H_l \cdot \gamma + 2 \cdot \frac{H_l}{B_f} \left(H_l \cdot \gamma + 2 \cdot p_o' \right) \cdot K_s \cdot \tan\phi \cdot A_f \tag{5.27}$$

$F_{v,b}$ is determined assuming the foundation bears on the surface of the lower liquefied layer,
A_f is foundation area,
H_l is distance from a foundation level to the level of soft layer below,
γ is unit weight of top soil,
p'_o is effective overburden stress at the foundation depth,
K_s is the coefficient of punching shear that is calculated from the equation

$$K_s \cdot \tan\phi = \frac{3 \cdot c_u}{B_f \cdot \gamma} \tag{5.28}$$

ϕ is friction angle of top layer,
c_u is undrained shear strength of soil in the bottom layer,
B_f is diameter of an equivalent circular foundation
$F_{v,b} = (c_u N_c + p') A_f$, where $N_c = 5.14$,
p' effective overburden stress at the top of soft bottom layer.

The ultimate sliding capacity is $\alpha c_u A'$, where

α is a mobilisation factor of the undrained shear strength c_u
A' is the foundation effective area $= B' L'$, $B' = B - 2\,M/V$, $L' = L - 2\,M/V$, M is the overturning moment, V is the vertical force

- **In drained conditions in coarse grained and fine grained soil (long-term)**
  According to the general bearing capacity equation, which was proposed by Meyerhof (1963) as a way to address issues in Terzaghi's earlier equation, the ultimate bearing stress q_f may be estimated as:

$$\begin{aligned} q_f = & 0.5 \cdot \gamma \cdot B' \cdot N_\gamma \cdot s_\gamma \cdot i_\gamma \cdot d_\gamma \cdot b_\gamma \cdot g_\gamma + \\ & c' \cdot N_c \cdot s_c \cdot i_c \cdot d_c \cdot b_c \cdot g_c + \\ & \gamma \cdot D_f \cdot N_q \cdot s_q \cdot i_q \cdot d_q \cdot b_q \cdot g_q \end{aligned} \tag{5.29}$$

γ is the bulk unit weight of soil beneath wall base or it is the submerged unit weight if the ground water level is at or above the soil surface,
B' is effective base width $=B - 2\,M/V$, M is the overturning moment, V is the vertical force
c' is soil cohesion
D_f is the foundation depth below the ground surface

The factors N, s, i, d, b, g according to different sources are given in Table 5.2 for silica and quartz minerals.

Equation (5.29) does not take into account the inertial forces in soil beneath foundation. The effect can be taken into account by considering additional angle of foundation base of ground surface inclination of arctan[0.65 a_h/g] (radians), g is the gravitational acceleration, a_h is design peak horizontal acceleration, 0.65 is factor for an effective value of acceleration. Informative Annex F of EN 1998-5 (2004) provides a general expression for checking of stability against seismic bearing capacity for a strip foundation at the surface of homogeneous soil.

The ultimate sliding capacity is $\zeta A'[c' + (q_f - u)\tan\phi']$, where

ζ is friction mobilisation factor, ~2/3 for smooth steel and precast concrete, =1 for cast in place concrete and brick
A' is the foundation effective area $= B' L'$, $B' = B - 2\,M/V$, $L' = L - 2\,M/V$, M is the overturning moment, V is the vertical force
q_f is the ultimate bearing pressure
u is the water pressure at the level of foundation underside
ϕ is soil friction angle

Table 5.2 Bearing capacity *N*, shape *s*, load inclination *i*, base depth *d*, base inclination *b* and ground surface inclination *g* factors for calculation of bearing capacity of shallow footings in terms of effective stresses in drained conditions

	EN 1997-1 (2004)	(Hansen 1970)
N_q	$e^{\pi \tan\phi}\tan^2(45° + \phi'/2)$	$e^{\pi \tan\phi}\tan^2(45° + \phi'/2)$
N_γ	$2(N_q - 1)\tan\phi'$	$1.5(N_q - 1)\tan\phi'$
N_c	$(N_q - 1)\cot\phi'$	$(N_q - 1)\cot\phi'$
s_q	$1 + B'/L'\sin\phi'$	$1 + B'/L' i_q \sin\phi'$
s_γ	$1 - 0.3B'/L'$	$1 - 0.4\, i_\gamma\, B'/L'$
s_c	$(s_q N_q - 1)/(N_q - 1)$	$0.2\,(1 - 2\, i_c)\, B'/L$
i_q	$[1 - H/(V+ A'c'\cot\phi')]^m$	$[1 - 0.5H/(V+ A'c'\cot\phi')]^5$
i_γ	$[1 - H/(V+ A'c'\cot\phi')]^{m+1}$	$[1 - 0.7H/(V+ A'c'\cot\phi')]^5$
i_c	$i_q - (1 - i_q)/(N_c\tan\phi')$	$0.5 - 0.5[1 - H/(A'c')]^{0.5}$
d_q	–	$1 + 2\tan\phi'(1\text{-}\sin\phi')^2 D_f/B'$ for $D_f < B'$ $1 + 2\tan\phi'(1\text{-}\sin\phi')^2\arctan(D_f/B')$ for $D_f > B'$
d_γ	–	1
d_c	–	$0.4D_f/B'$ for $D_f < B'$ $0.4D_f/B'\arctan(D_f/B')$ for $D_f > B'$
b_q	$(1\text{-}\alpha \tan\phi')^2$	$e^{-2\,\alpha\,\tan\phi'}$
b_γ	$(1\text{-}\alpha \tan\phi')^2$	$e^{-2.7\,\alpha\,\tan\phi'}$
b_c	$b_q - (1 - b_q)/(N_c\tan\phi')$	$\alpha/2.565$
g_q	–	$(1 - 0.5\tan\beta)^5$
g_γ	–	$(1 - 0.5\tan\beta)^5$
g_c	–	$\beta/2.565$
M	$(2 + B'/L')/(1 + B'/L')$ for H in B′ direction $(2 + L'/B')/(1 + L'/B')$ for H in L′ direction	–

Notes:
H, *V* are the resultant horizontal and vertical forces on loaded area,
ϕ' is soil friction angle in static or cyclic condition
α,β are base and ground surface inclination to the horizontal in radians. Positive α upwards in direction of *H* force, positive β downwards
c' is soil cohesion in static or cyclic condition
A' is the effective foundation area $= B' \times L'$.. $B' = B - 2\,M/V$, $L' = L - 2\,M/V$, *B* is foundation width, *L* is foundation length, *M* is overturning moment, *V* is vertical force

5.2.2.2 Shallow Depth in Rock

Informative Annex G of EN 1997-1 (2004) contains graphs with presumed bearing resistance for square pad foundations bearing on rock (for settlements not exceeding 0.5 % of the foundation width) based on uniaxial compressive strength and discontinuity spacing.

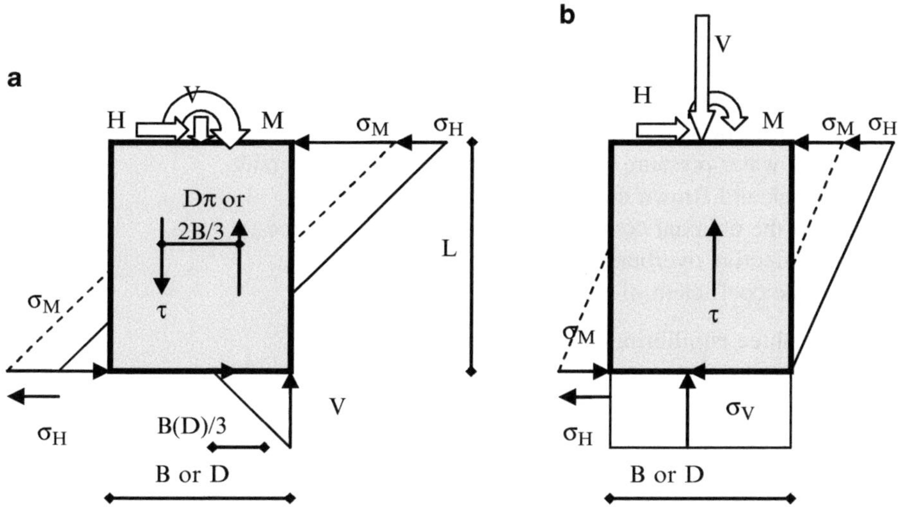

Fig. 5.17 (**a**) Case the vertical displacement caused by rotation is greater than settlement (the effect of V force on the σ_v at the bottom is ignored), (**b**) case settlement greater than the vertical displacement caused by rotation (the effect of moment M on the σ_v at the bottom is ignored)

Kulhaway and Carter (1994) suggested that the ultimate bearing capacity q_f of rock mass can be calculated as:

$$q_f = UCS \cdot \left(\sqrt{s} + \sqrt{m \cdot \sqrt{s} + s}\right) \tag{5.30}$$

UCS is the uniaxial compressive strength of rock pieces
m, s are Hoek and Brown coefficients from Table 3.10.

5.2.2.3 Medium Depth in Soil and Rock

The bearing capacity of box/block shaped foundation is affected by both side and toe bearing resistances. Two principal modes of response exist:

- **Rotation induced vertical displacement is greater than the foundation settlement under the vertical load** (Fig. 5.17a). Five unknown stresses σ_M, σ_H, τ, $\sigma_{V1,2}$ cannot be calculated from the three equations of equilibrium of overturning moments and forces in two perpendicular directions alone. If the ultimate (or allowable) axial and shear stresses are considered so that $\sigma_V = q_f$ and $\tau = \alpha c_u$ or $\zeta[c' + (q_f - u)\tan\phi']$ in soil or minimum 63(UCS in kPa/100)$^{0.5}$ (kPa) for rock masses according to Rowe and Armitage (1984), $\sigma_M = \sigma_H = \sigma'_v K_p + 2\,c K_p^{1/2}$ ($c = c_u$ or c') or $s^{1/2}$UCS according to Kulhaway and Carter (1994) for rock masses then the equilibrium equations will have negative values when the ultimate (allowable) stresses at the bottom and sides have not been exceeded.

q_f is the ultimate bearing capacity,
α is mobilisation factor of the undrained shear strength c_u,
ζ is friction mobilisation factor,
c' is soil cohesion,
ϕ is soil friction angle,
u is the water pressure at the level of foundation underside,
s is Hoek and Brown coefficient from Table 3.10,
UCS is the uniaxial compressive strength of rock pieces,
σ'_v is effective overburden stress,
K_p is the coefficient of passive soils resistance

The three equilibrium equations are:

$$\begin{aligned}\Sigma M_{at\,centroit} = {} & M + H \cdot L/2 - \sigma_M \cdot (D\,or\,W) \cdot L/2 \cdot 2/3 \cdot L \\ & - \tau \cdot \left(D/2 \cdot \pi \cdot D/\pi\,or\,B \cdot W \cdot 2B/3\right) \cdot L \\ & - \tau \cdot \left(D^2 \cdot \pi/8\,or\,B/2 \cdot W\right) \cdot L/2 - \sigma_v \cdot \left(2/3 \cdot D^2/4 \cdot D/3\right. \\ & \left. or\ B/4 \cdot W \cdot B/3\right) = 0 \\ \Sigma H = {} & H - \sigma_H \cdot (D\,or\,W) \cdot L + \tau \cdot \left(D^2 \cdot \pi/8\,or\,B/4 \cdot W\right) = 0 \\ \Sigma V = {} & V - \sigma_V \cdot \left(2/3 \cdot D^2/4\,or\,B/4 \cdot W\right) = 0\end{aligned} \tag{5.31}$$

M is overturning moment
H is horizontal force
L is foundation depth
σ_M is the horizontal axial stress from M
D is diameter for circular shapes
W is width for rectangular shapes
τ is shear stress
B is breadth for rectangular shapes
σ_V is the vertical axial stress
σ_H is the horizontal axial stress from H
V is the vertical force

- **Rotation induced vertical displacement is smaller than the foundation settlement under the vertical load** (Fig. 5.17b). The three equilibrium equations are:

$$\begin{aligned}\Sigma M_{at\,centroit} &= M + H \cdot L/2 - \sigma_M \cdot (D\,or\,W) \cdot L/2 \cdot 2/3 \cdot L \\ &\quad + \tau \cdot \left(D^2 \cdot \pi/4\,or\,B \cdot W\right) \cdot L/2 = 0 \\ \Sigma H &= H - \sigma_H \cdot (D\,or\,W) \cdot L - \tau \cdot \left(D^2 \cdot \pi/4\,or\,B \cdot W\right) = 0 \\ \Sigma V &= V - \sigma_V \cdot \left(D^2 \cdot \pi/4\,or\,B \cdot W\right) - \tau \cdot (D \cdot \pi\,or\,2 \cdot B \cdot W) \cdot L = 0\end{aligned} \tag{5.32}$$

M is overturning moment
H is horizontal force
L is foundation depth
σ_M is the horizontal axial stress from M
D is diameter for circular shapes
W is width for rectangular shapes
τ is shear stress
B is breadth for rectangular shapes
σ_V is the vertical axial stress
σ_H is the horizontal axial stress from H
V is the vertical force

5.2.2.4 Deep During Static and Cyclic Load

- **Single pile (barrette) in the vertical direction**

  In coarse grained soil without cohesion and fine grained soil in the long term (i.e. with zero shear strength at zero effective stress), the force along pile shaft is commonly calculated as:

$$\Sigma\sigma'_{v,avr} \cdot K_s \cdot \tan\delta_\phi \cdot D_p \cdot \pi \cdot L_p \tag{5.33}$$

$\sigma'_{v,avr}$ is an average effective overburden pressure in the middle of a soil layer
L_p pile length within a soil layer,
K_s is the coefficient of lateral effective stress acting on pile shaft that varies in the rage from 1 to 2 of K_o for large displacement driven piles or from 0.75 to 1.75 of K_o for small displacement driven piles in silica and quartz minerals (from Table 7.1 in Tomlinson 2001). Large displacement piles are for example sand compaction and Franki piles, which cause lateral compaction of adjacent soil due to insertion of a pile during pile driving. Lower values of K_s correspond to loose to medium dense and upper values of K_s to medium dense to dense soil. K_o is the coefficient of soil lateral effective stress. For piles in chalk, the unit shaft resistance is given in Table 5.3 from Lord et al. (2002).
δ_ϕ is friction angle between ground and pile shaft (usually assumed equal to about 2/3 of the ground friction angle ϕ for steel and pre-cast concrete driven piles and equal to ϕ for cast in place concrete piles),
D_p is pile diameter. For sheet pile walls, $D_p\ 2\ L_p$ is used instead of $D_p\ \pi\ L_p$ in Eq. (5.33), where D_p is the width of the wall member being driven by a hammer.

The force at the toe of a plugged pile is commonly calculated in non-cohesive ground as:

$$\sigma'_v \cdot N_q \cdot \frac{D_p{}^2 \cdot \pi}{4} \tag{5.34}$$

Table 5.3 Values of unit shaft friction and unit end bearing for piles in chalk with application limits (<) (Lord et al. 2002)

Pile type	Unit shaft friction	Unit end bearing
Bored in low to medium dense chalk and driven	$0.8\sigma'_v < 320$kPa	$<200N_{SPT}$
Bored in high density chalk	0.1UCS	
Continuous flight auger (CFA)	$0.45\sigma'_v < 100$kPa	
Small displacement piles (H, tubular, sheet) in not high density chalk	<20kPa	$<300N_{SPT}$
Small displacement piles (H, tubular, etc.) in high density chalk	<120kPa	

Notes:
σ'_v is the effective overburden stress in chalk disregarding overburden
UCS is the unconfined compressive strength of chalk pieces
N_{SPT} is the number of blow counts of the standard penetration test $= 2.5q_{c,CPT}$ where q_c is the cone resistance of cone penetration tests
Low density chalk dry density $\gamma_d < 1.55$ Mg/m^3
Medium density chalk dry density $\gamma_d = 1.55$ to 1.70 Mg/m^3
High density chalk dry density $\gamma_d = 1.70$ to 1.95 Mg/m^3
Very high density chalk dry density $\gamma_d > 1.95$ Mg/m^3

or of an unplugged pile as:

$$\sigma'_v \cdot N_q \cdot (D_p - d_p) \cdot \pi \cdot d_p \tag{5.35}$$

in which case the internal shaft friction will be taken into account

σ'_v is effective overburden pressure at the level of pile toe,

N_q is ground bearing capacity factor, ~0.196 exp(0.171ϕ) according to Berezantsev et al. (1961), ϕ is ground friction angle in degrees in the range $25° < \phi < 42°$ for silica and quartz minerals. For piles in chalk, the unit end bearing is given in Table 5.3 from Lord et al. (2002).

D_p is external pile diameter,

d_p is pile wall thickness for hollow piles. For sheet pile walls, the toe force is much smaller in comparison with the side force and can be ignored.

In fine grained soil with undrained shear strength in fully saturated and undrained condition (i.e. when there is no time for soil consolidation to take place under applied load), the force along pile shaft is commonly calculated as:

$$\Sigma\alpha_p \cdot c_{u,avr} \cdot D_p \cdot \pi \cdot L_p \tag{5.36}$$

α_p is ground cohesion mobilization factor that can be obtained from Fig. 5.18, which is based on Tomlinson (2001).

$c_{u,avr}$ is average undrained shear strength below pile tip

L_p is pile length within a soil layer

D_p is pile diameter. For sheet pile walls, $D_p\ 2\ L_p$ is used instead of $D_p\ \pi\ L_p$ in Eq. (5.36), where D_p is the width of the wall member being driven by a hammer.

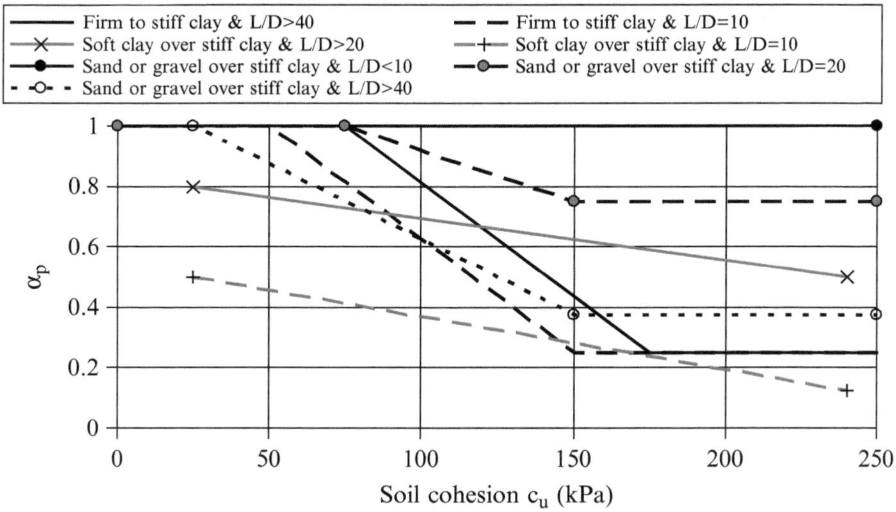

Fig. 5.18 Ground undrained shear strength c_u mobilisation factor α_p

Heaving of soil in near ground surface layers will cause different pile shaft forces from the forces according to Eq. (5.36).

If very soft to soft clay exist near the ground surface then the settlement of near top clay layer relative to the pile due to excessive water pressure dissipation or creep will cause negative shaft friction i.e. additional load on the pile.

The force at the toe of a plugged pile is commonly calculated in cohesive ground as:

$$9 \cdot c_u \cdot \frac{{D_p}^2 \cdot \pi}{4} \tag{5.37}$$

or of an unplugged pile as:

$$9 \cdot c_u \cdot (D_p - d_p) \cdot \pi \cdot d_p \tag{5.38}$$

in which case the internal shaft friction will be taken into account

c_u is undrained shear strength of ground under/around pile tip.

D_p is external pile diameter,

d_p is pile wall thickness for hollow piles For sheet pile walls, the toe force is much smaller in comparison with the side force and will be ignored.

Cone penetrometer is a scaled model of a pile. Forces acting along a pile shaft and at the pile toe can be calculated by applying the scaling factor of $D/3.57$ to the force acting along the shaft of cone penetrometer and $0.25*\pi D^2\ (10)^{-1}$ to the force acting at cone penetrometer tip, where the pile diameter D is in cm and pile squared diameter D^2 is in cm^2, the standard cone diameter is 3.57 cm and the cross sectional area is 10 cm^2. Jardine et al. (2005) published an elaborated

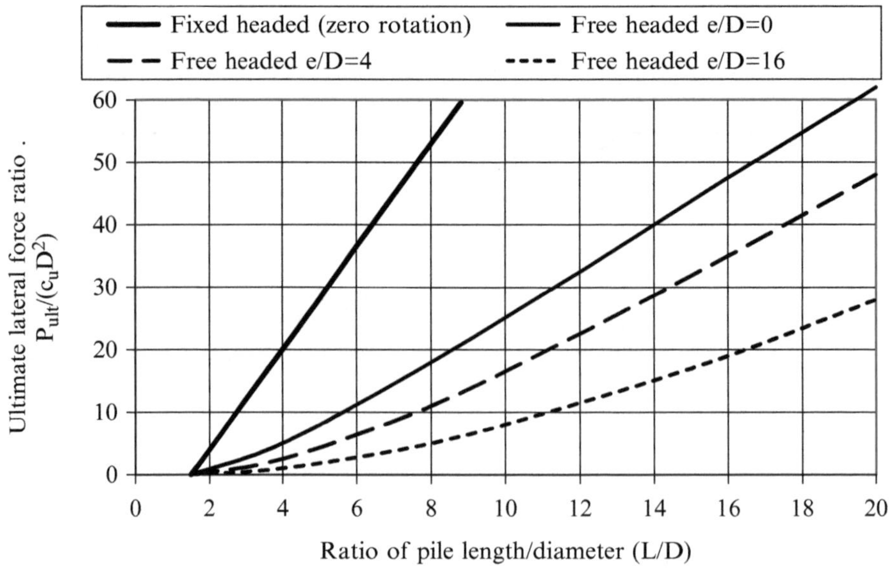

Fig. 5.19 Ultimate lateral force ratio for short piles (L/D < 6) in fine grained soil with undrained shear strength c_u for different eccentricities e above ground surface of the horizontal force on pile

method for calculation of capacity of driven piles in sand and clay based on cone penetrometer test results.

In **rock mass** Fig. E.1. of EN 1997-2 (2007) shows unit shaft resistance for axially loaded piles. The maximum value shown is 300 kPa. Table E.4 of EN 1997-2 (2007) shows unit end bearing for weathered rock. The maximum value could exceed 4.5 MPa. Rowe and Armitage (1984) propose that an average unit shaft friction in rock is $142(\text{UCS in kPa}/100)^{0.5}$ (kPa) and a lower bound unit shaft friction in rock is $63(\text{UCS in kPa}/100)^{0.5}$ in kPa, (UCS is the unconfined compressive strength of rock pieces) while the lower bound end bearing is calculated according to Eq. (5.30).

- **Single pile (barrette) in the horizontal direction**

The ultimate lateral force ratios for short and long piles in fine and coarse grained soil adopted from Broms (1964a, b) are shown in Figs. 5.19, 5.20, 5.21, and 5.22.

For rock mass, the limited lateral pressure is $s^{1/2}$UCS according to Kulhaway and Carter (1994), where s is Hoek and Brown coefficient from Table 3.10 and UCS is the unconfined compressive strength of rock pieces.

If flow type slope instability occurs in a liquefied soil layer then JRASHB (2002) suggests that the maximum lateral earth pressure on pile above liquefied layer equals to the maximum passive pressure and 0.3 times the total overburden pressure in liquefied layer.

If sliding type slope instability occurs then the lateral force R_x applied to a pile can be determined from Eq. (5.4) for $F_s = 1$ and circular cylindrical slip surfaces or based on Fig. 5.9 and description in Sect. 5.1.2.4 for $F_s = 1$ and prismatic slip surfaces.

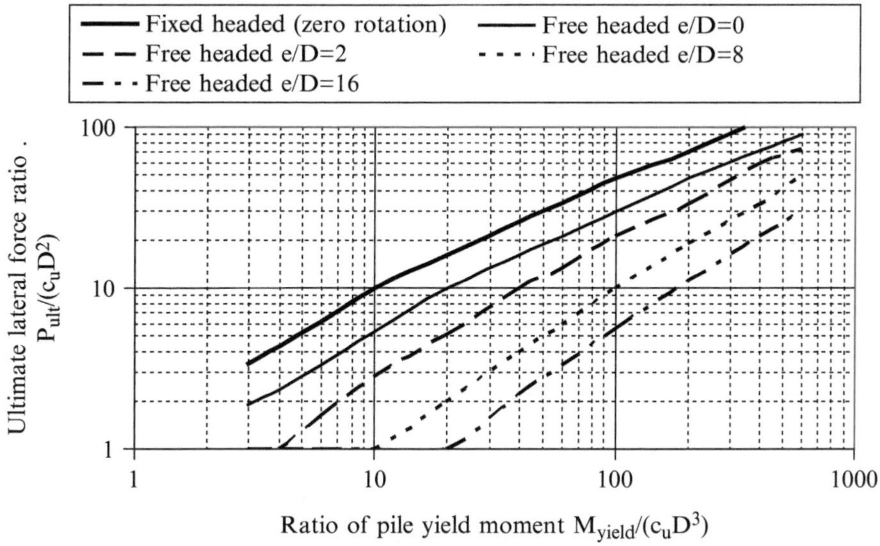

Fig. 5.20 Ultimate lateral force ratio for long piles (length/diameter D > 6) in fine grained soil with undrained shear strength c_u for different eccentricities e above ground surface of the horizontal force on pile

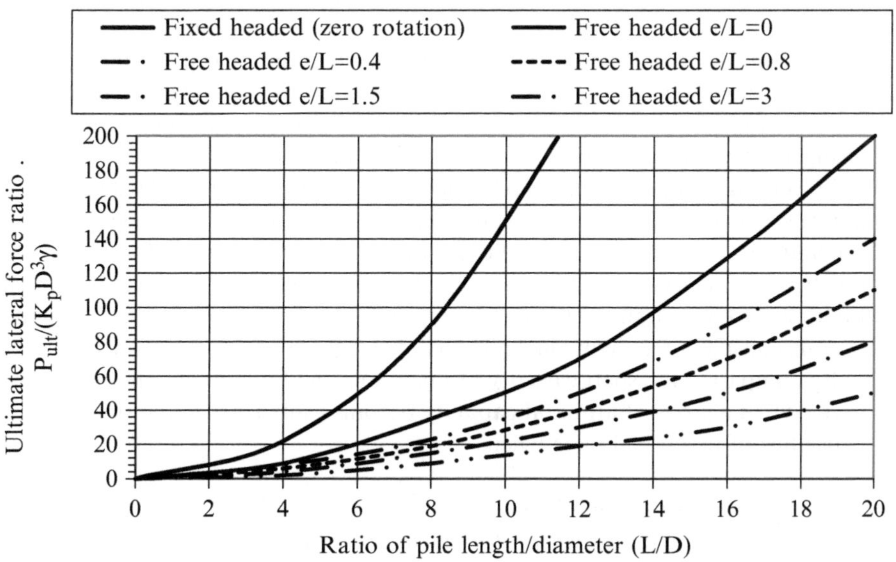

Fig. 5.21 Ultimate lateral force ratio for short piles (L/D < 6) in coarse grained and fine grained soil (long term) with the coefficient of passive resistance K_p and unit weight γ for different eccentricities e above ground surface of the horizontal force on pile

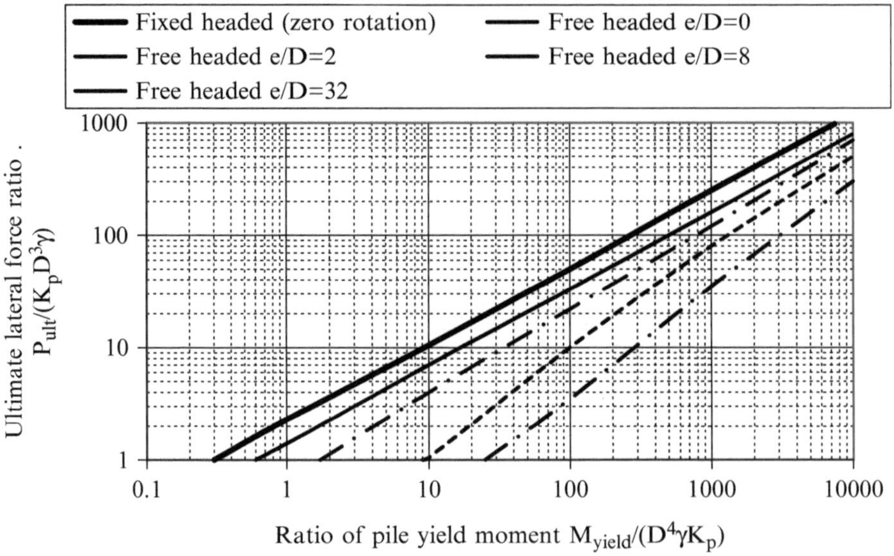

Fig. 5.22 Ultimate lateral force ratio for long piles (length/diameter D > 6) in coarse grained and fine grained soil (long term) with the coefficient of passive resistance K_p and unit weight γ for different eccentricities e above ground surface of the horizontal force on pile

- **Pile (barrette) group in static and cyclic condition except earthquakes**

  Terzaghi and Peck (1948) suggested that the group axial capacity is the lesser of the sum of the ultimate axial capacities of the individual piles in the group or the bearing capacity for block failure of the group. Whitaker (1957) found that there was a critical value of pile spacing at which the mechanism of failure changed from block failure to individual pile failure.

  The **end bearing capacity** of piles should not be much decreased with decrease in pile spacing based on the assumed mechanism of failure of soil under pile base according to Berezantzev et al. (1961). This statement is not valid for the case of pile bases in coarse grained layer overlying fine grained soft soil layer for which Eq. (5.27) for punch through type failure can be applied for consideration of end bearing capacity of an individual pile and of an equivalent block failure.

  By comparing the shaft capacity of pile groups and individual piles, it can be derived that a group **shaft axial capacity** becomes smaller than the sum of shaft capacities of individual piles when:

$$\begin{aligned} &Pile\,centre\,to\,centre\,spacing < \alpha_p \cdot \pi \cdot D\,or \\ &Pile\,centre\,to\,centre\,spacing < {}^{K_s}/_{K_o} \cdot \pi \cdot D \end{aligned} \tag{5.39}$$

α_p is ground cohesion mobilization factor that can be obtained from Fig. 5.18, which is based on Tomlinson (2001).

D is pile diameter

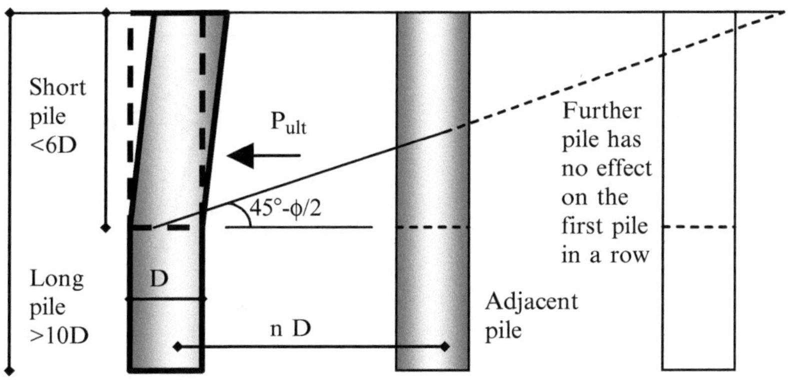

Fig. 5.23 Sketch of the effect of an adjacent pile on the ultimate lateral force P_{ult} of the first pile

K_s is the coefficient of lateral effective stress acting on pile shaft that varies in the rage from 1 to 2 of K_o for large displacement driven piles or from 0.75 to 1.75 of K_o for small displacement driven piles (from Table 7.1 in Tomlinson 2001). Large displacement piles are for example sand compaction and Franki piles, which cause lateral compaction of adjacent soil due to insertion of a pile during pile driving. Lower values of K_s correspond to loose to medium dense and upper values of K_s to medium dense to dense soil.

K_o is the coefficient of soil lateral effective stress

For a single row of piles, the pile centre to centre spacing in Eq. (5.39) is halved.

For the **lateral pile capacity** and considered linear surface of the passive wedge by Broms (1964b) shown in Fig. 5.23, it follows that the ultimate lateral capacity P_{ult} of a pile adjacent to another pile reduces approximately with the ratio of the passive wedge area between two adjacent piles and the total passive wedge area:

$$Part\ of\ P_{ult} = \frac{12 \cdot n \cdot \tan\left(45^\circ + \phi/2\right) - n^2}{36 \cdot \tan^2\left(45^\circ + \phi/2\right)} \tag{5.40}$$

ϕ is soil friction angle

n is the number of pile diameters of the pile centre to centre spacing between two adjacent piles

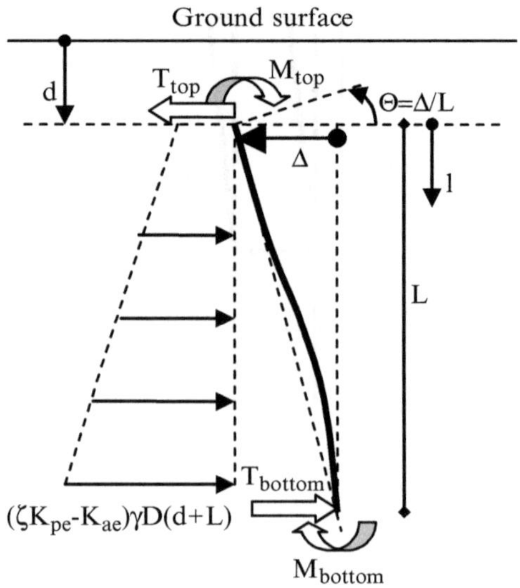

Fig. 5.24 Calculation scheme for the kinematic effect on shear forces T and bending moments M in a pile with diameter D embedded within a soil layer when subjected to horizontal pile top displacement Δ and pile top rotation Θ during an earthquake

5.2.2.5 Pile (Barrettes) Group During Earthquakes

- **Kinematic interaction**

  EN 1998-5 (2004), in clause 5.4.2(6)P, specifies that bending moments developing due to kinematic interaction shall be computed only when all of the following conditions occur simultaneously:

1. The ground profile is of type *D* (Fig. 2.8), S_1 or S_2 and contains consecutive layers of sharply differing stiffness, (S_1 – Deposits consisting, or containing a layer of at least 10 m thick, of soft clay/silt with plasticity index >40 % and high water content, with transversal wave velocity <100 m/s and undrained shear strength range from 10 to 20 kPa, S_2 – Deposits of liquefiable soil, of sensitive clay, or any other soil profile not included in types *A* – *E* or S_1)
2. The zone is of moderate or high seismicity, i.e. the product a_gS exceeds 0.10 g, (i.e. exceeds 0.98 m/s^2), and the supported structure is of importance class III or IV.

EN 1998-5 (2004) does not specify how to perform the computation. Figure 5.24 shows a possible calculation scheme (adopted from Srbulov 2010b) for rather homogeneous soil conditions, without presence of soil layers with very different stiffness as in the case of liquefied layers. Liyanapathirana and Poulos (2005) described a pseudo static approach for seismic analysis of piles in liquefying soil.

Using the expressions for engineering beams subjected to the end displacement and rotation (Jenkins 1989) and linearly increasing loading (Young and

Budynas 2002), the bending moments M and shear forces T are for a long pile with $L > 10D$:

$$M_{top} = \frac{6 \cdot E \cdot I \cdot \Delta}{L^2} - \frac{4 \cdot E \cdot I \cdot \Theta}{L} + \left(\varsigma \cdot K_{pe} - K_{ae}\right) \cdot \gamma \cdot D \cdot \left(\frac{L^3}{30} + \frac{d \cdot L^2}{12}\right)$$

$$M_{bottom} = \frac{6 \cdot E \cdot I \cdot \Delta}{L^2} - \frac{2 \cdot E \cdot I \cdot \Theta}{L} - \left(\varsigma \cdot K_{pe} - K_{ae}\right) \cdot \gamma \cdot D \cdot \left(\frac{L^3}{20} + \frac{d \cdot L^2}{12}\right)$$

$$T_{top} = \frac{12 \cdot E \cdot I \cdot \Delta}{L^3} - \frac{6 \cdot E \cdot I \cdot \Theta}{L^2} + \left(\varsigma \cdot K_{pe} - K_{ae}\right) \cdot \gamma \cdot D \cdot \left(\frac{3 \cdot L^2}{20} + \frac{d \cdot L}{2}\right)$$

$$T_{bottom} = \frac{12 \cdot E \cdot I \cdot \Delta}{L^3} - \frac{6 \cdot E \cdot I \cdot \Theta}{L^2} + \left(\varsigma \cdot K_{pe} - K_{ae}\right) \cdot \gamma \cdot D \cdot \left(\frac{7 \cdot L^2}{20} + \frac{d \cdot L}{2}\right) \tag{5.41}$$

For a short pile with $L < 6D$:

$$M_{top} = \frac{3 \cdot E \cdot I \cdot \Delta}{L^2} - \frac{3 \cdot E \cdot I \cdot \Theta}{L} + \left(\varsigma \cdot K_{pe} - K_{ae}\right) \cdot \gamma \cdot D \cdot \left(\frac{7 \cdot L^3}{120} + \frac{d \cdot L^2}{8}\right)$$

$$M_{bottom} = 0$$

$$T_{top} = \frac{3 \cdot E \cdot I \cdot \Delta}{L^3} - \frac{3 \cdot E \cdot I \cdot \Theta}{L^2} + \left(\varsigma \cdot K_{pe} - K_{ae}\right) \cdot \gamma \cdot D \cdot \left(\frac{9 \cdot L^2}{40} + \frac{3 \cdot d \cdot L}{8}\right)$$

$$T_{bottom} = \frac{3 \cdot E \cdot I \cdot \Delta}{L^3} - \frac{3 \cdot E \cdot I \cdot \Theta}{L^2} + \left(\varsigma \cdot K_{pe} - K_{ae}\right) \cdot \gamma \cdot D \cdot \left(\frac{11 \cdot L^2}{40} + \frac{5 \cdot d \cdot L}{8}\right) \tag{5.42}$$

For an intermediate pile length with $10D < L < 6D$, a linearly interpolated values can be used proportionally to the pile length.

The bending moments and shear forces at a depth l are:

$$M_l = M_{top} - T_{top} \cdot d + \left(\varsigma \cdot K_{pe} - K_{ae}\right) \cdot \gamma \cdot D \cdot \left(\frac{l^3}{6} - \frac{d \cdot l^2}{2}\right)$$
$$T_l = T_{top} - \left(\varsigma \cdot K_{pe} - K_{ae}\right) \cdot \gamma \cdot D \cdot \left(\frac{l^2}{2} - d \cdot l\right) \tag{5.43}$$

E is Young's modulus of pile material
I is the second moment of cross sectional area of pile $= D^4\pi/64$ for diameter D, $B^4/12$ for a square cross section with side length B
Δ is the maximum differential displacement between pile top and bottom $= d_g$ (Eq. 4.35) for piles in soil or $d_{g,soil} + d_{g,rock}$ for piles socketed into rock. The sum is for the unfavourable case of opposite direction of soil and rock motion
Θ is the rotation of pile top, if any, $=\Delta/L$. For piles under columns of long superstructure it may be that $\Theta \sim 0$ if the structure oscillates in phase with the ground
L is pile length
D is pile diameter
d is the depth of pile top below the ground surface
ζ is the coefficient of mobilisation of passive earth pressure in coarse grained soil $\sim 0.25 + 0.75\ (0.07\Delta\ L^{-1}$ to $0.25\Delta\ L^{-1})^{0.5}$ based on Fig. C.3 and Table C.2 in EN 1997-1 (2004) in loose soil for which the kinematic interaction needs to be analysed
K_{pe} is the coefficient of passive soil pressure according to Eq. (E.4) in EN 1998-5 (2004) i.e. Mononobe-Okabe Eq. (5.65). For piles placed behind other pile, the reduction coefficient of K_{pe} according to Eq. (5.40) applies for the friction angle ϕ in cyclic condition
K_{ae} is the coefficient of active soil pressure according to Eqs. (E.2 or E.3) in EN 1998-5 (2004) i.e. Mononobe-Okabe Eq. (5.64). For piles placed in front of other pile, the reduction coefficient of K_{ae} according to Eq. (5.44) applies

$$Part\ \ of\ \ K_{ae} = \frac{12 \cdot n \cdot \tan\left(45° - {}^{\phi}\!/_{2}\right) - n^2}{36 \cdot \tan^2\left(45° - {}^{\phi}\!/_{2}\right)} \tag{5.44}$$

ϕ is soil friction angle in cyclic condition
n is the number of pile diameters of the pile centre to centre spacing between two adjacent piles
γ is soil unit weight (=total unit weight if Mononobe-Okabe Eqs. (5.64) and (5.65) are applied i.e. according to Eqs. (E.5), (E.12) or (E.15) in EN 1998-5, 2004)

Nikolau et al. (2001) suggested the following formula obtained by data fitting for bending moment M in a pile due to kinematic soil pile interaction caused by different pile and soil stiffness

$$M \cong 0.042 \cdot a_{p,h} \cdot \rho_1 \cdot H_1 \cdot D^3 \cdot \left(\frac{L}{D}\right)^{0.30} \cdot \left(\frac{E_p}{E_1}\right)^{0.65} \cdot \left(\frac{V_{s2}}{V_{s1}}\right)^{0.50} \tag{5.45}$$

$a_{p,h}$ is the peak horizontal acceleration at the soil surface in the free field, ρ_1 is unit density of top soil layer,
H_1 is the thickness of the top soil layer,
L is pile length,
D is pile diameter,
E_p is pile Young modulus,
E_1 is Young modulus of top soil layer,
$V_{s2,1}$ are transversal wave velocities at the bottom and at the top of soil layer respectively

Examples of back analyses of the kinematic interactions are provided by (Srbulov 2011).

- **Inertial interaction**
  EN 1998-5 (2004), in clause 6, specifies that the effects of dynamic soil-structure interaction shall be taken into account in:

1. Structures where P–δ (2nd order) effect is significant
2. Structures with massive or deep seated foundations, such as bridge piers, offshore caissons, and silos,
3. Slender tall structures, such as towers and chimneys,
4. Structures supported on very soft soil, with average shear wave velocity in the top 30 m les than 100 m/s, such as those soil in ground type S_1

According to EN 1998-5 (2004) clause 5.4.2(3)P, analyses to determine the inertial forces along the pile, as well as deflection and rotation at the pile head, shall be based on discrete or continuum models that can realistically (even if approximately) reproduce:

1. The flexural stiffness of the pile (E I)
2. The soil reactions along the pile, with due consideration to the effects of cyclic loading and the magnitude of strains in soil (K_{dv})
3. The pile-to-pile dynamic interaction effects (also called dynamic "pile-group" effects) (α_v)
4. The degree of freedom of the rotation at/of the pile cap, or of the connection between the pile and the structure (β)

Figure 5.25 shows a possible calculation scheme (adopted from Srbulov 2010b).

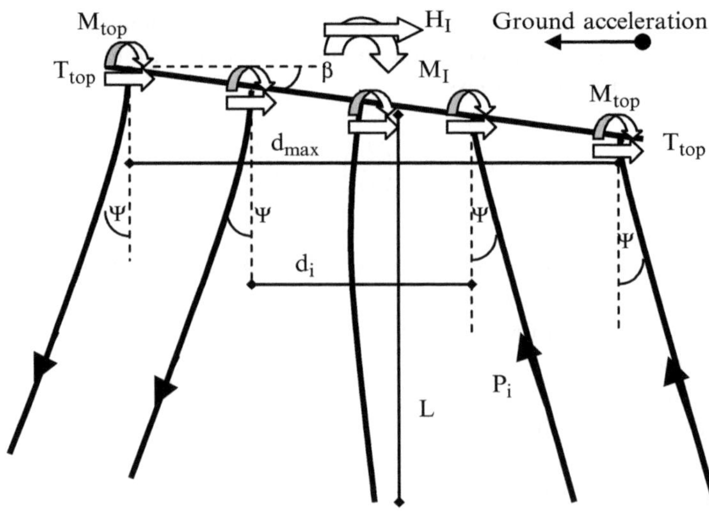

Fig. 5.25 Calculation scheme of the shear forces T_{top} and bending moments M_{top} at the tops of piles inclined for angle Ψ to the vertical due to the inertial effect (H_I, M_I) during an earthquake

$$\begin{aligned} P_i &= \frac{M_I \cdot K_{dv} \cdot d_{\max} \cdot L \cdot {}^{d_i}/_{d_{\max}}}{N_{piles} \cdot 7 \cdot E \cdot I \cdot \cos\psi + K_{dv} \cdot L \cdot \sum d_i^2 \cdot \cos\psi} \\ M_{top} &= \frac{7 \cdot P_{\max} \cdot E \cdot I \cdot \cos\psi}{K_{dv} \cdot d_{\max} \cdot L} \\ \beta &= \frac{M_{top} \cdot L}{3.5 \cdot E \cdot I} \\ T_{top} &= -\frac{H_I}{N_{piles}} \end{aligned} \tag{5.46}$$

P_i is the axial force in a pile spaced at the horizontal distance d_i
M_I is the overturning moment of inertial forces acting on a pile group
K_{dv} is pile dynamic axial stiffness $= (K_{sv}\ k_{dv} + i\omega\ C_v)\alpha_v$ with the coefficients given in Table 5.4 based on Gazetas (1991). Absolute value of the complex dynamic axial stiffness $= (K_{sv}{}^2\ k_{dv}{}^2 + \omega^2\ C_v{}^2)^{0.5}\alpha_v$. Alternatively, K_{dv} can be determined from pile tests.
d_{max} is the maximum horizontal distance between edge piles in a row
L is pile length
d_i is the horizontal spacing between two piles located symmetrically from the middle of a pile group
N_{piles} is the number of piles in a group
E is Young's modulus of pile material
I is the second moment of cross sectional area of pile $= D^4\pi/64$ for diameter D, $B^4/12$ for a square cross section with side length B

Table 5.4 Static axial stiffness K_{sv} and dynamic coefficients for long piles with $L > l_c$ when bedrock is located at depth $H \sim 2L$

Type	Linear increase of soil modulus with depth z	Parabolic increase of soil modulus with depth z	Constant soil modulus
	$E_s = E_{so}\, z/D$	$E_s = E_{so}\, (z/D)^{0.5}$	E_s
Active length l_c	$l_c \approx 2 \cdot D \cdot \left(\frac{E_p}{E_{so}}\right)^{0.2}$	$l_c \approx 2 \cdot D \cdot \left(\frac{E_p}{E_{so}}\right)^{0.22}$	$l_c \approx 2 \cdot D \cdot \left(\frac{E_p}{E_s}\right)^{0.25}$
Static axial stiffness K_{sv}	$K_{sv} \approx 1.8 \cdot E_{sL} \cdot D \cdot \left(\frac{L}{D}\right)^{0.55} \cdot$ $\cdot \left(\frac{E_p}{E_{sL}}\right)^{-(L/D)\cdot(E_p/E_{sL})}$ $E_{sL} = E_{so} \cdot {}^{L}/_{D}$	$K_{sv} \approx 1.9 \cdot E_{sL} \cdot D \cdot \left(\frac{L}{D}\right)^{0.6} \cdot$ $\cdot \left(\frac{E_p}{E_{sL}}\right)^{-(L/D)\cdot(E_p/E_{sL})}$ $E_{sL} = E_{so} \cdot \sqrt{{}^{L}/_{D}}$	$K_{sv} \approx 1.9 \cdot E_s \cdot D \cdot \left(\frac{L}{D}\right)^{2/3} \cdot$ $\cdot \left(\frac{E_p}{E_s}\right)^{-(L/D)\cdot(E_p/E_s)}$
Axial dynamic stiffness coefficient k_{dv}	$k_{dv} \approx 1 \; for \; \omega \cdot D/V_{sL} < 0.5$	$k_{dv} \approx 1 \; for \; L/D < 20$ $k_{dv} \approx 1 + {}^{1}/_{3}\sqrt{{}^{\omega \cdot D}/_{V_{sL}}}$ $for \; L/D \geq 50$	$k_{vd} \approx 1 \; for \; L/D < 15$ $k_{dv} \approx 1 + \sqrt{{}^{\omega \cdot D}/_{V_{sL}}}$ $for \; L/D \geq 50$
Axial damping coefficient C_v	$C_v = \frac{2}{3} \cdot \left(\frac{\omega \cdot D}{V_{sL}}\right)^{-1/3} \cdot$ $\cdot \rho \cdot V_{sL} \cdot \pi \cdot D \cdot L \cdot r_d$ $for \; f > 1.5 \cdot \frac{V_{s,average}}{4 \cdot H}$ $r_d \approx 1 - e^{-2\cdot(E_p/E_{sL})\cdot(L/D)^{-2}}$ $C_v \approx 0 \; for \; f \leq \frac{V_{s,average}}{4 \cdot H}$ *Linearly* int *erpolated for other f*	$C_v \approx \frac{3}{4} \cdot \left(\frac{\omega \cdot D}{V_{sL}}\right)^{-1/4} \cdot$ $\cdot \rho \cdot V_{sL} \cdot \pi \cdot D \cdot L \cdot r_d$ $for \; f > 1.5 \cdot \frac{V_{s,average}}{4 \cdot H}$ $r_d \approx 1 - e^{-1.5\cdot(E_p/E_{sL})\cdot(L/D)^{-2}}$ $C_v \approx 0 \; for \; f \leq \frac{V_{s,average}}{4 \cdot H}$ *Linearly* int *erpolated for other f*	$C_v \approx \left(\frac{\omega \cdot D}{V_s}\right)^{-1/5} \cdot$ $\cdot \rho \cdot V_s \cdot \pi \cdot D \cdot L \cdot r_d$ $for \; f > 1.5 \cdot \frac{V_{s,average}}{4 \cdot H}$ $r_d \approx 1 - e^{-\cdot(E_p/E_{sL})\cdot(L/D)^{-2}}$ $C_v \approx 0 \; for \; f \leq \frac{V_{s,average}}{4 \cdot H}$ *Linearly* int *erpolated for other f*

(continued)

Table 5.4 (continued)

Type	Linear increase of soil modulus with depth z	Parabolic increase of soil modulus with depth z	Constant soil modulus
	$E_s = E_{so}\, z/D$	$E_s = E_{so}\, (z/D)^{0.5}$	E_s
Interaction factor α_v for axial in-phase oscillations of two adjacent piles	$\alpha_v \approx \sqrt{2} \cdot \left(\frac{d_p}{D}\right)^{-3/4} \cdot \cdot e^{-0.5 \cdot \beta \cdot \omega \cdot d_p / V_{sL}} \cdot e^{-i \cdot \omega \cdot \sqrt{2} \cdot d_p / V_{sL}}$	$\alpha_v \approx \sqrt{2} \cdot \left(\frac{d_p}{D}\right)^{-2/3} \cdot \cdot e^{-2/3 \cdot \beta \cdot \omega \cdot d_p / V_{sL}} \cdot e^{-i \cdot \omega \cdot \sqrt{2} \cdot d_p / V_{sL}}$	$\alpha_v \approx \sqrt{2} \cdot \left(\frac{d_p}{D}\right)^{-1/2} \cdot \cdot e^{-\beta \cdot \omega \cdot d_p / V_s} \cdot e^{-i \cdot \omega \cdot \cdot d_p / V_s}$

Notes:
L is pile length
H is bedrock depth or 2 L
D is pile diameter
E_p is pile Young's modulus
$E_{s(o)}$ is soil (basic) axial modulus reduced depending on the value of peak horizontal ground acceleration at the surface according to Table 4.1 in EN 1998-5 (2004) as plotted in Fig. 3.18 for G/G_{max}
$\omega = 2\pi f$ where f is frequency of pile vibration ~ $V_{s,average}/(4H)$
V_{sL} is soil transversal wave velocity at a depth L
V_s is soil transversal wave velocity at a depth L/2
ρ is soil unit density
$V_{s,average}$ is average soil transversal wave velocity over depth H
d_p is the horizontal distance between two piles
i is an imaginary number $(-1)^{0.5}$

Absolute value of the complex interaction factor for a constant E_s for example is

$$\alpha_v \approx \sqrt{2} \cdot \left(\frac{d_p}{D}\right)^{-1/2} \cdot \cdot e^{-\left[\left(\beta \cdot \omega \cdot d_p / V_s\right)^2 + \left(\omega \cdot \sqrt{2} \cdot d_p / V_s\right)^2\right]^{0.5}}$$

Ψ is the pile inclination to the vertical
P_{max} is the axial force in the piles spaced at d_{max}
H_I is the horizontal inertial force acting on a pile group
– sign is introduced in the case of superposition of the shear forces from kinematic and inertial effects

Examples of back analyses of the inertial interactions are provided by Srbulov (2011).

5.2.3 *Movement*

5.2.3.1 Shallow Depth Under Static Load

- **In undrained conditions (short-term) in fine grained soil**
  Immediate average settlement s_i of uniformly loaded flexible areas is:

$$s_i = \frac{\Delta\sigma_v \cdot B \cdot \mu_1 \cdot \mu_o}{E} \tag{5.47}$$

$\Delta\sigma_v$ is load acting at depth D of a foundation
B is foundation width (or diameter)
E is an average soil stiffness modulus over depth range D to $2B$ for circles and squares and D to $4B$ for strip foundations
μ_I is the coefficient dependent on the thickness of soil layer H between the foundation depth D and the maximum depth of $2B$ or $4B$ depending on the shape of foundation
μ_o is the coefficient dependent on depth D to width B ratio and foundation shape L/B, where L is foundation length

The coefficients μ_I, μ_o are shown in Fig. 5.26 adopted from Janbu et al. (1956)

- **In drained conditions in coarse grained soil and rock and fine grained soil (long-term)**
  The settlement of **coarse grained soil** can be calculated according to de Beer and Martens (1957):

$$s_d = \sum \frac{\Delta d}{C} \cdot \ln_e \frac{\sigma_v{}' + \Delta\sigma_v}{\sigma_v{}'} \tag{5.48}$$

Σ is the sum over depth range between a foundation depth and the depth where $\Delta\sigma_v = 0.2\sigma_v{}'$ (EN 1997-1 2004)
Δd is the thickness of a sub layer
$C = 1.9\, q_c / \sigma_v{}'$ is the compressibility coefficient proposed by Meyerhof (1965)
q_c is cone penetrometer test point resistance
$\sigma_v{}'$ is the effective vertical stress in the middle of a sub layer with thickness Δd

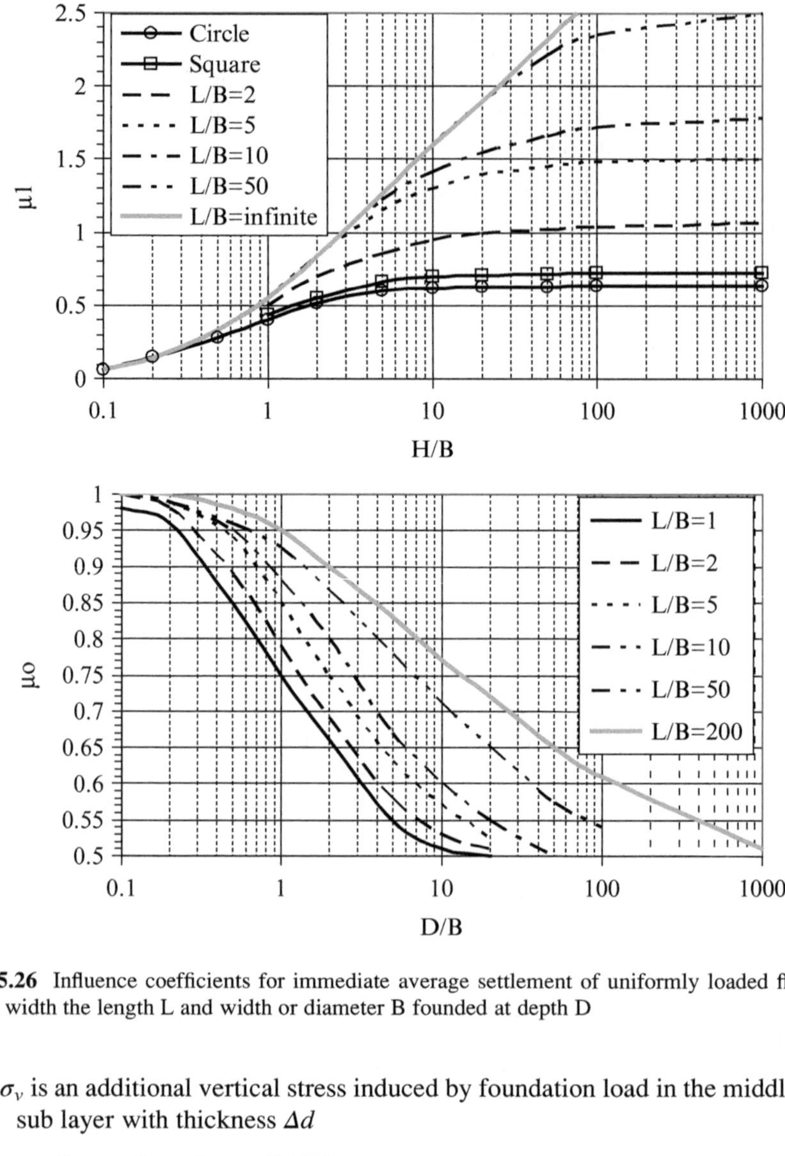

Fig. 5.26 Influence coefficients for immediate average settlement of uniformly loaded flexible areas width the length L and width or diameter B founded at depth D

$\Delta\sigma_v$ is an additional vertical stress induced by foundation load in the middle of a sub layer with thickness Δd

According to Boussinesq (1885)

$$\Delta\sigma_v = \frac{3 \cdot F \cdot d^3}{2 \cdot \pi \cdot R^5}$$
$$R = \sqrt{r^2 + d^2} > 0 \tag{5.49}$$

F is the force acting at the foundation depth
d is depth to a middle of a sub layer with thickness Δd

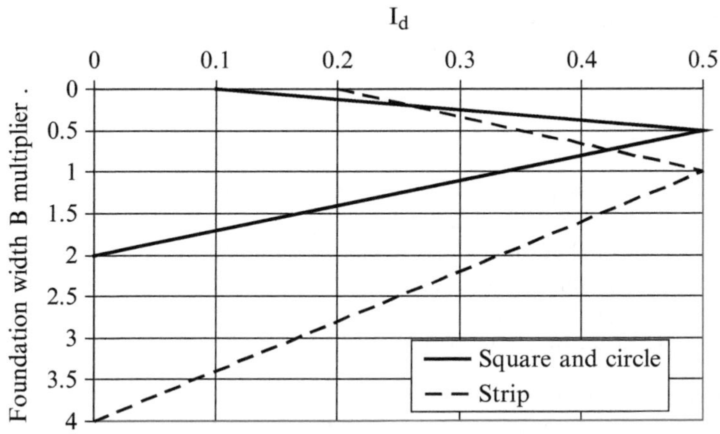

Fig. 5.27 Strain influence factor I_d diagram

r is the horizontal distance between the place of application of the force F and the place where $\Delta\sigma_v$ is calculated (r = 0 for single foundation) Schmertmann (1970) and Schmertmann et al. (1978) proposed the following formula for calculation of settlement of **coarse grained soil**

$$
\begin{aligned}
s_d &= C_1 \cdot C_2 \cdot (\Delta\sigma_{vo} - \sigma_{vo}') \cdot \sum_0^{d_{\max}} \frac{I_d}{C_3 \cdot E} \cdot \Delta d \\
C_1 &= 1 - 0.5 \cdot \frac{\sigma_{vo}'}{q - \sigma_{vo}'} \\
C_2 &= 1.2 + 0.2 \cdot \log_{10} t \\
C_3 &= 1.25 \quad for \quad square \quad foundation \\
C_3 &= 1.75 \quad for \quad strip \quad foundation
\end{aligned}
\tag{5.50}
$$

$\Delta\sigma_{vo}$ is additional vertical stress at the level of a foundation
σ_{vo}' is the initial effective vertical stress at the level of a foundation
t is time in years
I_d is a strain influence factor according to Fig. 5.27
E is soil Young's modulus = 2.5 q_c for circular and square foundations,
= 3.5 q_c for strip foundation
q_c is cone penetrometer test point resistance
Δd is the thickness of a sub layer
d_{max} is 2 times width of square and circular foundation or 4 times width of a strip foundation

The settlement of **rock mass** can be calculated as:

$$s_d = \sum \frac{\Delta\sigma_v}{E} \Delta d \tag{5.51}$$

Σ is the sum over depth range between a foundation depth and the depth where $\Delta\sigma_v = 0.2\sigma_v'$ (EN 1997-1 2004)
$\Delta\sigma_v$ is an additional vertical stress induced by foundation load in the middle of a sub layer with thickness Δd according to Eq. (5.49)
E is rock mass stiffness modulus
Δd is the thickness of a sub layer

Burland and Burbridge (1985) proposed a method to calculate settlement of coarse grained soil under spread foundations based on the results of standard penetration tests

The settlement of **fine grained soil (long-term)** is

$$s_d = \Sigma \frac{C_r \cdot \Delta d}{1 + e_o} \cdot \log_{10} \frac{p_c'}{\sigma_v'} + \Sigma \frac{C_c \cdot \Delta d}{1 + e_o} \cdot \log_{10} \frac{\sigma_v' + \Delta\sigma_v}{p_c'} \tag{5.52}$$

C_r is recompression index in case of over consolidated soil otherwise $= 0$
Δd is the thickness of a sub layer
e_o is the initial void ratio
p_c' is the effective preconsolidation pressure in over consolidated soil, otherwise $= \sigma_v'$
σ_v' is the effective vertical stress in the middle of a sub layer with thickness Δd
C_c is compression index
$\Delta\sigma_v$ is an additional vertical stress induced by foundation load in the middle of a sub layer with thickness Δd according to Eq. (5.49)

Time t necessary to achieve certain **degree of consolidation U** can be calculated for saturated soil, constant soil properties, small strain and one dimensional water flow:

$$t = \frac{T_v \cdot H^2}{c_v} \tag{5.53}$$

T_v is the time factor
H is the thickness of consolidated layer
c_v is the coefficient of consolidation

The correlation between U and T_v are shown in Fig. 5.28 adopted from Janbu et al. (1956)

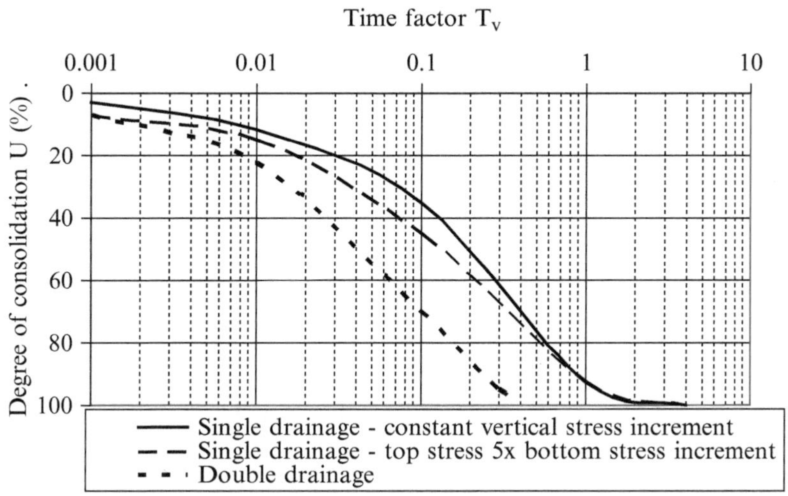

Fig. 5.28 Relationship between degree of consolidation and time factor

Examples of more detailed approach and the case of use of pressure relief wells are presented by Srbulov (2008).

- **Creep settlement in loose sand and rock fill, highly plastic or soft clay and peat**

$$s_s = d \cdot C_\alpha \cdot \log_{10}\frac{t_2}{t_1} \tag{5.54}$$

d is layer thickness
C_a is the secondary compression index
t_2 is the time where secondary settlement is calculated
t_l is the time at the end of primary consolidation from Eq. (5.53) for the time factor corresponding to 100 % degree of consolidation

5.2.3.2 Medium Depth Under Static Load

In the vertical direction, the settlement is simply a ratio between applied vertical load and static stiffness K_v given in Table 5.5 from Gazetas (1991). In the horizontal direction and rotation, Fig. 5.29,

Table 5.5 Static stiffness coefficients for a medium depth foundation

Direction	Coefficient
Vertical	$K_v = \frac{2 \cdot G \cdot L}{1-\nu} \cdot \left(0.73 + 1.54 \cdot \left(\frac{A_b}{4 \cdot L^2}\right)^{0.75}\right) \cdot \left(1 + \frac{D}{21 \cdot B} \cdot \left(1 + 1.3 \cdot \frac{A_b}{4 \cdot L^2}\right)\right) \cdot \left(1 + 0.2 \cdot \left(\frac{A_w}{A_b}\right)^{2/3}\right)$
Horizontal – lateral	$K_{h,B} = \frac{2 \cdot G \cdot L}{2-\nu} \cdot \left(2 + 2.5 \cdot \left(\frac{A_b}{4 \cdot L^2}\right)^{0.85}\right) \cdot \left(1 + 0.15 \cdot \sqrt{\frac{D}{B}}\right) \cdot \left(1 + 0.52 \cdot \left(\frac{h_c \cdot A_w}{B \cdot L^2}\right)^{0.4}\right)$
Horizontal – longitudinal	$K_{h,L} = \left(K_{h,B} - \frac{0.2}{0.75-\nu} \cdot G \cdot L \cdot \left(1 - \frac{B}{L}\right)\right) \cdot \left(1 + 0.15 \cdot \sqrt{\frac{D}{B}}\right) \cdot \left(1 + 0.52 \cdot \left(\frac{h_c \cdot A_w}{B \cdot L^2}\right)^{0.4}\right)$
Rotating around longitudinal axis	$K_{r,B} = \frac{G}{1-\nu} \cdot {I_B}^{0.75} \cdot \left(\frac{L}{B}\right)^{0.25} \cdot \left(2.4 + 0.5 \cdot \frac{B}{L}\right) \cdot \left\{1 + 1.26 \cdot \frac{d}{B} \cdot \left[1 + \frac{d}{B} \cdot \left(\frac{d}{D}\right)^{-0.2} \sqrt{\frac{B}{L}}\right]\right\}$
Rotating around lateral axis	$K_{r,L} = \frac{G}{1-\nu} \cdot {I_L}^{0.75} \cdot 3 \cdot \left(\frac{L}{B}\right)^{0.15} \cdot \left\{1 + 0.92 \cdot \left(\frac{d}{L}\right)^{0.6} \cdot \left[1.5 + \left(\frac{d}{L}\right)^{1.9} \cdot \left(\frac{d}{D}\right)^{-0.6}\right]\right\}$

(continued)

Table 5.5 (continued)

Direction	Coefficient
Coupled horizontal & rotational	$K_{hr,B} \approx \frac{1}{3} \cdot d \cdot K_{h,B}$ $K_{hr,L} \approx \frac{1}{3} \cdot d \cdot K_{h,L}$

Notes:
G is shear modulus
L is a half of the foundation length
ν is Poisson's ratio
A_b is the base area of a foundation
D is foundation depth below ground surface
B is a half of foundation width $B < L$
A_w is actual sidewall-soil contact area
d is actual sidewall-soil contact depth
h is the height of foundation centre of gravity above its base
I_B is the second moment of area of foundation around the longitudinal axis of the foundation
I_L is the second moment of area of foundation around the transversal axis of the foundation

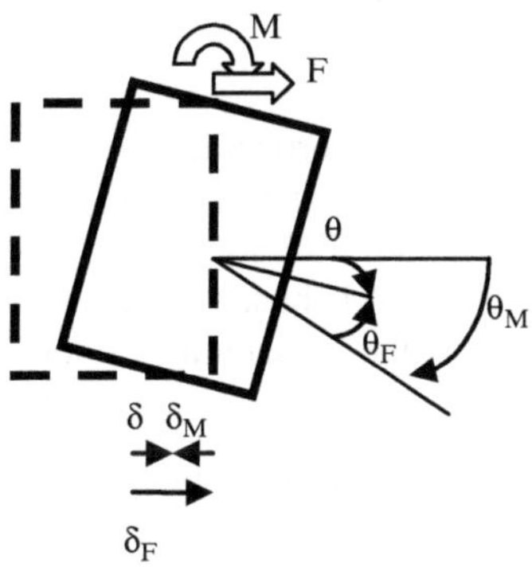

Fig. 5.29 Applied actions to (Moment, Force) and responses ($\delta = \delta_F - \delta_M$, $\theta = \theta_M - \theta_F$) of a medium depth block

$$
\begin{aligned}
\delta &= \frac{B_{22} \cdot F - B_{12} \cdot M}{B_{11} \cdot B_{22} - B_{12}^2} \\
\vartheta &= \frac{B_{11} \cdot M - B_{12} \cdot F}{B_{11} \cdot B_{22} - B_{12}^2} \\
B_{11} &= K_h \\
B_{12} &= K_{hr} - K_h \cdot h_c \\
B_{22} &= K_r + K_h \cdot h_c^2 - 2 \cdot K_{hr} \cdot h_c
\end{aligned} \tag{5.55}
$$

δ is the horizontal displacement at the centre of gravity of a foundation
ϑ is the rotation of a box or block
F is the horizontal force maximum amplitude
M is the rotating moment maximum amplitude
K_h, K_{hr}, K_r are the coefficients given in Table 5.5 from Gazetas (1991)
h_c is the height of centroid of a moving block above its foundation level

5.2.3.3 Deep Under Static Load

- **Single pile (barrette) in the vertical direction**
  The settlement is a ratio between applied axial force and the static axial stiffness K_{sv} from Table 5.4.
- **Single pile (barrette) in the horizontal direction and rotation**
  The horizontal displacement and rotation at the top are the ratios between applied horizontal force and moment at the top and the static stiffness in horizontal direction and rotation given in Table 5.6 from Gazetas (1991) and Table C.1 of EN 1998-5 (2004).
- **Group of piles (barrettes)**
  Poulos and Davis (1980) provided influence coefficients for pile groups loaded in axial and lateral direction. Alternatively, an equivalent medium depth foundation method described in Sect. 5.2.3.2 can be used instead. For a constant stiffness of soil, the effect of load at the location of one pile on displacement at the location of other pile can be estimated using Midlin (1936) formulas for homogeneous half space.

$$s = \frac{V \cdot (1 - \nu)}{2 \cdot \pi \cdot G \cdot r} \tag{5.56}$$

s is the vertical displacement at a pile top
V is the vertical force (=axial pile force)
ν is soil Poisson's ratio
G is soil shear stiffness modulus
r is the horizontal distance between the place where force V is acting and the place where s is calculated

For the horizontal load H at a pile top and a constant stiffness of soil, the effect of load at the location of one pile on displacement at the location of other pile can be estimated using Cerrutti (1882) formulas for homogeneous half space.

$$\begin{aligned} u_x &= \frac{H}{2 \cdot \pi \cdot G \cdot R} \cdot \left(1 - \nu + \nu \cdot \frac{x^2}{R^2}\right) \\ u_y &= \frac{H \cdot \nu}{2 \cdot \pi \cdot G} \cdot \frac{x \cdot y}{R^3} \end{aligned} \tag{5.57}$$

u is the horizontal displacement at a pile top
H is the horizontal force at a pile top
ν is soil Poisson's ratio
G is soil shear stiffness modulus
$R = (x^2 + y^2)^{1/2}$
x is the horizontal distance between the place where force H is acting and the place where u_x is calculated in the same direction of force H

Table 5.6 Static stiffness coefficients for piles (barrettes) for the horizontal and rotational modes of motion

Type	Linear increase of soil modulus with depth z $E_s = E_{so}\, z/D$	Parabolic increase of soil modulus with depth z $E_s = E_{so}\, (z/D)^{0.5}$	Constant soil modulus E_s
Horizontal	$K_h = 0.6 \cdot D \cdot E_{so} \cdot \left(\frac{E_p}{E_{so}}\right)^{0.35}$	$K_h = 0.8 \cdot D \cdot E_{so} \cdot \left(\frac{E_p}{E_{so}}\right)^{0.28}$	$K_h = D \cdot E_s \cdot \left(\frac{E_p}{E_s}\right)^{0.21}$
Rotational	$K_r = 0.15 \cdot D^3 \cdot E_{so} \cdot \left(\frac{E_p}{E_{so}}\right)^{0.80}$	$K_r = 0.15 \cdot D^3 \cdot E_{so} \cdot \left(\frac{E_p}{E_{so}}\right)^{0.77}$	$K_r = 0.15 \cdot D^3 \cdot E_s \cdot \left(\frac{E_p}{E_s}\right)^{0.75}$
Coupled horizontal and rotational	$K_{hr} = -0.17 \cdot D^2 \cdot E_{so} \cdot \left(\frac{E_p}{E_{so}}\right)^{0.60}$	$K_{hr} = -0.24 \cdot D^2 \cdot E_{so} \cdot \left(\frac{E_p}{E_{so}}\right)^{0.53}$	$K_{hr} = -0.22 \cdot D^2 \cdot E_s \cdot \left(\frac{E_p}{E_s}\right)^{0.50}$

Notes:
D is pile diameter
E_p is pile Young's modulus
$E_{s(o)}$ is soil (basic) axial modulus

y is the horizontal distance between the place where force H is acting and the place where u_y is calculated in the perpendicular direction to force H

The above formulas were derived for a half space without presence of piles.

5.2.3.4 Shallow, Medium Depth and Deep Under Cyclic Load

Gazetas (1991) provided dynamic stiffness and damping coefficients for all depths of foundations.

- **For shallow depth foundations,** Wolf (1994) formulated the rocking and horizontal equations of motions for harmonic rotational loading with circular frequency $\omega = 2\pi f$, f is frequency of vibration, as:

$$\begin{aligned} &\left[-\omega^2 \cdot (I + \Delta M_\theta) + S_\theta\right] \cdot \theta - S_\omega \cdot e \cdot u = M_\omega \\ &-\omega^2 \cdot m \cdot e \cdot \theta + \left[-\omega^2 \cdot m + S_\omega\right] \cdot u = H_\omega \end{aligned} \tag{5.58}$$

m is the mass of foundation,
e is the eccentricity of m with respect to the underside of foundation, $I = m/12^*(h^2 + w^2)$,
h is the foundation height
w is the foundation width
$\Delta M_\theta = 0.3^*\pi^*(\nu - 1/3)^*\rho^* r^5$ when $1/3 < \nu < 1/2$ otherwise $\Delta M_\theta = 0$,
ν is Poisson's ratio of soil,
ρ is soil unit density,
r is the equivalent radius of foundation $= (4^*J/\pi)^{1/4}$,
J is the second moment of area of a foundation,
$S_\omega = K_o{}^*(k_o + i^*b_o)$,
$K_o = r_o{}^*V_t{}^{2*}A/z_o$,
$r_o = (A/\pi)^{1/2}$,
A is the foundation area,
V_t is the transversal wave velocity in soil,
$z_o = r_o{}^*\pi / 8^*(2 - \nu)$,
$b_o = \omega^* z/V_t$,
$k_o = 1 - \mu_o / \pi * r_o / z_o{}^* b_o{}^2$,
$\mu_o = 2.4^*\pi^*(\nu - 1/3)$ for $\nu > 1/3$ otherwise $\mu_o = 0$,
$S_\theta = K_\theta{}^*(k_\theta + i^* b_o{}^* c_\theta)$,
$K_\theta = 3^*\rho^* V^{2*}\pi^* r^4/4/z$, $V = V_l$ for $\nu < 1/3$ and $V = 2^*V_t$ for $1/3 < \nu < 1/2$,
V_l is the longitudinal wave velocity in soil
$z = 9^*r^*\pi / 32^*(1 - \nu)^*(V/V_t)^2$,
$k_\theta = 1 - 4^*\mu / 3 /\pi^* r/z * b_o{}^2 - 1/3^*b_o{}^2/(1 + b_o{}^2)$,
$\mu = 0.3^*\pi^*(\nu - 1/3)$ for $\nu > 1/3$ otherwise $\mu = 0$,
$c_\theta = 1/3 {}^*b_o{}^2/(1 + b_o{}^2)$
$i = (-1)^{0.5}$

The solution of the above two linear equations of motions provides the values of horizontal displacement u and rotation θ at any time depending on M_ω and H_ω.

Examples of more detailed analyses of vibration of shallow foundations including the use of rubber bearings and viscoelastic dampers are provided by Srbulov (2010c, 2011).

- **For deep foundations in the axial direction,** the settlement is a ratio between applied axial load and K_{dv}, which is pile dynamic axial stiffness $= (K_{sv}\, k_{dv} + i\omega\, C_v)\alpha_v$ with the coefficients given in Table 5.4 based on Gazetas (1991). Absolute value of the complex dynamic axial stiffness $= (K_{sv}^{\ 2}\, k_{dv}^{\ 2} + \omega^2\, C_v^{\ 2})^{0.5}\alpha_v$, where α_v is pile group effect coefficient given in Table 5.4, $\omega = 2\pi f$, f is vibration frequency. Alternatively, K_{dv} can be determined from pile tests.
- **For deep foundations in the horizontal direction and rotation**, the horizontal displacement and rotation at the top are the ratios between applied horizontal force and the moment at the top and dynamic horizontal, rotational and coupled stiffness of a pile $K_{h,r,hr} = (K_{h,r,hr}^{\ \ 2} + \omega^2 C_{h,r,rh})^{0.5}$. The values of $K_{h,r,hr}$ are given in Table 5.6 (dynamic stiffness coefficients $k_{h,r,rh} = 1$), $C_{h,r,rh}$ are given in Table 5.7, $\omega = 2\pi f$, f is vibration frequency.

Causes of foundation movements due to hydraulic failure, erosion, liquefaction, rock cavitations, soil structure collapse, heave, slope instability, vibration, earthquakes, and ice formation are described in Sect. 4.

5.2.4 Execution

5.2.4.1 Shallow and Medium Depth

- It is important to protect base soil under foundation upon excavation to the foundation level by placing a layer of lean concrete immediately after excavation to minimize the effect of physical weathering.
- Minimum foundation depth depends on the depth of local soil freezing, depth of significant moisture change in swelling and shrinkage soil and on scour depth when water flow exists
- For medium depth foundation relying on ground side resistance, it is important to ensure that such resistance is not compromised by opening of gaps in shrinkage soil, during cyclic loading in fine grained soil or other reasons.
- When medium depth foundation is not constructed in an open excavation due to limited space then lateral soil pressure distribution on flexible supports is shown in Fig. 5.30 based on Peck (1969) and Goldberg et al. (1976).

Table 5.7 Damping coefficients for piles (barrettes) for the horizontal and rotational modes of motion

Type	Linear increase of soil modulus with depth z	Parabolic increase of soil modulus with depth z	Constant soil modulus
	$E_s = E_{so}\, z/D$	$E_s = E_{so}\,(z/D)^{0.5}$	E_s
Natural frequency	$f_s = 0.19 \cdot V_{tH}/H$	$f_s = 0.223 \cdot V_{tH}/H$	$f_s = 0.25 \cdot V_t/H$
Horizontal	$C_h = 2 \cdot K_h \cdot D_h/\omega$ $D_h = 0.6 \cdot \beta + 1.8 \cdot f \cdot D/V_{to}$, $f > f_s$ $D_h = 0.6 \cdot \beta, f \leq f_s$	$C_h = 2 \cdot K_h \cdot D_h/\omega$ $D_h = 0.7 \cdot \beta + 1.2 \cdot f \cdot D \cdot \left(\frac{E_p}{E_{so}}\right)^{0.08} / V_{to}$ $f > f_s$ $D_h = 0.7 \cdot \beta, f \leq f_s$	$C_h = 2 \cdot K_h \cdot D_h/\omega$ $D_h = 0.8 \cdot \beta + 1.1 \cdot f \cdot D \cdot \left(\frac{E_p}{E_s}\right)^{0.17} / V_t$ $f > f_s$ $D_h = 0.8 \cdot \beta, f \leq f_s$
Rotational	$C_r = 2 \cdot K_r \cdot D_r/\omega$ $D_r = 0.2 \cdot \beta + 0.4 \cdot f \cdot D/V_{to}$, $f > f_s$ $D_r = 0.2 \cdot \beta, f \leq f_s$	$C_r = 2 \cdot K_r \cdot D_r/\omega$ $D_r = 0.22 \cdot \beta + 0.35 \cdot f \cdot D \cdot \left(\frac{E_p}{E_{so}}\right)^{0.1} / V_{to}$ $f > f_s$ $D_r = 0.22 \cdot \beta, f \leq f_s$	$C_r = 2 \cdot K_r \cdot D_r/\omega$ $D_r = 0.35 \cdot \beta + 0.35 \cdot f \cdot D \cdot \left(\frac{E_p}{E_s}\right)^{0.1} / V_t$ $f > f_s$ $D_r = 0.25 \cdot \beta, f \leq f_s$
Coupled horizontal and rotational	$C_{rh} = 2 \cdot K_{rh} \cdot D_{rh}/\omega$ $D_{rh} = 0.3 \cdot \beta + f \cdot D/V_{to}$, $f > f_s$ $D_{rh} = 0.3 \cdot \beta, f \leq f_s$	$C_{rh} = 2 \cdot K_{rh} \cdot D_{rh}/\omega$ $D_{rh} = 0.6 \cdot \beta + 0.7 \cdot f \cdot D \cdot \left(\frac{E_p}{E_{so}}\right)^{0.05} / V_{to}$, $f > f_s$ $D_{rh} = 0.35 \cdot \beta, f \leq f_s$	$C_{rh} = 2 \cdot K_{rh} \cdot D_{rh}/\omega$ $D_{rh} = 0.8 \cdot \beta + 0.85 \cdot f \cdot D \cdot \left(\frac{E_p}{E_s}\right)^{0.18} / V_t$, $f > f_s$ $D_{rh} = 0.5 \cdot \beta, f \leq f_s$

Interaction factor for in-phase oscillations of two adjacent piles	Not well defined	Not well defined	$\alpha_{h,90^o} \approx 3/4 \cdot \alpha_v$ $\alpha_{h,0^o} \approx 0.5 \cdot \left(\frac{D}{d_p}\right)^{0.5} \cdot e^{-\beta \cdot \omega \cdot d_p / V_t} \cdot e^{-i \cdot \omega \cdot d_p / V_t}$

Notes:
V_{tH} is the transversal wave velocity at the bottom of a soil layer
H is soil layer thickness
K_h, K_r, K_{rh} are given in Table 5.6
$\omega = 2\pi f$
f is the frequency of vibration
β is soil hysteretic damping ~0.05
V_{to} is soil transversal velocity at a depth D
D is pile diameter
d_p is pile length
E_p is pile Young's modulus
$E_{s(o)}$ is soil (basic) axial modulus
α_v is given in Table 5.4

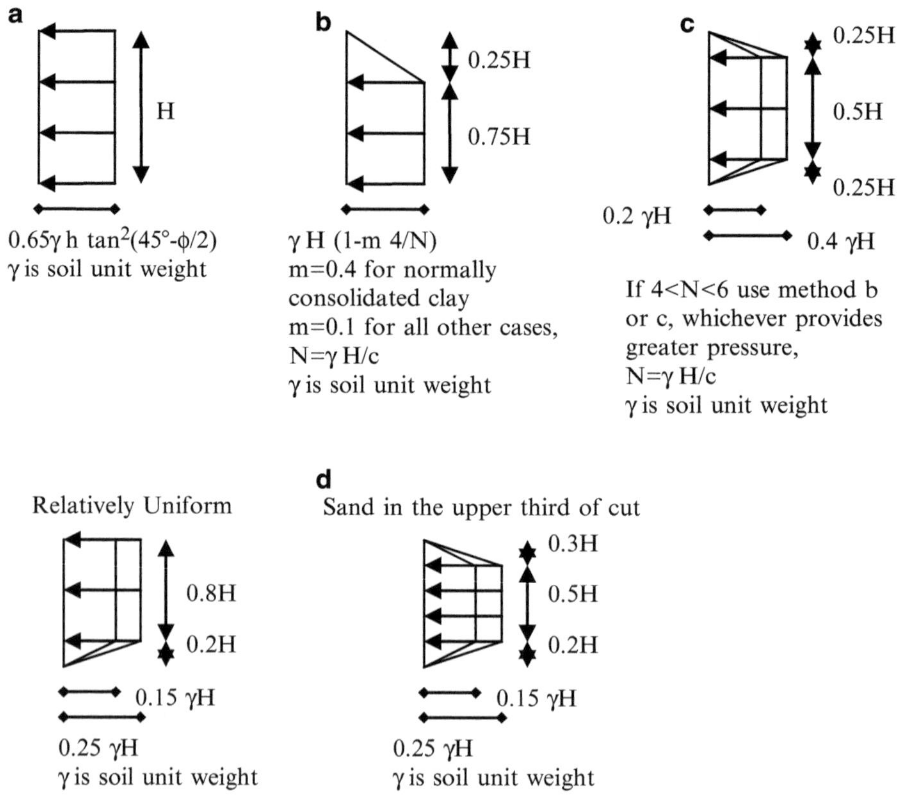

Fig. 5.30 Pressure distribution behind temporary (and flexible) bracings along excavation depth H, (**a**) sand, (**b**) soft clay with $N > 6$, (**c**) stiff clay with $N < 4$, (**d**) dense cohesive sand or very stiff sandy clay

5.2.4.2 Deep

- The following codes contain useful details: EN 1536 (2010), EN 12699 (2001), EN 1993-5 (2007), EN 14199 (2005), EN ISO/DIS 22477-1, ASTM D1143, D3689, D3966, D4945, D5882, D6760, D7383.
- Osterberg cell test is not standardized but is used for testing of pile shaft capacity by applying hydraulic pressure inside a flat hollow steel cell located at the pile toe.
- The Elliot method uses inflatable cylinder(s) located near the pile top to remove concrete mixed with spoil instead of using pneumatic hammers that cause vibration disease to the operatives. Spoil may originate from bentonite mixture used to stabilize excavation before concreting or from excavated soil debris.
- Airlift is used to clean spoil accumulated at the pile toe with ground removal from the void. Failure to remove spoil at the pile toe can result in excessive settlement.

Table 5.8 Values of hammer coefficient K (BSP Pocket Book, 1969)

Hammer	K
Drop hammer, winch operated	0.8
Drop hammer, trigger release	1
Single acting hammer	0.9
BSP double acting hammer	1
McKiernan-Terry diesel hammer	1

- Pile driving refusal criteria are defined in clause 12.5.6 of API RP2A-WSD (2007) for example. In most cases, pile penetration of less than 1 mm under a single hammer stroke is considered as pile driving refusal.
- The use of pile driving formulas for estimation of pile axial capacity is allowed only after the calibration with the results of static load tests performed on the same type of pile, of similar length and cross section, and in similar ground condition (e.g. clause 7.6.2.5(2) of EN 1997-1: (2004))
- Hiley (1925) formula for an ultimate axial pile capacity

$$Q_u = \frac{K \cdot W \cdot H \cdot \eta}{s + c/2} \tag{5.59}$$

K is hammer coefficient given in Table 5.8
W is hammer ram weight
H is the maximum stroke of a hammer ram
η is hammer efficiency obtainable from a hammer manufacturer
s is final set (penetration/blow) obtainable from dynamic pile testing e.g. using CAPWAP method
c is the sum of the temporary elastic compression of the pile (difference between the top and bottom pile settlement) obtainable from dynamic pile testing e.g. using CAPWAP method

- Janbu (1953) formula for an ultimate axial pile capacity

$$Q_u = \frac{1}{K_u} \cdot \frac{\eta \cdot W \cdot H}{s} \tag{5.60}$$

W is hammer ram weight
H is the maximum stroke of a hammer ram
η is hammer efficiency =0.7 for good, =0.55 for average, =0.4 for difficult or bad driving conditions
s is final set (penetration/blow) obtainable from dynamic pile testing e.g. using CAPWAP method

$$K_u = C_d \cdot \left[1 + \left(1 + \frac{\lambda_e}{C_d} \right)^{1/2} \right]$$
$$C_d = 0.75 + 0.15 \cdot \frac{W_p}{W} \tag{5.61}$$
$$\lambda_e = \frac{\eta \cdot W \cdot H \cdot L}{A \cdot E \cdot s^2}$$

L is length of pile
A is pile cross sectional area
E is modulus of elasticity of pile material
W is weight of hammer
W_p weight of pile
H is the maximum stroke of hammer ram
s is final set (penetration/blow) obtainable from dynamic pile testing e.g. using CAPWAP method
η is hammer efficiency =0.7 for good, =0.55 for average, =0.4 for difficult or bad driving conditions

- Poulos (2005) considered some of the consequences of geological and construction imperfections on pile behaviour in practice.
- Thrornburn and Thornburn (1977) listed the following problems during construction of **cast-in-place concrete piles**:

  1. Over break (excavation instability, cavitations) in unstable mainly water bearing strata results in a significant excess volume of spoil being excavated. The installation of the temporary steel casing within a pile bore, either during or on completion of the boring operation, conceals the cavities which remain outside the casing and which may be full of water. This may happen if the casing depth has not been maintained ahead of the bottom of the pile bore. Depending on the workability of placed concrete, pile reinforcement may be left exposed and smaller (in ground water) or larger (above ground water) diameter pile may be cast.
  2. Debris in the form of small block like portions of rock and soil (but also boring equipment parts, footwear and cement bags) may accumulate at the bases of the bores made for piles. Once pile shaft capacity is exceeded, large settlement follows further increase in pile load when such debris is present.
  3. Extraction of double temporary casing have caused shaft defects in instances where the outer larger diameter temporary casing was extracted before the inner smaller diameter casing in very fine grained soil of low permeability. The formation of discontinuous water filled annular spaces between the outer surface of the inner casing and the surrounding soil is most likely the reason for the problem to occur.
  4. Extraction of single temporary casing of bored and driven cast-in-place piles caused several types of defects related to the workability and head of concrete placed within the temporary steel linear prior to their

extraction. Such defects are wasting, local reduction in shaft diameter, necking. Complete loss of workability of fresh concrete from premature stiffening ('flash set') may be due to water absorption into dry aggregates, the use of too fine cement, delay between mixing and placing the concrete, the use of sulphate resistant cement with sea dredged aggregates and high pressure at the base of a long column of concrete. Problems of extraction can be accentuated when dented and dirty casing are used.

5. Excessive amount of steel reinforcement prevented free flow of concrete due to aggregate interlocking, arching and adherence to even clean temporary casing. The heavy reinforcement replacements used are heavy steel sections consisting of steel rails, tubes or rolled hollow section. If the viscous flow of fresh concrete causes reinforcement twisting and sinking, additional bars should not be pushed into fresh concrete as the additional rods can bend outwards into soil.
6. The use of low slump concrete mix can result in partial penetration of reinforcement cage with almost complete separation of the pile shaft during extraction of temporary casing. The use of low slump mix may result in occasional penetration of groundwater into pile shaft and seriously dilute the cement paste. The use of high water content mixes and the self-compaction of a highly workable concrete can result in excessive bleeding from the exposed concrete surface in contact with soil.
7. Placing of concrete in dry conditions can result in formation of weak and partially segregated concrete at a pile base. A small volume of vermiculite or other suitable buoyant material should be used as an initial separation layer between the first batch of concrete in the tremie tube (125 and 200 mm diameter) and the water in the open-ended tremie. The continuous flow of concrete and extraction of tremie tube as well as temporary casing must not be interrupted and restarted.
8. The use of poker vibrator to compact concrete instead of self-compaction resulted in difficulties or impossibilities of extraction of temporary casing and concrete segregation at the top portion of pile.
9. Rapid ground water flow along steep interfaces between different ground types resulted in leaching out of cement from concrete mixture and washing of aggregates leading to pile shaft collapse and exposure of steel reinforcement upon subsequent excavation. Provision of permanent linear is the only safe solution in such situations.
10. Inadequate site investigation work, fraudulent workmanship and inadequate site supervision are also sources of problems for deep foundations.

- Healy and Weltman (1980) listed problems associated with **steel pile driving** as:

1. Damaged pile top (head) (e.g. buckling, longitudinal cracking, distortion) due to unsuitable hammer weight, incorrect use of dollies, helmets, packing, rough cutting of pile ends, overdriving.
2. Damaged pile shaft (e.g. twisting, crumpling, bending) due to unsuitable hammer weight, inadequate directional control, overdriving, obstructions.

3. Collapse of tubular piles due to insufficient thickness
4. Damaged pile toe (e.g. buckling, crumpling, turning up) due to overdriving, redriving, obstructions, difficulty toeing into rock particularly for inclined piles
5. Base plate rising relative to the casing, loss of plugs and shoe in cased piles due to poor welding, overdriving, incorrect use of concrete plugs.

For **concrete pile driving**:

1. Damaged pile head (e.g. shattering, cracking, spalling of concrete) due to unsuitable reinforcement detail, lack of reinforcement, protruding longitudinal bars, poor concrete, incorrect concrete cover, unsuitable hammer weight, incorrect use of dollies, helmets, packing, overdriving
2. Damaged pile shaft (e.g. fracture, cracking, spalling of concrete) due to excessive restraint on piles during driving, unsuitable hammer weight, poor concrete, incorrect concrete cover, obstructions, overdriving, incorrect distribution of driving stresses from use f incorrect dollies, helmets or packing.
3. Damaged pile toe (e. g collapsing, cracking, spalling of concrete) due to unsuitable hammer weight, overdriving, poor concrete, lack of reinforcement, incorrect concrete cover, obstructions, lack of rock shoe where applicable.

For **timber pile driving** as:

1. Damaged pile head (e.g. splitting, brooming) due to unsuitable hammer weight, incorrect use of dollies, helmets, packing, insufficient hoop reinforcement, overdriving.
2. Damaged pile toe (e.g. splitting, brooming) due to unsuitable hammer weight, lack of driving shoe, overdriving

In dense coarse-grained soil, it is possible to employ **jetting** by forcing water at high pressure in the vicinity of a pile toe to loosen and wash upward soil permitting easier driving. The jetting may cause a general instability, which could affect adjacent piles or structures.

Preboring may be used to overcome damage to piles at shallow depths. Driving a precast concrete pile through a prebored hole of slightly smaller diameter than the pile may cause damaging tensile stresses.

Redriving may be necessary after adding a section to a pile, after plant breakdown, or when only driving part lengths in a sequential operation but should be avoided whenever possible. Redriving should be carried out as soon as possible, preferably the same day as driving, as driven pile after a period of time may have develop set up in weak rock or stiff clay.

Soil movement caused by pile driving may cause damage to adjacent piles or structures nearby by uplifting and laterally displacing them in fine-grained soil or causing additional settlement in loose coarse-grained soil. **High pore water pressure develops** in fine-grained soil on pile driving. The pore water pressure dissipation after driving causes additional settlement and negative shaft friction.

Corrosion of buried steel pile onshore can be about 0.01 mm/year. Offshore, **steel piles** may be corroded by oxidation or electrolytic action in the splash zone and below the water level. Table 4–1 of EN 1993-5 contains guidance on corrosion rates for different conditions and time periods. Loss of steel thickness varies from 0.01 mm/year in undisturbed natural soil, to 0.06 mm/year in non-compacted and aggressive fills (ashes, slags, etc.). Corrosion may be minimised by application of protective coating, the use of cathodic protection or both. Coatings based on coal tar, vulcanised rubber membrane, hot dip galvanising, flame sprayed metals or epoxy based should have a long maintenance free life and resistant to abrasion by sand carried by currents and waves and damage during handling and pile driving. **Concrete piles** may suffer the reinforcement corrosion possibly accelerated by stray electrical currents, leaching of calcium from the cement matrix in splash zone, alkali-silica reaction of aggregates, poor workmanship and inadequate site supervision. **Timber piles** may suffer wood decay, marine borers (Limnoria, Teredo) within the tidal zone, fungus attack at pile head. The most common protection against Limnoria is to treat piles with arsenate and creosote and for Teredo with creosote coal tar solutions, or with combination.

5.3 Retaining Walls

5.3.1 Types

Two types are considered:

- **Massive**, which use their self weight to transfer nearly horizontal soil forces on the wall into nearly vertical forces on ground under the wall (e.g. brick work, reinforced concrete with counterforts also called cantilever or counterfort and not treated as massive although they engage the backfill mass, crib, gabion, reinforced soil with facing blocks, caisson or cellular cofferdams filled with sand, concrete blocks, Fig. 5.31)
- **Slender**, which use their flexural stiffness to transfer nearly horizontal active soil forces on the wall in nearly horizontal passive (resisting) forces onto the ground in front of the wall (e.g. cantilever, anchored, propped, Fig. 5.32).

5.3.2 Stability/Capacity

The main factors affecting retaining wall stability/capacity are:

- Addition of loading/fill at the back
- Large water pressure difference between the front and the back due to prevented drainage

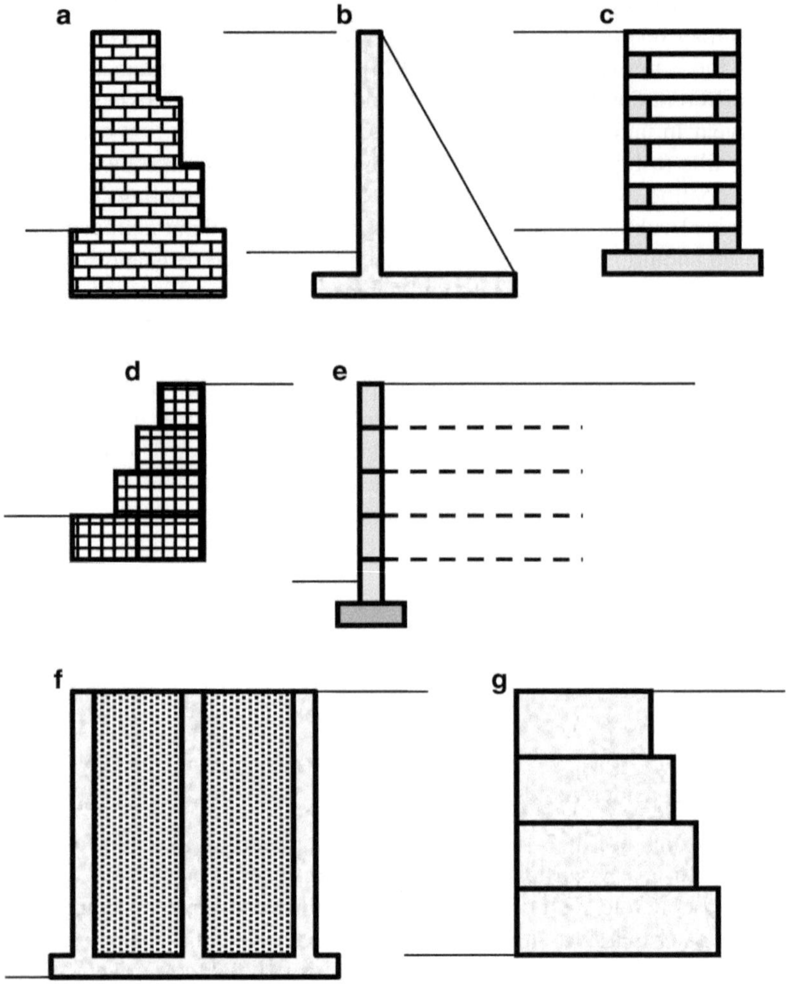

Fig. 5.31 Sketches of cross sections of massive retaining walls (**a**) brick work, (**b**) reinforced concrete with counterforts that engage backfill mass (also called a cantilever or counterfort wall and is not treated as a massive wall), (**c**) crib, (**d**) gabion (metal mesh basket with rock fill), (**e**) reinforced soil with facing blocks, (**f**) caisson (or cellular cofferdam) filled with sand, (**g**) concrete blocks

- Earthquake acceleration
- Soil shear strength decrease in cyclic conditions
- Excavation in front of the wall
- Liquefaction and/or fluidization of ground in front (and beneath massive walls)
- Liquefaction (fluidization) of the backfill, increasing active force
- Resonant response of wall during cyclic load
- Ground freezing at the back of wall (and beneath massive walls)
- Ground swelling behind a wall

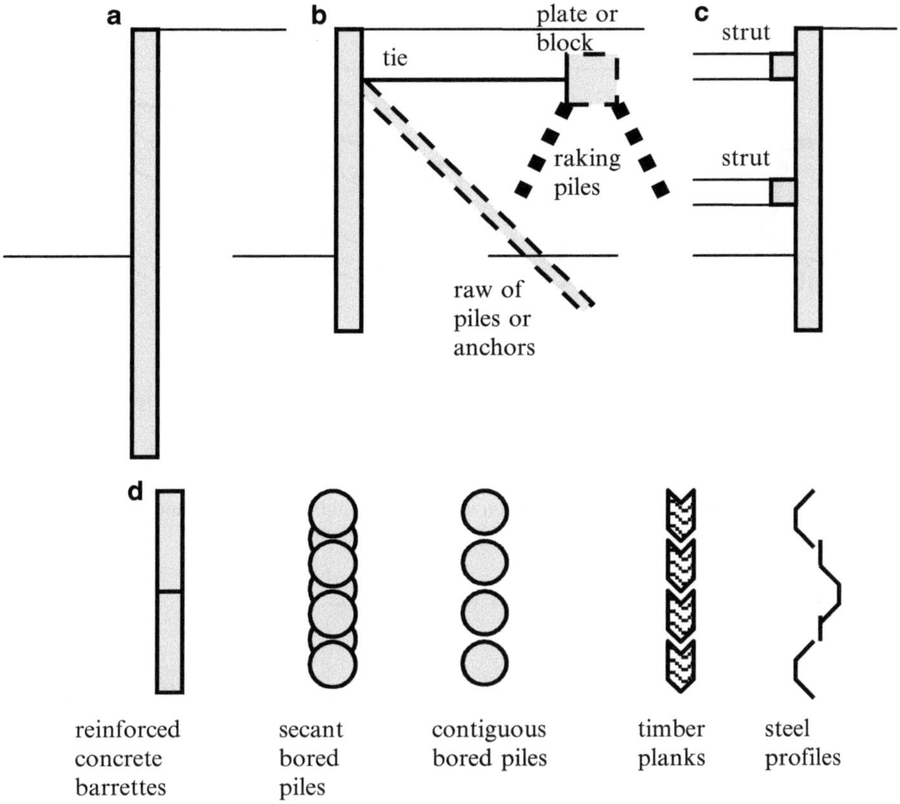

Fig. 5.32 Sections through (**a**) cantilever, (**b**) anchored, (**c**) propped, (**d**) different materials for slender retaining walls

5.3.2.1 Lateral Forces

- **Active force at the back of a wall in static condition** E_{as} according to Coulomb (1776)

$$E_{as} = \cdot K_{as} \cdot \left(\gamma \cdot \frac{{h_w}^2}{2} + q \cdot h_w \right) - 2 \cdot c \cdot \sqrt{K_{as}} \cdot h_w$$
$$K_{as} = \frac{\cos^2(\phi - o)}{\cos^2 o \cdot \cos(\delta_b + o) \cdot \left[1 + \sqrt{\frac{\sin(\delta_b + \phi) \cdot \sin(\phi - \eta_w)}{\cos(\delta_b + o) \cdot \cos(\eta_w - o)}} \right]^2} \tag{5.62}$$

γ is the unit weight of soil behind a wall
h_w is the height of a wall
q is uniform surcharge behind a wall

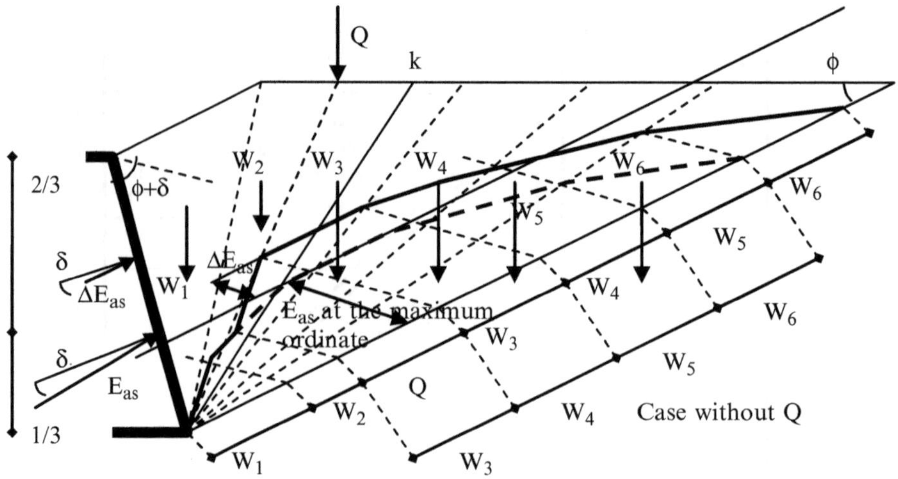

Fig. 5.33 Active force E_{as} in static condition due to the weights $W_{1...6}$ of ground wedges 1–6 and additional force ΔE_{as} due to linear load Q according to Culmann (1875)

δ_b is friction angle at the back soil-wall interface
ϕ is soil friction angle
c is soil cohesion
o is the back of wall inclination to the vertical (positive from the back to the front of the wall)
η_w is inclination to the horizontal of soil surface at the back of a wall (positive upwards)

The force is located at 1/3 of wall height from the bottom and is inclined downwards at an angle δ_b with respect to the normal to a wall back.

For varying inclination of the ground surface behind a wall and/or linear loads, the active force can be determined using graphical method proposed by Culmann (1875) as shown in Fig. 5.33. The location of additional active force on massive walls is shown in Fig. 5.34. For flexural walls Williams and Waite proposed pressure distribution shown in Fig. 5.35.

Full active forces are achievable at a relatively small rotations and translations of a wall. For loose soil, the horizontal displacement to wall height ratio range is 0.2 % for translation to 1 % for rotation around a wall top and for dense soil the range is from 0.05 % for translation to 0.5 % for rotation around a wall top according to Table C.1 in EN 1997 (2004).

- **Passive force in the front of a wall in static condition E_{ps}**

$$E_{ps} = K_{ps} \cdot \left(\gamma \cdot \frac{h_w{}^2}{2} + q \cdot h_w \right) + 2 \cdot c \cdot \sqrt{K_{ps}} \cdot h_w \qquad (5.63)$$

γ is the unit weight of soil behind a wall
h_w is the height of a wall
q is uniform surcharge behind a wall

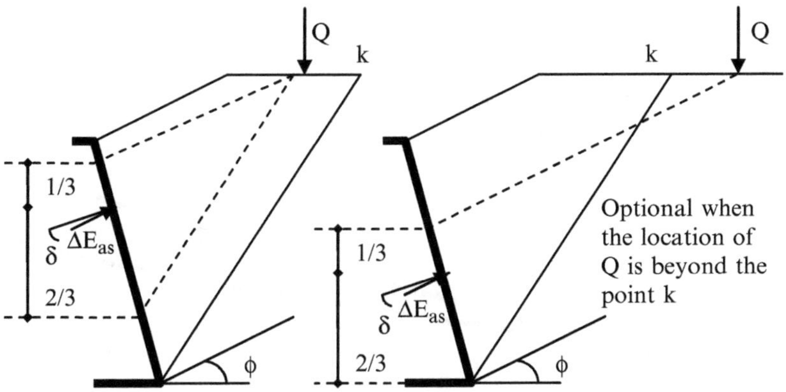

Fig. 5.34 The location of additional force ΔE_{as} due to linear load Q according to the empirical method after Terzaghi (the location of point k is from Fig. 5.32; the line is passing through the location of the maximum ordinate for E_{as})

Fig. 5.35 Pressure diagram for a line load acting on a flexural wall Williams and Waite (1993)

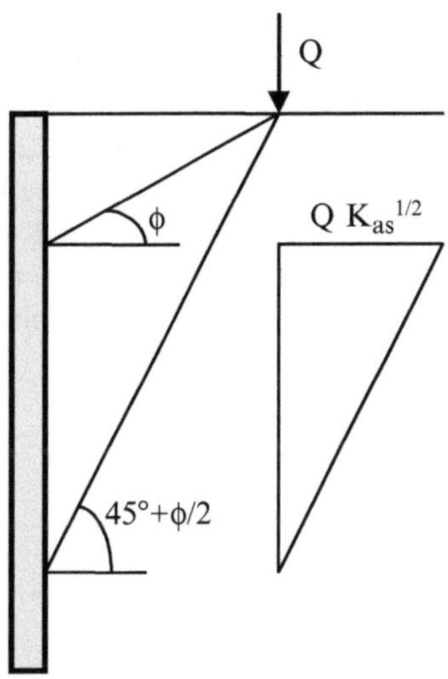

The assumption of a passive wedge with planar base according to Coulomb (1776) is not on the safe side for large soil friction angles as the actual base of a passive wedge has a concave shape. Caquot and Kerisel (1948) provided values of the coefficient of passive resistance K_{ps} using a shape of logarithmic spiral instead of planar shape. The coefficients are shown in Fig. 5.36 from EN 1997-1 (2004).

The force is located at 1/3 of wall height from the bottom and is inclined upwards at an angle δ_b with respect to the normal to a wall front.

Significant ground displacements are necessary for full activation of passive forces. For loose soil, the horizontal displacement to wall height ratio range is 5 % for translation to 25 % for rotation around a wall bottom and for dense soil the range is from 3 % for translation to 10 % for rotation around a wall bottom according to Table C.2 in EN 1997 (2004). For the horizontal displacements

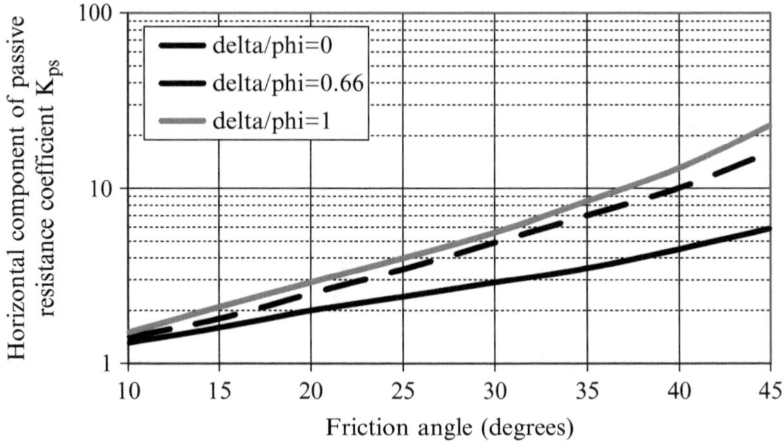

Fig. 5.36 Horizontal component of the coefficient of passive ground resistance for horizontal ground behind a wall

smaller than the maximum ones, the coefficient of mobilisation of passive earth pressure in coarse grained soil ~0.25 + 0.75 ($0.07\Delta\, h_w^{-1}$ to $0.25\Delta\, h_w^{-1})^{0.5}$ based on Fig. C.3 and Table C.2 in EN 1997-1 (2004), where Δ is actual horizontal displacement and h_w is the height of a wall.

- **Active force at the back of a wall in seismic condition** is calculated using Mononobe-Okabe (M–O) method following Okabe (1926) and Mononobe and Matsuo (1929) for non-cohesive soil.

$$E_{ae} = \frac{1}{2} \cdot K_a \cdot \gamma \cdot {h_w}^2$$

$$K_{ae} = \frac{\cos^2(\phi - o - \psi)}{\cos\psi \cdot \cos^2 o \cdot \cos(\delta_b + o + \psi) \cdot \left[1 + \sqrt{\frac{\sin(\delta_b+\phi)\cdot\sin(\phi-\eta_w-\psi)}{\cos(\delta_b+o+\psi)\cdot\cos(\eta_w-o)}}\right]^2} \tag{5.64}$$

$$\psi = \arctan\left(\frac{a_h}{g}\right)$$

γ is the total unit weight of soil behind a wall,
h_w is wall height,
a_h is the horizontal acceleration (approximately 0.65 of the peak value for an equivalent harmonic motion),
δ_b is friction angle between wall back and soil,
ϕ is soil friction angle in cyclic condition,
o is the back of wall inclination to the vertical (positive from back to front),
η_w is inclination to the horizontal of soil surface at the back of a wall (positive upwards).

When the difference $\phi - \eta_w - \psi$ in Eq. (5.64) becomes less than zero then the sine function becomes negative and the square root an imaginative number. This represents the condition when the slip of soil at the back of a wall occurs. However, retaining wall itself may have reserve resistance against sliding and can prevent slip of soil at the back of wall so that the lateral force can continue to build up. EN 1998-5 (2004) in its normative annex E chooses to ignore the whole term in brackets [ ] in Eq. (5.64) when $\phi - \eta_w - \psi < 0$. The point of application of the M–O active lateral force is assumed to be at one third above the base of the wall.

- **Passive force in the front of a wall in seismic condition** is according to Mononobe-Okabe method for non-cohesive soil

$$E_{pe} = \frac{1}{2} \cdot K_p \cdot \gamma \cdot {h_w}^2$$

$$K_{pe} = \frac{\cos^2(\phi + o - \psi)}{\cos\psi \cdot \cos^2 o \cdot \cos(\delta_b - o + \psi) \cdot \left[1 - \sqrt{\frac{\sin(\delta_b+\phi)\cdot\sin(\phi+\eta_w-\psi)}{\cos(\delta_b-o+\psi)\cdot\cos(\eta_w-o)}}\right]^2} \tag{5.65}$$

$$\psi = \arctan\left(\frac{a_h}{g}\right)$$

γ is the total unit weight of soil at the back of a wall,
h_w is wall height,
a_h is the horizontal acceleration (approximately 0.65 of the peak value for an equivalent harmonic motion),
δ_b is friction angle between wall back and soil,
ϕ is soil friction angle in cyclic condition,
o is the back of wall inclination to the vertical (positive from back to front),
η_w is inclination to the horizontal of soil surface at the back of a wall (positive upwards).

The point of application of the M–O passive lateral force is assumed to be at one third above the base of the wall.

When the difference $\phi + \eta_w - \psi$ in Eq. (5.65) becomes less than zero then the sine function becomes negative and the square root an imaginative number. This represents the condition when the slip of soil in front of a wall occurs. EN 1998-5 (2004) in its normative annex E does not provide recommendation for this situation but with analogy to the active force the whole term in brackets [ ] in Eq. (5.65) can be ignored when $\phi + \eta_w - \psi < 0$.

- **Seismic increment of active static force**

Seed and Whitman (1970) performed a parametric study using M–O method to evaluate the effects of various input parameters on the magnitude of dynamic earth pressures. They observed that the maximum total earth pressure acting on a retaining wall can be divided into two components: the initial static pressure E_{as} and the dynamic increment ΔE_{ad} due to the base motion. They recommended that the dynamic increment acts at approximately 0.6 of wall height from its base. Towhata and Islam (1987) stated that total passive lateral force can also be divided into static and dynamic components in which case the dynamic component acts in the opposite direction of the static component.

Following Richards et al. (1999) idea, it is assumed in the simple model by Srbulov (2011) that the value of dynamic increment of active soil lateral force ΔE_{ad} in the horizontal direction for a vertical wall ($\varrho = 0$) with horizontal soil surface in the back ($\eta_w = 0$) is proportional to the sum (integral) of inertial forces acting on an active zone behind wall because the wall prevents free ground movements and is subjected to the lateral forces.

$$\Delta E_{ad} = 0.65 \cdot a_{top} \cdot \rho \cdot \tan\left(45^\circ - \frac{\phi}{2} + \frac{\psi}{2}\right) \cdot \frac{h_w{}^2}{2} \tag{5.66}$$

a_{top} is the peak horizontal acceleration at the top of a wall
ρ is unit soil density behind a wall
ϕ is soil friction angle in cyclic condition behind a wall
$\psi = \arctan(0.65\ a_h/g)$,
a_h is the peak horizontal ground acceleration behind a wall,
h_w is wall height

The location of ΔE_{ad} force is at 2/3 of wall height from its base. The resultant of the static E_{as} and of dynamic increment ΔE_{ad} is located between 1/3 and 2/3 of wall height from its base depending on their relative values.

- **Lateral dynamic water force in the front of a wall** (and hydrostatic water force decrease) according to Westergaard (1931)

$$P_w = \frac{7}{12} \cdot \frac{a_h}{g} \cdot \gamma_w \cdot h^2 \tag{5.67}$$

a_h is the horizontal ground acceleration
g is the gravitational acceleration
γ_w is the unit weight of water
h is depth of water

The force acts at 0.4 h distance from the bottom.

- **Swelling, ice forming and vibration effect** pressures are described in Sects. 4.7.3.2, 4.13.2 and 4.10.1.

5.3.2.2 Massive Walls

Basic mechanisms of failures of massive retaining walls are shown in Fig. 5.37.

- Bearing capacity and sliding are checked according to Sect. 5.2.2.1 and 5.2.2.2
- Overturning around the external edge is checked according to Sect. 5.1.2.1
- Global failure according to Sect. 5.1.2.3 and 5.1.2.4

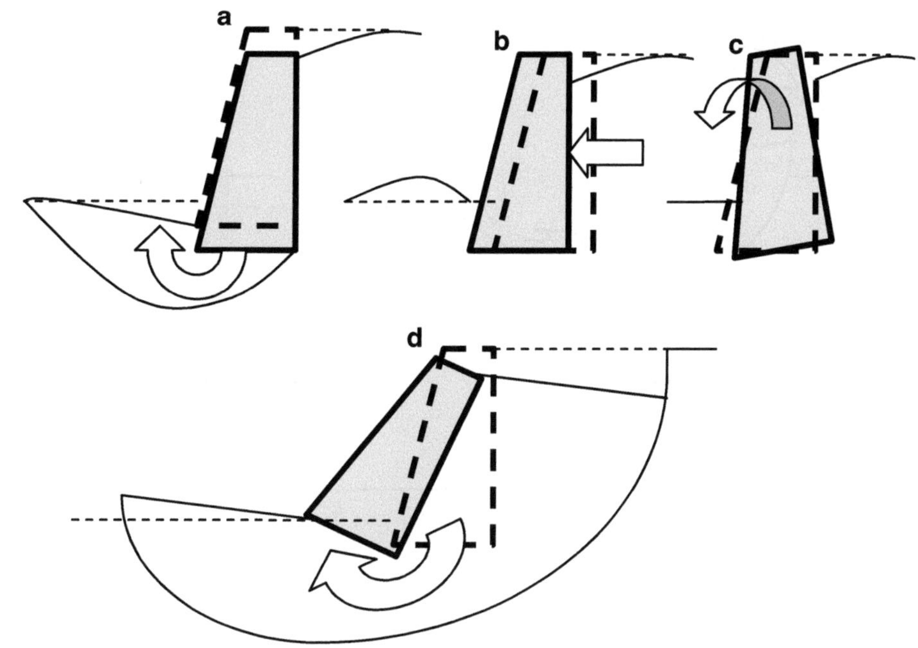

Fig. 5.37 Failure mechanisms of massive retaining walls (**a**) exceedance of base bearing, (**b**) sliding along toe, (**c**) overturning around external edge, (**d**) deep seated soil slip

5.3.2.3 Slender Walls

Basic mechanisms of failures of slender retaining walls are shown in Fig. 5.38.

- **Stability against forward rotation**
  The stability is expressed in terms of the factor of safety which is a ratio between resisting and driving rotational moments acting on a wall. Driving moments are caused by lateral active force at the back of a wall and resisting moments are caused by lateral passive force in the front of a wall as well as by struts and ties if present.
- **Stability against backward rotation**
  It is checked according to Sect. 5.1.2.1.
- **Flexural capacity**
  The flexural capacity is checked by calculating the ratio between acting bending moment at a particular cross section of wall and the second moment of cross sectional area of a wall times a half of wall thickness. The calculated maximum bending stress must be smaller than allowable bending stress for wall material considered.
- **Capacity of strut or tie against break**
  The capacity against break is calculated as the ratio between acting force and cross sectional area of a strut/tie, which must be smaller than allowable stress for

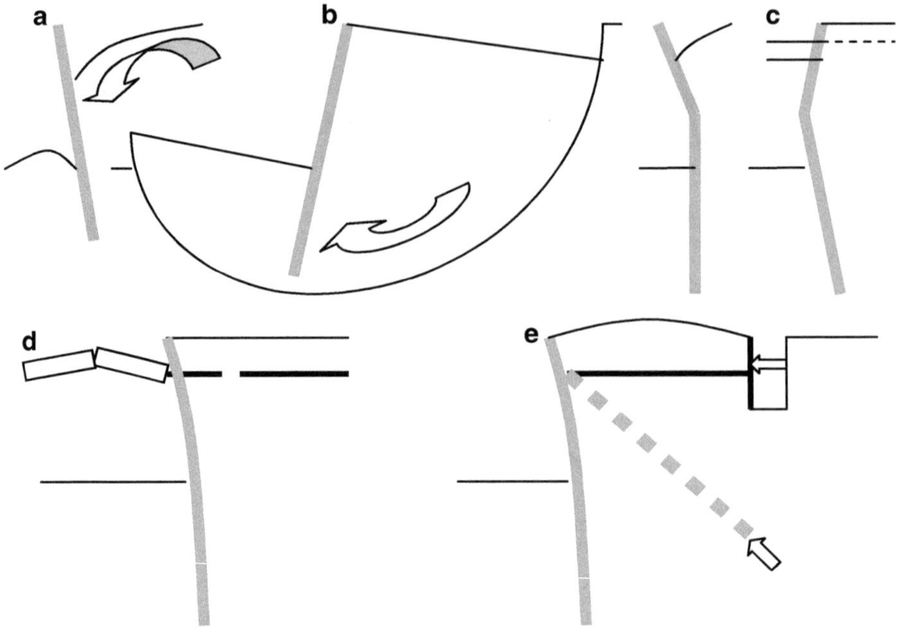

Fig. 5.38 Types of failures of slender walls (**a**) forward rotational, (**b**) backward rotational, (**c**) flexural, (**d**) strut or tie break, (**e**) pull-out of tie or anchor

strut/tie material considered. For struts in compression, capacity against buckling is also checked using Euler equation:

$$P_{critical} = \pi^2 \cdot \frac{E \cdot I}{l_k^2} \tag{5.68}$$

E is Young's modulus of a strut,
I is the second moment of a cross sectional area of a strut,
l_k is critical length of a strut, for both ends hinged = strut length, for one end hinged and the other with prevented rotation = 0.707 of strut length, for both ends with prevented rotation = 0.5 of strut length.

- **Pull out capacity of tie or anchor**
  The capacity can be obtained from field pull out test or it can alternatively be estimated by applying soil mechanics theory. The theory provides acceptable location and dimensions of a tie plate/block within the passive zone to avoid its influence on the active zone behind wall. Figure 5.39 shows a cross section for the case of a wall with vertical back and the horizontal backfill.
- **Equilibrium of acting forces**
  Several methods exist for consideration of wall equilibrium, and bending moment distribution along its length. Burland et al. (1981) summarised and commented on the available methods, Fig. 5.40, as follows:

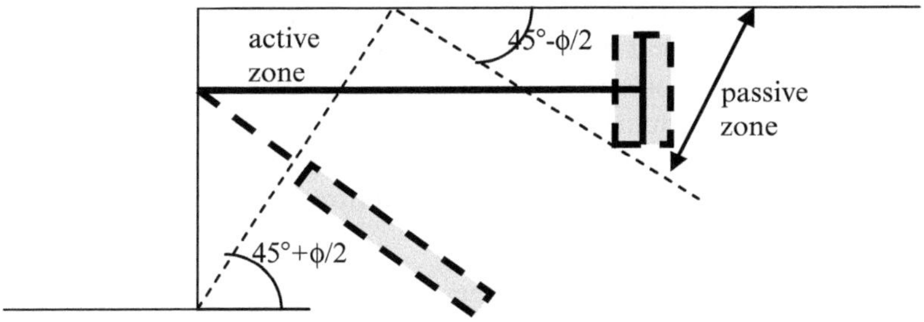

Fig. 5.39 Acceptable location of a tie plate/block and anchor behind a wall with vertical back and the horizontal backfill to prevent the influence of the passive zone on the active zone behind wall

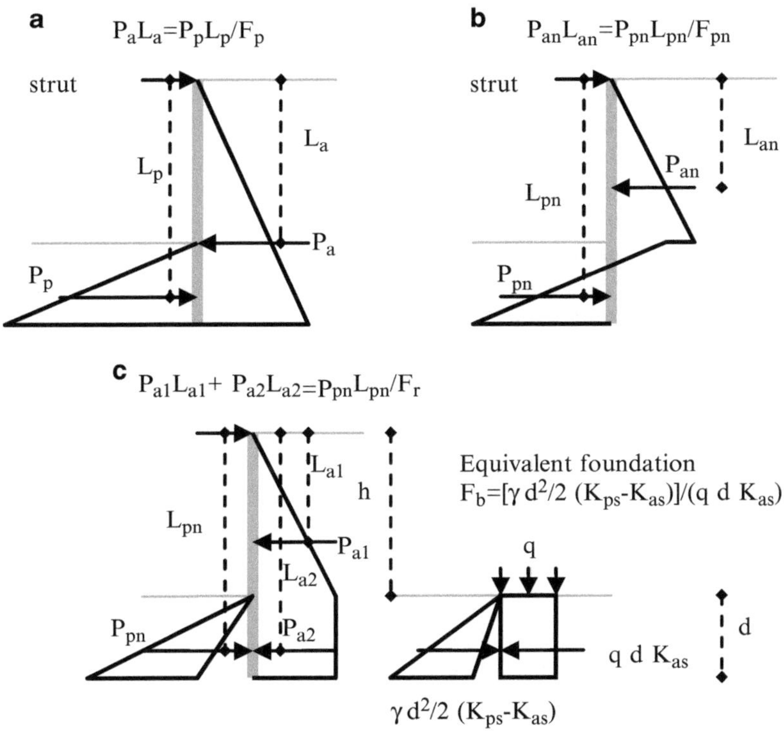

Fig. 5.40 Different methods for consideration of slender wall equilibrium, (**a**) gross lateral force, (**b**) net lateral force, (**c**) bearing capacity analogy

1. Gross lateral forces moment equilibrium uses a factor of safety F_p with respect to the total passive resistance of the toe and could be regarded as a load factor. Values of between 1.5 and 3 are usually used.
2. Net lateral force moment equilibrium uses a load factor of safety F_{np} of the moment of the net passive force. A value of 2 is normally adopted.

3. The method uses gross lateral forces moment equilibrium but multiplies the minimum embedment depth by a factor of safety F_d. A value of 1.7–1.2 is used for granular soil and 2–1.4 for undrained cohesive soil.
4. The method determines the average shear strength required to achieve limiting equilibrium. The factors of safety F_s of available soil shear strength from 1.25 to 1.5 for soil friction angle and from 1.5 to 2 for soil cohesion are commonly used.

Burland et al. (1981) remarked that: *The use of both F_p and F_{np} can lead to very unsatisfactory results. There appears to be no logical or consistent relationship between F_p and F_s and its use can lead to very conservative values of wall penetration for drained conditions with ϕ' less than 25° and for undrained conditions. With regard to* F_{np}, *its use with currently recommended values of about 2 leads to F_s generally less than 1.1 for both undrained and drained conditions. It should, therefore, only be used with great caution and with much higher values, which are compatible with acceptable values of F_s. For drained conditions in uniform ground, the use of F_d appears to be entirely satisfactory. However, it should not be used for undrained conditions or where the strength properties of ground vary significantly with depth.*

The concept of factoring passive resistance of the toe of an embedded retaining wall is attractive in principle as the overall stability can be expressed as a single number. This is not the case for shear strength where the engineer may be faced with the possibility of a number of factors for strength and wall friction. Hence there are considerable benefits in developing a definition of factor of safety on passive resistance which and be shown to be logical and consistent.

Simple bearing capacity analogy is sketched in Fig. 5.40c **for frictional only soil above ground water level**.

$$\begin{aligned}
F_b &= \frac{0.5 \cdot \gamma \cdot d^2 \cdot (K_{ps} - K_{as})}{\gamma \cdot d \cdot K_{as}} \\
F_r &= \frac{P_{pn} \cdot L_{pn}}{P_{a1} \cdot L_{a1} + P_{a2} \cdot L_{a2}} \\
P_{pn} &= 0.5 \cdot \gamma \cdot d^2 \cdot (K_{ps} - K_{as}) \\
P_{a1} &= 0.5 \cdot \gamma \cdot h^2 \cdot K_{as} \\
P_{a2} &= \gamma \cdot h \cdot K_{as} \cdot d \\
L_{pn} &= h + 2/3 \cdot d \\
L_{a1} &= 2/3 \cdot h \\
L_{a2} &= h + 0.5 \cdot d
\end{aligned} \tag{5.69}$$

γ is soil unit weight,
d is embedment depth,
h is excavation depth in front of the wall
K_{as} is given in Eq. (5.62)
K_{ps} is given in Fig. 5.36

For soil with cohesion c' and friction ϕ' in terms of effective stresses:

$$
\begin{aligned}
F_b &= \frac{0.5 \cdot \gamma \cdot d^2 \cdot (K_{ps} - K_{as}) + 2 \cdot c' \cdot d \cdot (K_{ac} + K_{pc})}{\gamma \cdot h \cdot K_{as} \cdot d} \\
K_{ac} &= \sqrt{K_{as}}, K_{pc} = \sqrt{K_{ps}} \\
F_r &= \frac{P_{pn1} \cdot L_{pn1} + P_{pn2} \cdot L_{pn2}}{P_{a1} \cdot L_{a1} + P_{a2} \cdot L_{a2}} \\
P_{n1} &= 0.5 \cdot \gamma \cdot d^2 \cdot (K_{ps} - K_{as}) \\
P_{n2} &= 2 \cdot c' \cdot d \cdot (K_{pc} + K_{ac}) \\
P_{a1} &= 0.5 \cdot (K_{as} \cdot h \cdot \gamma - 2 \cdot c' \cdot K_{ac}) \cdot \left[h - 2 \cdot c' \cdot {}^{K_{ac}}/_{(\gamma \cdot K_{as})}\right] \\
P_{a2} &= K_{as} \cdot h \cdot \gamma \cdot d \\
L_{pn1} &= h + {}^2/_3 \cdot d \\
L_{pn2} &= h + 0.5 \cdot d \\
L_{a1} &= {}^1/_3 \cdot \left[2 \cdot h - 2 \cdot c' \cdot {}^{K_{ac}}/_{(\gamma \cdot K_{as})}\right] \\
L_{a2} &= h + 0.5 \cdot d
\end{aligned}
\tag{5.70}
$$

γ is soil unit weight,
d is embedment depth,
h is excavation depth in front of the wall,
c' is soil cohesion in terms of effective stresses,
q is surcharge at the excavation level,
$K_{()s}$ are coefficients of soil lateral active (a) and passive (p) pressures according Eq. (5.62) and Fig. 5.36 respectively.

For **cohesive soil in undrained conditions and in terms of total stresses:**

$$
\begin{aligned}
F_r &= \frac{4 \cdot c_u \cdot d \cdot L_{pn}}{0.5 \cdot (\gamma \cdot h - 2 \cdot c_u)^2 \cdot {}^{L_{a1}}/_{\gamma} + \gamma \cdot h \cdot d \cdot L_{a2}} \\
&for \gamma \cdot h \geq 2 \cdot c_u or \ \ ignore \ \ it \\
L_{pn} &= h + 0.5 \cdot d \\
L_{a1} &= {}^1/_3 \cdot \left(2 \cdot h - 2 \cdot {}^{c_u}/_{\gamma}\right) \\
L_{a2} &= h + 0.5 \cdot d
\end{aligned}
\tag{5.71}
$$

γ is soil unit weight,
d is embedment depth,
h is excavation depth in front of the wall,
c_u is soil undrained shear strength

When **surcharge (s) acts at the top and (p) at the excavation level in front of a wall**:

$$\begin{aligned}
F_b &= \frac{p \cdot d \cdot (K_{ps} - K_{as}) + 0.5 \cdot \gamma \cdot d^2 \cdot (K_{ps} - K_{as})}{(\gamma \cdot h - p) \cdot d \cdot K_{as}} \\
F_r &= \frac{P_{pn1} \cdot L_{pn1} + P_{pn2} \cdot L_{pn2}}{P_{a1} \cdot L_{a1} + P_{a2} \cdot L_{a2} + P_{a3} \cdot L_{a3}} \\
P_{pn1} &= 0.5 \cdot \gamma \cdot d^2 \cdot (K_{ps} - K_{as}) \\
P_{pn2} &= p \cdot d \cdot (K_{ps} - K_{as}) \\
P_{a1} &= s \cdot K_{as} \cdot h \\
P_{a2} &= 0.5 \cdot \gamma \cdot h^2 \cdot K_{as} \\
P_{a3} &= (s + \gamma \cdot h - p) \cdot d \cdot K_{as} \\
L_{pn1} &= h + 2/3 \cdot d \\
L_{pn2} &= h + 0.5 \cdot d \\
L_{a1} &= 0.5 \cdot h \\
L_{a2} &= 2/3 \cdot h \\
L_{a3} &= h + 0.5 \cdot d
\end{aligned} \tag{5.72}$$

γ is soil unit weight,
d is embedment depth,
h is excavation depth in front of the wall
K_{as} is given in Eq. (5.62)
K_{ps} is given in Fig. 5.36

For a **ground water level at depth *j* below the ground surface behind a wall**:

$$\begin{aligned}
F_r &= \frac{P_{pn} \cdot L_{pn}}{P_{a1} \cdot L_{a1} + P_{a2} \cdot L_{a2} + P_{a3} \cdot L_{a3} + P_{w1} \cdot L_{w1} + P_{w2} \cdot L_{w2}} \\
P_{pn} &= 0.5 \cdot d \cdot \left\{ \begin{array}{l} (K_{ps} - K_{as}) \cdot d \cdot \gamma - \dfrac{2 \cdot \gamma_w \cdot d}{2 \cdot d + h - j} \cdot [(K_{ps} - K_{as}) \cdot \\ (h + d - j) + K_{as} \cdot (h - j)] \end{array} \right\} \\
P_{a1} &= 0.5 \cdot \gamma \cdot j^2 \cdot K_{as} \\
P_{a2} &= 0.5 \cdot (h - j) \cdot \gamma \cdot K_{as} \cdot \left[j + h - \frac{2 \cdot \gamma_w \cdot (h - j) \cdot d}{\gamma \cdot (2 \cdot d + h - j)} \right] \\
P_{a3} &= d \cdot \gamma \cdot K_{as} \cdot \left[h - \frac{2 \cdot \gamma_w \cdot (h - j) \cdot d}{\gamma \cdot (2 \cdot d + h - j)} \right] \\
L_{pn} &= h + 2/3 \cdot d L_{a1} = 2/3 \cdot j \\
L_{a2} &\sim 0.5 \cdot (j + h) \\
L_{a3} &= h + 0.5 \cdot d
\end{aligned} \tag{5.73}$$

γ is soil unit weight,
d is embedment depth,
h is excavation depth in front of the wall
K_{as} is given in Eq. (5.62)
K_{ps} is given in Fig. 5.36

For **walls without struts and in a basic case** like the one shown in Fig. 5.40c,

$$
\begin{aligned}
F_r &= \frac{P_{pn} \cdot L_{pn}}{P_{a1} \cdot L_{aa1} + P_{a2} \cdot L_{a2}} \\
P_{pn} &= 0.5 \cdot \gamma \cdot d^2 \cdot (K_{ps} - K_{as}) \\
P_{a1} &= 0.5 \cdot \gamma \cdot h^2 \cdot K_{as} \\
P_{a2} &= \gamma \cdot h \cdot K_{as} \cdot d \\
L_{pn} &= 1/3 \cdot d \\
L_{a1} &= d + 1/3 \cdot h \\
L_{a2} &= 0.5 \cdot d
\end{aligned}
\tag{5.74}
$$

γ is soil unit weight,
d is embedment depth,
h is excavation depth in front of the wall
K_{as} is given in Eq. (5.62)
K_{ps} is given in Fig. 5.36

Water pressure, surcharges and cohesion may be included as in the case of a propped wall. The value of F_r between 1.5 and 2 would normally be appropriate according to Burland et al. (1981).

5.3.3 *Movement*

5.3.3.1 Massive Walls

- Settlement in static condition is calculated according to Sect. 5.2.3.1
- Horizontal displacement and rotation in static condition is calculated according to Sect. 5.2.3.2
- Permanent displacement during earthquake is calculated according to Sect. 5.1.3.4 for an equivalent sliding block shown in Fig. 5.41
- Permanent displacement after an earthquake if the factors of safety against sliding is reduced below one is calculated according to Sect. 5.1.3.4 for equivalent two blocks shown in Fig. 5.42. An example of the calculation of permanent displacements during and after earthquake of quay walls at Kobe port in Japan is provided by Srbulov (2011).

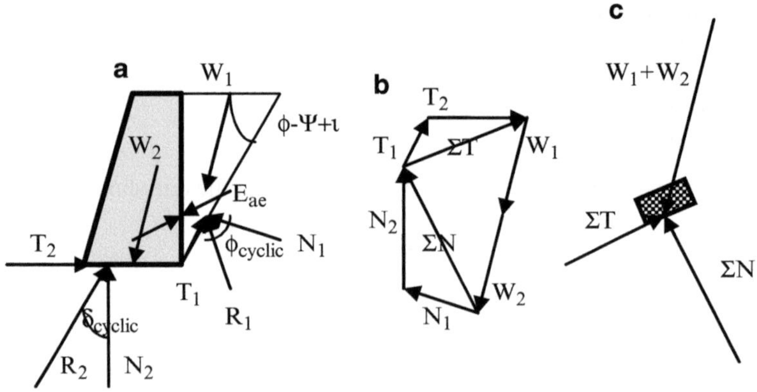

Fig. 5.41 Sketch of (**a**) cross section through a massive wall with acting forces during an earthquake, (**b**) polygon of axial and transversal forces acting on the slip surfaces, (**c**) an equivalent block for co-seismic sliding

Fig. 5.42 Sketch of the cross section through a massive wall and equivalent two blocks for analysis of post-seismic sliding

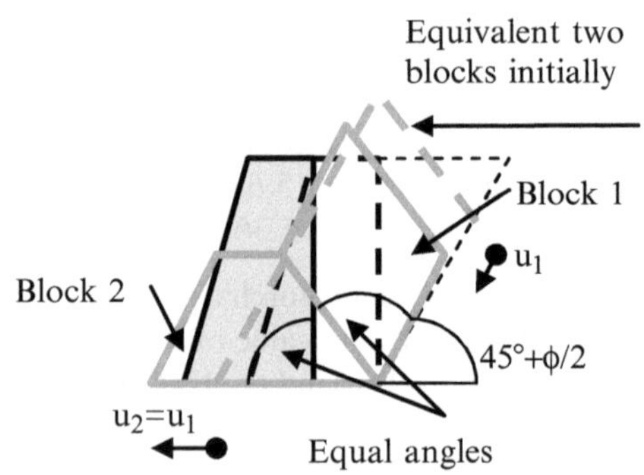

- Displacements due to swelling, ice forming and vibrations are described in Sects. 4.7.3.2, 4.13.2 and 4.10.3.

5.3.3.2 Slender Walls

- Clough et al. (1989) provided graphs for maximum lateral wall movement versus system stiffness and Gaba et al. (2003) provided graphs of horizontal movement and settlement of ground surface behind walls due to excavation in front of walls in stiff clay and sand, Fig. 5.43.
- Displacements due to swelling, ice forming and vibrations are described in Sects. 4.7.3.2, 4.13.2 and 4.10.3.

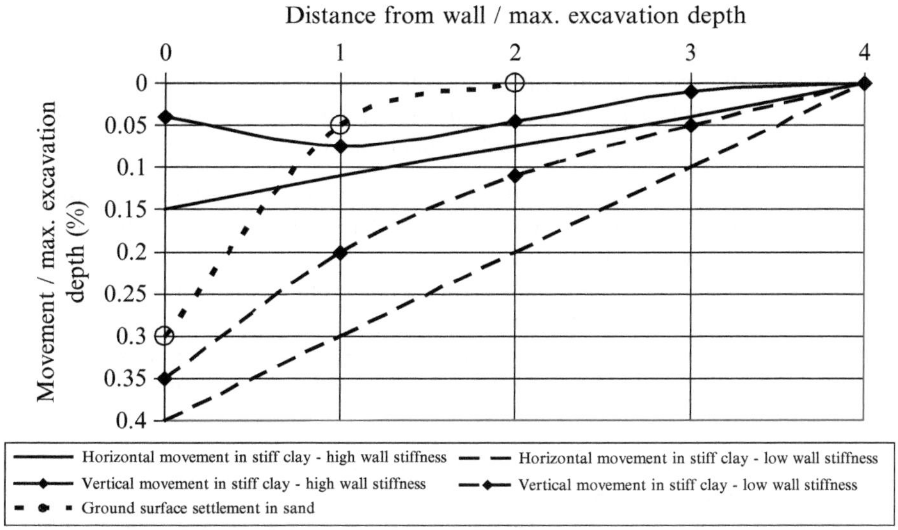

Fig. 5.43 Ground surface movements due to excavation in front of slender walls

5.3.4 Execution

- Overdig in front of a wall of 10 % wall height but maximum 0.5 m is commonly considered in design.
- Construction surcharge load of at least 10 kPa behind a wall is commonly considered in design
- Vertical construction joints are positioned at approximately 10 m spacing
- Compaction behind massive wall induces additional lateral stresses σ_h, which increase linearly from zero to a depth z_{cr} and remain constant to a depth h_c according to Ingold (1979)

$$h_c = \frac{1}{K} \cdot \sqrt{\frac{2 \cdot P}{\pi \cdot \gamma}}$$

$$\sigma_h = \sqrt{\frac{2 \cdot P \cdot \gamma}{\pi}} \tag{5.75}$$

$$z_{cr} = K \cdot \sqrt{\frac{2 \cdot P}{\pi \cdot \gamma}}$$

K is the coefficient of active or at rest ground pressure (for basement walls) behind a wall

P is the effective line load per metre of compaction roller

γ is unit density of compacted material behind a wall

- Useful information is provided in EN 1538 (2010) and EN 12063 (1999).
- Problems associated with construction of concrete bored pile and diaphragm walls typically relate to difficulties in concreting causing insufficient cover to reinforcement and lack of water tightness at joints. Gaba et al. (2003) list the following reasons for problems with slender walls:

  1. Inadequate understanding of the geological and hydro geological conditions
  2. Poor design and construction details and poor standard of workmanship, particularly of support system
  3. Construction operations and sequences that result in earth pressure different from those assumed in design
  4. Inadequate control of construction operations, e.g. over excavation of berms and formation, excessive surcharge loads from soil heaps and construction equipment.

- The execution problems mentioned for cast-in-place concrete piles can occur with concreting slender retaining walls.
- Temporary works are usually design by contractors but wall designer should be involved as well once preferable construction method is chosen
- Use of the observational method may result in the most cost effective walls

5.4 Anchors, Bolts and Nails

5.4.1 Types

The following types are usually used:

- Soil and rock anchors passive or prestressed
- Rock bolts (tendons grouted into a hole)
- Soil nails (bars driven or grouted in boreholes)

5.4.2 Stability/Capacity

Anchor, bolts and nails used to stabilise tunnels, slopes, retaining walls and foundation (against uplift only) are best placed in direction of intended movements of geo-structures when possible because the tendons have high tensile but low bending and shear capacity. Bending and shear stresses must be checked for the elements that can be subjected to bending and shearing. For rotational type movement, the tendons should be placed as far as possible from the centre of rotation to generate maximum resisting moments.

EN 1997-1 (2004) contains brief requirements for design and constructions of anchorages. The following limit states need to be considered for anchorages:

- Failure of the tendon (snap) or anchor head under load
- Corrosion induced failure
- Pull out due to the slip at the interface between anchorage and ground
- Pull out due to the slip at the interface between tendon and anchored length
- Pull out due to failure of ground containing anchored length
- Prestress force decrease due to creep and relaxation of ground
- Prestress force causing excessive deformation or failure of parts of the structure
- Global failure through and outside anchoraged/bolted/nailed zone

Shaft friction along anchored length can be calculated as for piles but a sufficient number of field tests are require confirming anchorage capacity (e.g. ASTM D4435, D4436)

Group of anchors even when installed at spacing not less than 1.5 m mentioned in EN 1997-1 (2004) could cause block type failure of ground if the ground resistance around the group perimeter is not greater than the resultant pull out force. For the vertical anchors, the weight of ground zone of conical shape engaged around an anchor is checked against pull-out force. The angle of the cone inclination to the vertical in rock is considered to be 45° and in soil equal to the angle of dilatation of soil Ψ. Shear stress at the interface between the cone and the surrounding ground is considered to be zero. According to Jewell (1992):

$$\Psi \approx C \cdot [D_r \cdot (10 - \ln_e \sigma') - 1] \tag{5.76}$$

Ψ is the angle (degrees) of dilatation of soil
C is a constant =6.25 for plane strain, =3.75 for triaxial compression
D_r is soil relative density from Eq. (3.14)
σ' the mean effective stress (kPa)

For steep slopes, acting force per a tendon head and on a rigid or flexible facing between such tendons is at least equal to the local active ground force (Eq. 5.62 or Fig. 5.33) as well as hydrostatic, swelling, ice forming, liquefaction and other forces defined in Chap. 4 if they can exist within the area covered by a tendon (min. 1.5^2 m^2). Force per a tendon can be greater than the forces at tendon head if greater force is necessary to maintain the global stability of slope. For gently inclined slopes, the tendon forces are determined from the analyses of global stability using one of the methods described in Sect. 5.1.2. Full tendon force can be considered in global stability analysis only if a critical slip surface passes in front of the anchored length of tendons as shown in Fig. 5.39.

The tensile force Z per unit width of a flexible facing can be determined as:

$$Z = \frac{p}{\alpha} \cdot \frac{e - w}{2} \tag{5.77}$$

Z is tensile force per unit width of a flexible facing
p is the pressure acting on a facing
e is the spacing of tendons between which facing exist

Table 5.9 Displacements at the top of steep soil nailed structures

	Weathered rock/stiff soil	Sandy soil	Clayey soil (not high or very high plasticity clay)
Horizontal & vertical displacement	H/1000	2H/1000	3H/1000
Coefficient k	0.80	1.25	1.50

Note: H is slope height

w is the width of plate under anchor head
α is the angle (in radians) of facing inclination with respect to its undeformed state determined by trial and error from the following formula (Srbulov 2001a)

$$p \cdot (e - w) \cdot \left(2 - \frac{\alpha^2}{2}\right) \cdot \left(1 - \frac{\alpha^2}{2}\right) - 2 \cdot \alpha^3 \cdot E = 0 \tag{5.78}$$

p is the pressure acting on a facing
e is the spacing of tendons between which facing exist
w is the width of plate under anchor head
E is modulus of deformation of a flexible facing (kN/m or equivalent units)

5.4.3 *Movement*

Axial extension of tendons can be calculated or measured from calculated or measured axial strain and length over which such strain exist.

Displacements at the top of steep soil nailed slopes are given in Table 5.9 from Clouterre Cloueterre (1991). Schlosser and Unterreiner (1991) suggest for steep hard faced slopes the vertical and horizontal deflection at the top of the slopes are

$$\delta_o = k \cdot (1 - \tan \phi) \cdot H \tag{5.79}$$

k coefficient is given in Table 5.9
ϕ is soil friction angle
H is slope height

5.4.4 *Execution*

- Useful information is provided in EN 1537 (2000) and EN 14490 (2010).
- Activation of tensile force in tendon inevitably leads to formation of cracks in the material surrounding tendon unless such material is flexible (Epoxy but

expensive) or repeated grouting (not cheap) of the annulus between a tendon and borehole wall is performed using tube-a-manchette method (grouting pipe with sleeves covering holes in the pipe to prevent back flow of grout when fresh and under pressure).

- Corrosion rates are highly dependable on local conditions. For U.K. conditions, atmospheric corrosion of steel is approximately 0.035 mm/year, within ground 0.015 mm/year, immersed in sea water and within tidal zone 0.035 mm/year, in water splash zones 0.075 mm/year (Clause 4.4.4.4.3 of BS 8002, 1994)
- Manufacturers of anchors/bolts provide details for their installation.
- Installed anchors/bolts/nails are tested in situ.

5.5 Reinforced Soil

5.5.1 Types

The following types are considered:

- Gently inclined soil slopes
- Steep soil slopes
- Retaining walls
- Unpaved roads
- Embankments on soft base

5.5.2 Stability/Capacity

Modes of failure of reinforced soil include:

- Sliding over reinforcement
- Reinforcement rupture
- Reinforcement pull-out
- Base bearing capacity exceedance
- Internal and external (global) instability
- Excessive deformation

5.5.2.1 Gently Inclined Soil Slopes

- The factor of safety of a reinforced slope for **translational slides** based on Eq. (5.2):

$$F_s = \frac{c + [\gamma \cdot d \cdot \cos\alpha \cdot (\cos\alpha - k_h \cdot \sin\alpha) - \gamma_w \cdot h + T \cdot \sin\alpha] \cdot \tan\phi}{\gamma \cdot d \cdot \cos\alpha \cdot (\sin\alpha + k_k \cdot \cos\alpha) + \gamma_w \cdot {}^{h}/\cos\alpha \cdot \tan\alpha - T \cdot \cos\alpha} \quad (5.80)$$

γ is ground unit weight
d is the vertical depth to slip surface
α is the inclination to the horizontal of the slip surface
γ_w is water unit weight
h is the piezometric height above the slip surface
k_h is the ratio between the horizontal and the gravitational acceleration , the vertical acceleration can be accounted for by multiplying γ by $(1 \pm k_v)$, where k_v is the ratio between the vertical and the gravitational acceleration
c is cohesion
ϕ is friction angle
T is the horizontal reinforcement force per metre length of slide

- For **rotational slides**, Eq. (5.4) can be used with R_x denoting reinforcement force.
- Reinforcement force is found for required factor of safety of slope stability of say 1.3. The **rupture strength of the reinforcement** is determined by multiplication of the required force with the coefficients. – Jewell (1996) suggests the use of coefficients for the rupture strength of reinforcement as: 1.2 for mechanical damage, 1.1 for environmental degradation, and 1.5 for material factor (due to extrapolation of test data to design life time) (in total 2.0). Koerner (1998) suggests the use of reduction coefficients for: installation damage of 1.1–1.4, creep 2–3, chemical degradation of 1.1–1.4, biological degradation of 1–1.2, seams of 1.33, and for holes in geotextiles of 1.11 (in total 2.4–10.4). BS 8006-1 (2010) provides only partial factors of 1.3 for sliding across surface of reinforcement and pull-out resistance of reinforcement for the ultimate limit state in addition to partial load factors.
- The **anchoring length or reinforcement** beyond the extent of a critical slip surface is determined from the results of pull-out tests on the reinforcement.
- The check of **sliding over reinforcement** is based on the shear strength of soil-reinforcement interface determined from direct shear tests.

5.5.2.2 Steep Soil Slopes

Jewell (1996) suggests the following design steps:

1. The required earth pressure coefficient K_{req} and the required reinforcement length L_r to slope height H ratio $(L_r/H)_{\text{overl}}$ and $(L_r/H)_{ds}$ may be determined from the charts shown in Fig. 5.44. The charts are valid for the coefficient of direct sliding $\alpha_{ds} = 0.8$. A correction may be applied in cases where $\alpha_{ds} < 0.8$.
2. Use greater of $(L_r/H)_{\text{overl}}$ and $(L_r/H)_{ds}$ to calculate the length of reinforcement L_r.

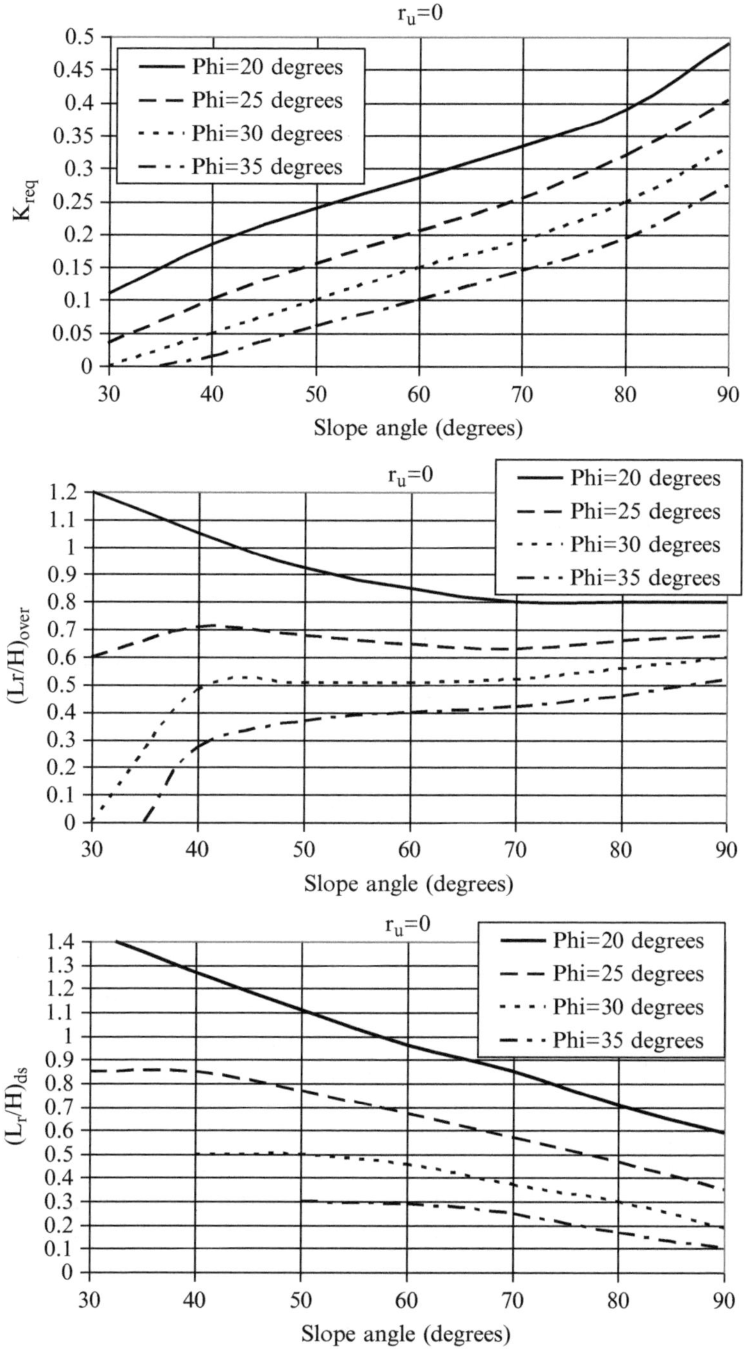

Fig. 5.44 Coefficients K_{req}, $(L_r/H)_{overl}$ and $(L_r/H)_{ds}$ for reinforced steep slopes depending on the coefficient of pore pressure (r_u) and backfill design friction angle (Phi) (Adopted from Jewell 1996)

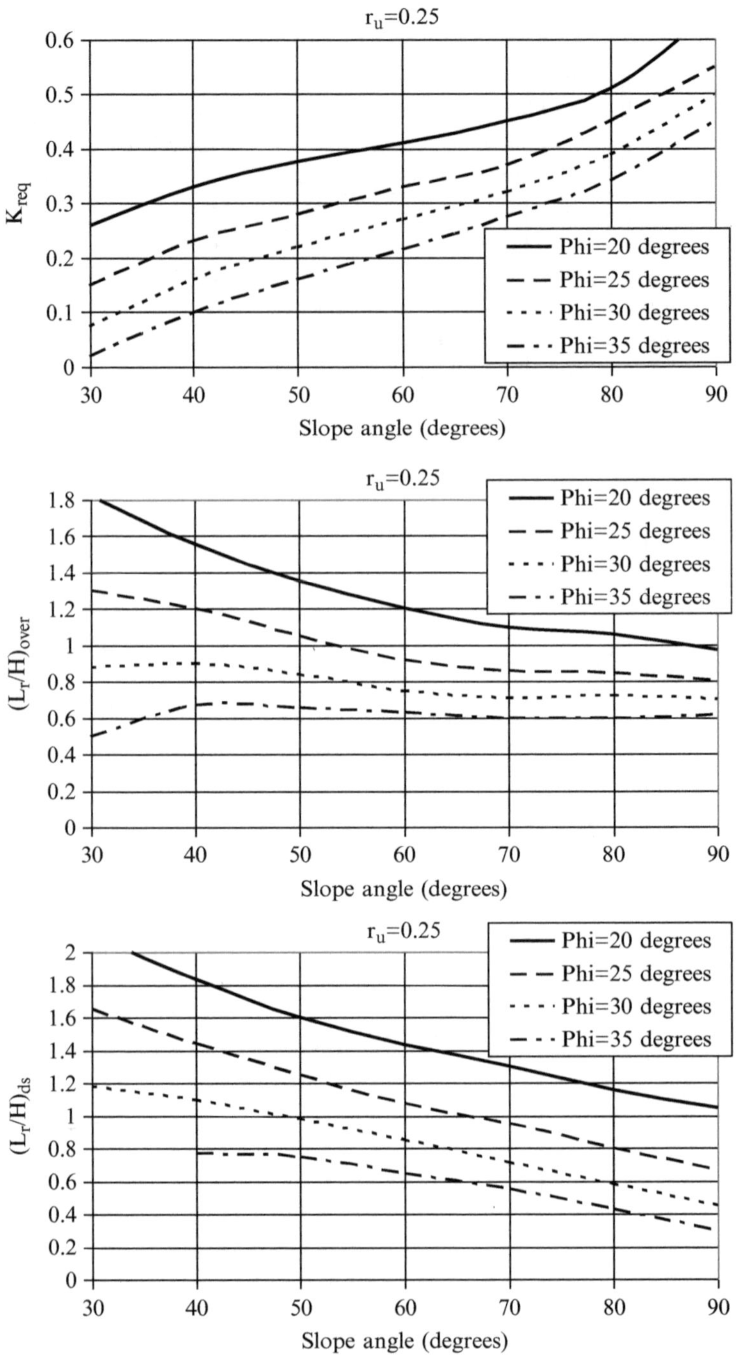

Fig. 5.44 (continued)

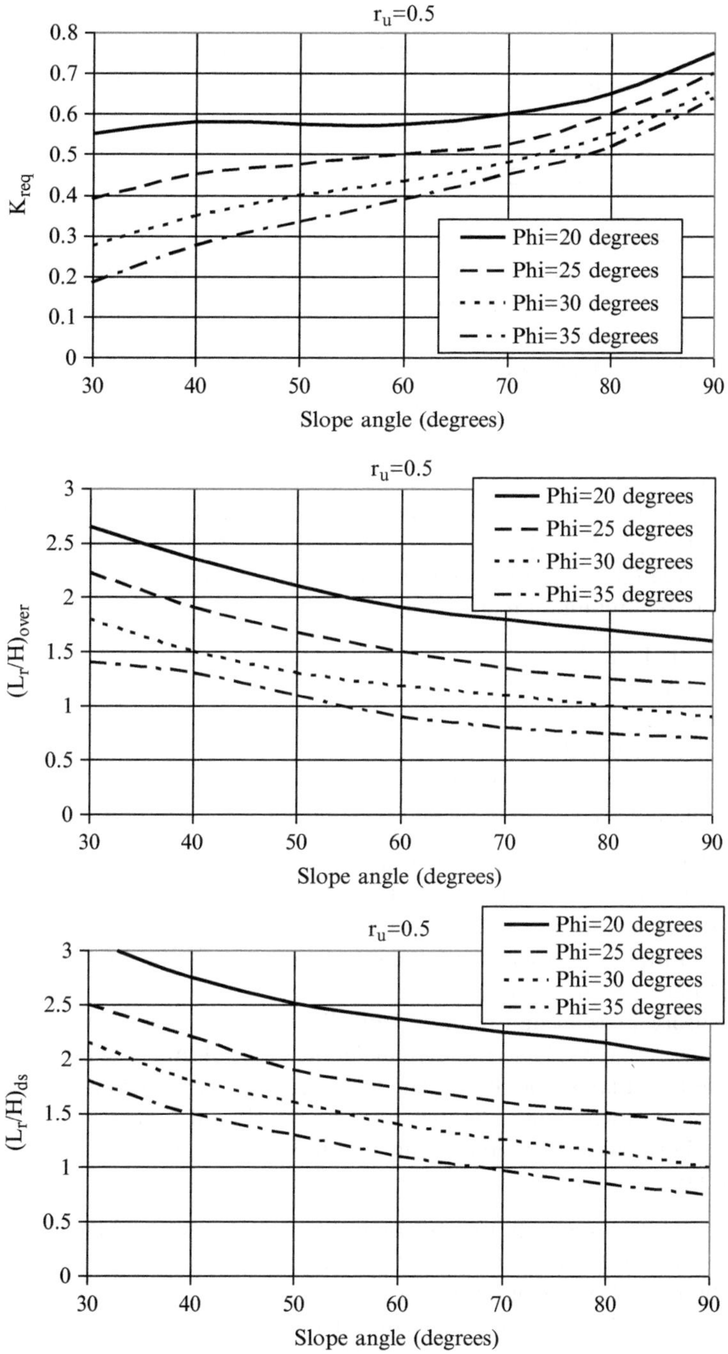

Fig. 5.44 (continued)

3. Calculate the maximum required stress at the base of the slope as $\sigma_{\mathrm{req}} = K_{req}\,\gamma\,H$ where γ is soil unit weight, H is slope height. Calculate the bond length L_b from the following formula

$$\frac{L_b}{H} = \frac{P_r}{2 \cdot W_r \cdot \gamma \cdot H^2} \cdot \frac{1}{1 - r_u} \cdot \frac{1}{\alpha_b \cdot \tan\phi'} \tag{5.81}$$

   H is the slope height
   P_r is the required force (kN/m) $= \sigma_{req}$ times the reinforcement vertical spacing
   W_r is the width of the reinforcement
   γ is soil unit weight
   r_u is the pore pressure coefficient (=pressure of water / total pressure)
   α_b is the bond shear coefficient from pull-out tests
   ϕ is soil friction angle

   From the values of L_r/H in step 3 and L_b/H in step 4, find the ratio L_b/L_r and the bond allowance $1 - L_b/L_r$. Find the design earth pressure coefficient, allowing for bond, $K_d = K_{req}/(1 - L_b/L_r)$. The design stress is $\sigma_d = K_d\ \gamma\ H$ and the design force $P_d = \sigma_d$ times the reinforcement vertical spacing.
4. Calculate the minimum required stress to allow for bond at the crest

   $\sigma_{\mathrm{min}} = \gamma\ \mathrm{H}\ \mathrm{L_b/L_r}\ \mathrm{K_{req}}$
   γ is soil unit weight
   H is the slope height
   L_b is the bond length
   L_r is the required reinforcement length
   K_{req} is the required earth pressure coefficient

   If calculated σ_{min} is smaller than the maximum stress induced by compaction (usually in the range from 10 to 30 kPa) then use the compaction stress instead. The greater of these two stresses σ can be used to calculate the design force $P_{d1} = \sigma$ times the vertical reinforcement spacing s_{v1} in the top part of the slope to depth $z_1 = P_{d1}/(s_{v1}\ \gamma\ K_d)$.
5. For reinforcement design force P_d (and P_{d1}) in kN/m multiplied by the coefficients for: damage 1.2, environmental degradation 1.1, material factor for design 1.5 (due to extrapolation of test data to design life time) calculate the number of layers of reinforcement from the slope height H (and z_1) and selected vertical spacing between reinforcement (min. 0.3 m for practical reasons of compaction of soil layers).

Koerner (1998) suggests the use of the same steps as Jewell (1996) but the coefficients for: installation damage of 1.1–1.4, creep 2–3, chemical degradation of 1.1–1.4, biological degradation of 1–1.2 (in total 2.4–7).

BS 8006-1 (2010) provides only partial factors of 1.3 for sliding across surface of reinforcement and pull-out resistance of reinforcement for the ultimate limit state in addition to partial load factors.

Overall stability of a slope for trial slip surfaces passing outside the reinforced zone is checked using the methods described in Sect. 5.1.2. Examples of back analyses of overall stability of geogrid reinforced steep slopes using limit equilibrium method are provided by Srbulov (2001b).

5.5.2.3 Retaining Walls

Reinforcement design is as for steep slopes. Jewell (1996) also states the following:

6. The effect of eventual surcharge q behind a wall is added to the stress caused by backfill as $K_a\, q$, where K_a is the coefficient of backfill active pressure (according to Eq. (5.62)).
7. The bond length L_b for the case of surcharge q is:

$$\frac{L_b}{H} = \frac{P_r}{2 \cdot W_r \cdot \gamma \cdot H^2 \cdot \left(1 + {}^{q}/_{\gamma}\right)} \cdot \frac{1}{1 - r_u} \cdot \frac{1}{\alpha_b \cdot \tan \phi'} \tag{5.82}$$

H is the slope height
P_r is the required force (kN/m) $= \sigma_{req}$ times the reinforcement vertical spacing
W_r is the width of the reinforcement
γ is soil unit weight
r_u is the pore pressure coefficient (=pressure of water / total pressure)
α_b is the bond shear coefficient from pull-out tests
ϕ is soil friction angle
q is surcharge at the top of backfill

8. The effect of eventual concentrated inclined load Q behind a wall is added to required reinforcement force as $Q_h + K_a^{1/2}\, Q_v$, where $_h$ and $_v$ refer to the horizontal and vertical component of the force Q, K_a is the coefficient of backfill active pressure according to Eq. (5.62). Additional required stress in the reinforcement due to a horizontal point load is $2\, Q_h/h_c$ at the top and zero at a depth $h_c = x / \tan(45° - \phi/2)$, where x is the horizontal distance of the force from the wall front. Additional required stress in the reinforcement due to a vertical point load is $Q_v K_a^{1/2}/h_c$ over the depth $h_c = x / \tan(45° - \phi/2)$, where x is the horizontal distance of the force from the wall front, and $Q_v K_a /[x + (B' + z)/2]$, where B' is the width over which the force Q_v is applied and z is depth below the backfill surface.
9. The minimum reinforcement length for the case of a wall foundation on a fine grained soil with the undrained shear strength c_u is:

$$\frac{L}{H} = \frac{K_{ab}}{\alpha_{ds}} \cdot \left[\frac{\gamma_b \cdot H}{2 \cdot c_u} + \frac{q}{c_u}\right] \tag{5.83}$$

L is reinforcement length
H is wall height
K_{ab} is the coefficient of active backfill pressure
α_{ds} is the coefficient of direct sliding within backfill
γ_b is the unit weight of backfill
q is surcharge on the backfill behind a wall

Koerner (1998) provides examples of calculations of stability of wrap around geotextile and geogrid with facing reinforced walls as for massive walls and remarks that geosynthetic manufacturers have their own design methods.

The bearing capacity of soil underneath a wall is checked according to Sect. 5.2.2.

For the check of external stability of a wall when trial slip surfaces pass through reinforced section, the contribution of reinforcement force on the overall stability can be considered for the reinforcement for which the length between the intersection with a trial slip surface and the end of the reinforcement is greater than the bond length L_b. Examples of back analyses of overall stability of geogrid reinforced retaining walls using limit equilibrium method are provided by Srbulov (2001b).

Jewell (1996) suggests the use of coefficients for the rupture strength of reinforcement as: 1.2 for mechanical damage, 1.1 for environmental degradation, and 1.5 for material factor (due to extrapolation of test data to design life time) (in total 2.0). Koerner (1998) suggests the use of the coefficients for: installation damage of 1.1–1.4, creep 2–3, chemical degradation of 1.1–1.4, biological degradation of 1–1.2 (in total 2.4–7) . BS 8006-1 (2010) provides only partial factors of 1.3 for sliding across surface of reinforcement and pull-out resistance of reinforcement for the ultimate limit state in addition to partial load factors.

5.5.2.4 Unpaved Roads

- Hammitt (1970) proposed an empirical equation for design of **unreinforced unpaved roads**:

$$\frac{D}{R} = [0.176 \cdot \log_{10}(N) + 0.12] \cdot \sqrt{1.687 \cdot {}^{p}/_{c_u} - 1} \tag{5.84}$$

D is the thickness of fill
R is the radius of loaded area
N is the number of load cycles
p is the vertical pressure on the fill surface
c_u is undrained shear strength of base soil

The formula has been derived for the standard correlation $CBR = c_u/30$
CBR is the California Bearing Ration (Sect. 2.1.6)

- Giroud and Noiray (1981) proposed the following equation for **static condition**:

$$\frac{p}{c_u} = N_c \left[1 + 2.29 \cdot \frac{D \cdot \tan\beta}{R} + 1.27 \cdot \left(\frac{D \cdot \tan\beta}{R} \right)^2 \right] \tag{5.85}$$

p is the vertical pressure on the fill surface
c_u is undrained shear strength of base soil
N_c is the bearing capacity factor $= \pi$ in the unreinforced case, $=2 + \pi$ in the reinforced case
D is the thickness of fill
β is the angle of load p spread through the fill $=31°$
R is the radius of loaded area

For traffic load, D from Hammitt (1970) formula for unreinforced fill is decreased for the thickness difference in static condition for unreinforced and reinforced cases from Giroud and Noiray (1981) formula

- Jewell (1996) provided the formula for **the maximum design tension force** P_r in the reinforcement:

$$P_r = p \cdot B \cdot \left[\frac{K_{as}}{\tan\beta} \cdot \ln_e\left(1 + {}^{D}/_{B} \cdot \tan\beta\right) - \tan\delta \right] + \left(K_{as} - \zeta \cdot K_{ps}\right) \cdot \gamma \cdot {}^{D^2}/_{2} \tag{5.86}$$

p is the vertical pressure on the fill surface
B is the side length of a square shaped loaded area
K_{as} is the coefficient of active pressure in the fill (Eq. 5.62)
β is the angle of load p spread through the fill $=31°$
D is the thickness of fill
$\tan\delta$ is the ratio between the horizontal and vertical load on the fill surface (from breaking or centrifuge force in a bend)
ζ is the coefficient of activation of the passive resistance in the fill (~2/3 or ½)
K_{Ps} is the coefficient of passive pressure in the fill (Fig. 5.36)
γ is the unit weight of fill

Koerner (1998) refers to Giroud and Noiray (1981) formula but considers reinforcement deflection for calculation of reinforcement force. Jewell (1996) suggests the use of coefficients for the rupture strength of reinforcement as: 1.2 for mechanical damage, 1.1 for environmental degradation, and 1.5 for material factor (due to extrapolation of test data to design life time) (in total 2.0). Koerner (1998) suggests the use of the coefficients for: installation damage of 1.1–1.4, creep 2–3, chemical degradation of 1.1–1.4, biological degradation of 1–1.2 (in total 2.4–7) . BS 8006-1 (2010) provides only partial factors of 1.3 for sliding across surface of reinforcement and pull-out resistance of reinforcement for the ultimate limit state in addition to partial load factors.

5.5.2.5 Embankments on Soft Base

- Jewell (1996) provides the expression for necessary slope inclination $\tan\beta = 1/n$ to prevent **sliding over basal reinforcement**:

$$n > \frac{K_{as}}{\alpha_{ds} \cdot \tan\phi'} \cdot \left[1 + \frac{2 \cdot q}{\gamma \cdot H}\right] \tag{5.87}$$

n is the horizontal length corresponding to 1 m vertical step along a slope
K_{as} is the coefficient of active pressure in an embankment according to Eq. (5.62)
α_{ds} is the coefficient for direct sliding along reinforcement-fill interface
ϕ' is the fill angle of internal friction
q is the surcharge on an embankment
γ is the unit weight of embankment fill
H is the embankment height

- The **overall factor of safety** F_s for reinforced embankment is according to Jewell (1996):

$$F_s = \frac{c_{uo}}{\gamma \cdot H} \cdot \left[4 + \rho \cdot n \cdot {}^{H}\!\Big/_{c_{uo}} + 2 \cdot \sqrt{2 \cdot (1 + \alpha) \cdot \rho \cdot n \cdot {}^{H}\!/_{c_{uo}}}\right] \tag{5.88}$$

c_{uo} is the undrained shear strength at the surface of soft clay under an embankment
γ is the unit weight of embankment fill
H is the embankment height
ρ is the rate of increase of undrained shear strength with depth
n is the horizontal distance corresponding to 1 m vertical step along a slope
α is the coefficient of application of shear stress to clay surface, $=-1$ for unreinforced case, $=0$ for no shear stress, $=1$ for full inward shear stress from reinforcement action. The value of α can be determined by pull-out tests

- The **maximum required design reinforcement force** P_r to maintain embankment equilibrium according to Jewell (1996):

$$P_r = \gamma \cdot H^2 \cdot \left[\alpha \cdot n \cdot \frac{c_{uo}}{F_s \cdot \gamma \cdot H} + {}^{K_{as}}\!/_{2}\right] \tag{5.89}$$

γ is the unit weight of embankment fill
H is the embankment height
α is the coefficient of application of shear stress to clay surface, $=-1$ for unreinforced case, $=0$ for no shear stress, $=1$ for full inward shear stress from reinforcement action. The value of α can be determined by pull-out tests

n is the horizontal distance corresponding to 1 m vertical step along a slope
c_{uo} is the undrained shear strength at the surface of soft clay under an embankment
F_s the factor of safety of the overall stability
K_{as} is the coefficient of active pressure in an embankment according to Eq. (5.62)

Jewell (1996) suggests the use of coefficients for the rupture strength of reinforcement as: 1.2 for mechanical damage, 1.1 for environmental degradation, and 1.5 for material factor (due to extrapolation of test data to design life time) (in total 2.0). Koerner (1998) suggests the use of the coefficients for: installation damage of 1.1–1.4, creep 2–3, chemical degradation of 1.1–1.4, biological degradation of 1–1.2 (in total 2.4–7) . BS 8006-1 (2010) provides only partial factors of 1.3 for sliding across surface of reinforcement and pull-out resistance of reinforcement for the ultimate limit state in addition to partial load factors.

- The **critical depth** z_{crit} for the overall sliding according to Jewell (1996):

$$z_{crit} = \sqrt{(1+\alpha)\cdot c_{uo}\cdot n\cdot\frac{H}{2\cdot\rho}} \tag{5.90}$$

α is the coefficient of application of shear stress to clay surface, =−1 for unreinforced case, =0 for no shear stress, =1 for full inward shear stress from reinforcement action. The value of α can be determined by pull-out tests
c_{uo} is the undrained shear strength at the surface of soft clay under an embankment
n is the horizontal distance corresponding to 1 m vertical step along a slope
H is the embankment height
ρ is the rate of increase of undrained shear strength with depth

Koerner (1998) used modified limit equilibrium analyses for embankments on soft clay with reinforcement placed at the top or several layers.

Examples of back analyses of the embankments over soft clay are provided by Srbulov (1999).

5.5.3 *Movement*

- Compacted fill settlement is small (of the order of 1 % of the fill thickness).
- The base of fill settlement due to fill weight is calculated according to Sect. 5.2.3.1.
- The horizontal movement is a product of reinforcement axial strain and the reinforcement length.

5.5.4 *Execution*

- Useful information is provided in EN 14475 (2006), by Jewell (1996) and Koerner (1998).
- Contractor must minimise the effect of installation damage and chemical degradation by careful handling and storing geosynthetic product away from ultraviolet radiation, heating and freezing prior to the installation.
- Manufacturers of reinforcement provide details for their installation.

5.6 Tunnels and Shafts

5.6.1 *Types*

The following types are considered:

- **Vertical shafts**
- **Shallow tunnels** constructed using open cut and cover method with the lateral ground support provided by diaphragm walls, contiguous or secant piles with anchors or props
- **Deep tunnels** in strong ground constructed using open machine excavation or blasting with (temporary) ground support provided by sprayed concrete, bolts and steel meshes or in week ground constructed using tunnel boring machines with ground support provided by precast reinforced concrete segments

5.6.2 *Stability/Capacity*

5.6.2.1 Vertical Shafts

- **In static condition**, Wong and Kaiser (1988) used the convergence – confinement method (usually applied to tunnels), which accounts for in situ stresses and ground properties, to define analytical predictive formulas for pressures $\sigma_{v,r,t}$ (vertical, radial, tangential) and soil horizontal deformation u_i around vertical shafts.

$$\sigma_v = \gamma \cdot h = p_o$$
$$\sigma_r = p_i$$
$$\sigma_t = 2 \cdot K_o \cdot p_o - p_i$$
$$For\,\phi \neq 0$$
$$K_o = (1 - \sin\phi) \cdot OCR^{\sin\phi}$$
$$K_p = \tan^2\left(45^o + \phi/2\right)$$

$$
\begin{aligned}
p_i &= \frac{2 \cdot K_o \cdot p_o}{K_p + 1}, K_o > K_{cr} \\
p_i &= \frac{p_o}{K_p}, K_o < K_{cr} \\
K_{cr} &= \frac{K_p + 1}{2 \cdot K_p} \\
&For\, c_u \neq 0 \\
p_i &= K_o \cdot p_o - c_u, 1 - {}^{c_u}/_{p_o} < K_o < {}^{c_u}/_{p_o} + 1 \\
p_i &= p_o - 2 \cdot c_u, K_o < 1 - {}^{c_u}/_{p_o}
\end{aligned} \tag{5.91}
$$

The radial horizontal displacement u_i before soil yield,

$$
u_i = r \cdot \frac{(K_o \cdot p_o - p_i) \cdot (1 + \nu)}{E} \tag{5.92}
$$

γ is soil unit weight
h is depth at which the stresses are calculated
ϕ is soil friction angle
OCR is the over consolidation ratio
c_u is soil undrained shear strength
r is the shaft radius,
E is Young modulus of soil,
v is Poisson's ratio of soil.

After soil yield, the value of displacement depends on the yield amount. Wong and Kaiser (1988) obtained a good agreement between the results obtained using Eqs. (5.91) and (5.92) and finite element analyses. They also concluded that the conventional design methods that provide the minimum support pressures required to maintain stability are not conservative. These pressures are generally less than those actually encountered if ground movements during construction are restricted with good ground control.

- **In seismic condition**, the additional lateral earth pressure on vertical shafts can be estimated using Eq. (5.66).

5.6.2.2 Shallow Tunnels Formed by Cut and Cover Method

- **In static condition**, Sect. 5.3.2 for flexural retaining walls is applicable.
- **In seismic condition**, Sect. 5.2.2.5 for kinematic interaction is applicable. An example of calculations for failed cut and cover Daikai station in Japan is provided by Srbulov (2011).

5.6.2.3 Deep Tunnels

- **In static condition**, Hoeg (1968) for the case of no slippage and at the location of tunnel extrados:

$$
\begin{aligned}
&\sigma_r = \frac{\sigma_v}{2} \cdot [(1+k) \cdot (1-a_1) - (1-k) \cdot (1 - 3 \cdot a_2 - 4 \cdot a_3) \cdot \cos(2 \cdot \alpha)] \\
&\sigma_\alpha = \frac{\sigma_v}{2} \cdot [(1+k) \cdot (1+a_1) + (1-k) \cdot (1 - 3 \cdot a_2) \cdot \cos(2 \cdot \alpha)] \\
&\tau_{r\alpha} = \frac{\sigma_v}{2} \cdot (1-k) \cdot (1 + 3 \cdot a_2 + 2 \cdot a_3) \cdot \sin(2 \cdot \alpha) \\
&s_r = \frac{\sigma_v}{2} \cdot \frac{R}{M_s} \cdot \begin{bmatrix} (1+k) \cdot (1-\nu_s) \cdot \left(1 + \dfrac{a_1}{1 - 2 \cdot \nu_s}\right) - \\ (1-k) \cdot \dfrac{1-\nu_s}{1-2 \cdot \nu_s} \cdot [(1 + a_2 + 4 \cdot (1-\nu_s) \cdot a_3] \cdot \cos(2 \cdot \alpha) \end{bmatrix} \\
&s_\alpha = \frac{\sigma_v}{2} \cdot \frac{R}{M_s} \cdot \frac{1-\nu_s}{1-2 \cdot \nu_s} \cdot (1-k) \cdot [(1 - a_2 + 2 \cdot (1 - 2 \cdot \nu_s) \cdot a_3] \cdot \sin(2 \cdot \alpha) \\
&a_1 = \frac{(1 - 2 \cdot \nu_s) \cdot (C-1)}{(1 - 2 \cdot \nu_s) \cdot C + 1} \\
&a_2 = \frac{(1 - 2 \cdot \nu_s) \cdot (1-C) \cdot F - 0.5 \cdot (1 - 2\nu_s)^2 \cdot C + 2}{[(3 - 2 \cdot \nu_s) + (1 - 2 \cdot \nu_s) \cdot C] \cdot F + (2.5 - 8 \cdot \nu_s + 6 \cdot \nu_s{}^2) \cdot C + 6 - 8 \cdot \nu_s} \\
&a_3 = \frac{[1 + (1 - 2 \cdot \nu_s) \cdot C] \cdot F - 0.5 \cdot (1 - 2 \cdot \nu_s) \cdot C - 2}{[(3 - 2 \cdot \nu_s) + (1 - 2 \cdot \nu_s) \cdot C] \cdot F + (2.5 - 8 \cdot \nu_s + 6 \cdot \nu_s{}^2) \cdot C + 6 - 8 \cdot \nu_s} \\
&C = 0.5 \cdot \frac{1}{1-\nu_s} \cdot \frac{M_s}{\frac{E}{1-\nu^2}} \cdot \frac{D}{t} \\
&F = 0.25 \cdot \frac{1 - 2 \cdot \nu_s}{1 - \nu_s} \cdot \frac{M_s}{\frac{E}{1-\nu^2}} \cdot \left(\frac{D}{t}\right)^3 \\
&M_s = \frac{E_s \cdot (1-\nu_s)}{(1+\nu_s) \cdot (1 - 2 \cdot \nu_s)} \\
&E_s = G_s \cdot 2 \cdot (1 + \nu_s)
\end{aligned}
\tag{5.93}
$$

σ_v is the vertical overburden stress,
k is the ratio between horizontal and vertical overburden stress,
R and D are the average radius and diameter of a cylinder,
t is cylinder thickness,
E is Young modulus,

ν is Poisson's ratio,
s subscript is for soil,
α angle is measured from the horizontal cylinder axis upwards,
s_r and s_α are the tangential and radial displacements respectively.

- **In seismic condition**, Penzien (2000) evaluated the racking deformation of rectangular and circular tunnel linings by soil structure interaction during earthquakes and provided formulae for calculation of sectional forces and bending moment in a circular lining without slippage between soil and lining:

$$
\begin{aligned}
P_\alpha &= \frac{-24 \cdot E \cdot J \cdot \Delta}{D^3 \cdot (1-\nu^2)} \cdot \cos\left(2 \cdot \alpha + \frac{\pi}{2}\right) \\
M_\alpha &= \frac{-6 \cdot E \cdot J \cdot \Delta}{D^2 \cdot (1-\nu^2)} \cdot \cos\left(2 \cdot \alpha + \frac{\pi}{2}\right) \\
T_\alpha &= \frac{-24 \cdot E \cdot J \cdot \Delta}{D^3 \cdot (1-\nu^2)} \cdot \sin\left(2 \cdot \alpha + \frac{\pi}{2}\right) \\
\Delta &= R \cdot \Delta_s \\
R &= \pm\frac{4 \cdot (1-\nu_s)}{(c+1)} \\
c &= \frac{24 \cdot (3 - 4 \cdot \nu_s) \cdot E \cdot J}{D^3 \cdot G_s \cdot (1-\nu^2)}
\end{aligned}
\tag{5.94}
$$

α angle is measured in a cross section from the horizontal tunnel axis downwards,
E is Young modulus of tunnel lining,
J is the second moment of circumferential cross section area,
D is the diameter of the middle of lining,
ν is Poisson's ratio,
s subscript is for soil,
G is soil shear modulus
Δ_s is the horizontal differential displacement over the tunnel height. It can be estimated from the horizontal displacement of ground surface d_g (Eq. 4.33) for design ground wave amplitude and from the predominant period of ground motion $> T_D$ in Fig. 4.32 as well as local transversal wave velocity to calculate the length of an assumed sinusoidal wave.

Positive axial force P_α is tensile, positive bending moment M_α is stretching internal side and compressing external side of lining, positive shear force T_α is oriented towards outside at the cross section inclined at an angle α to the horizontal. The formula is applicable to a uniform strain field around a tunnel, negligible inertial interaction between a tunnel and surrounding soil and sufficiently deep tunnels so that the free surface boundary conditions has little effect on the racking soil structure interaction.

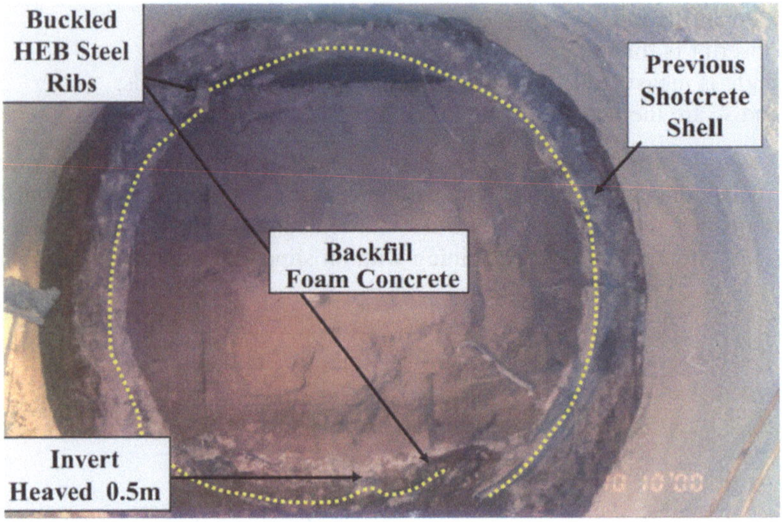

Fig. 5.45 Cross section of the bench pilot Bolu tunnel with marked locations of the maximum damage to the shotcrete lining (Kontoe 2009)

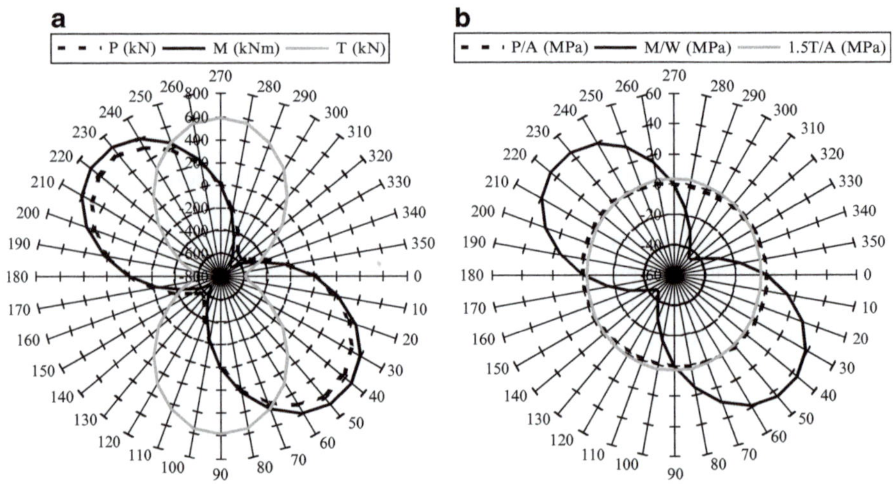

Fig. 5.46 (**a**) Axial P and shear T force and bending moment M, (**b**) Axial (P/A), maximum shear (1.5 T/A) and bending stress (M/W) in the shotcrete lining of the bench pilot Bolu tunnel due to differential horizontal ground displacement Δs of 0.015 m i.e. Δ of 0.038 m over the tunnel diameter of 5 m Srbulov (2011)

An example of the calculations performed for the bench pilot Bolu tunnel in Turkey, which deformed cross section is shown in Fig. 5.45, according to Eq. (5.94) is shown in Fig. 5.46 according to Srbulov (2011).

5.6.3 Movement

5.6.3.1 Vertical Shafts

- **In static condition**, the radial horizontal displacement u_i before soil yield can be estimated using Eq. (5.92).
- **In seismic condition,** the horizontal shaft movement can be estimated from the horizontal displacement of ground surface d_g (Eq. 4.33) for design ground wave amplitude and from the predominant period of ground motion $> T_D$ in Fig. 4.34 as well as local transversal wave velocity to calculate the length of an assumed sinusoidal wave.

5.6.3.2 Shallow Tunnels Formed by Cut and Cover Method

- **In static condition**, Sect. 5.3.3.2 is applicable.
- **In seismic condition**, the horizontal movement can be estimated from the horizontal displacement of ground surface d_g (Eq. 4.33) for design ground wave amplitude and from the predominant period of ground motion $> T_D$ in Fig. 4.34 as well as local transversal wave velocity to calculate the length of an assumed sinusoidal wave.

5.6.3.3 Deep Tunnels

- **In static condition**, the tangential and radial displacements s_r and s_α can be estimated from Eq. (5.93).
- **In seismic condition**, the horizontal differential displacement over the tunnel height can be estimated from the horizontal displacement of ground surface d_g (Eq. 4.35) for design ground wave amplitude and from the predominant period of ground motion $> T_D$ in Fig. 4.34 as well as local transversal wave velocity to calculate the length of an assumed sinusoidal wave.

5.6.4 Execution

- **Vertical shafts** and **shallow cut and cover tunnels**, the comments made in Sect. 5.2.4.2 are applicable.
- **Deep tunnels in week ground** are constructed using slurry shield or earth pressure balance tunnel boring machines of different design.
- **Deep tunnels in strong ground** are constructed using open face excavation machines of different design. Unsupported span stand up times depending on RMR (Sect. 3.2.2) are shown in Fig. 5.47 after Stacey and Page (1986) following Bieniawski (1976). Bolt covering area (Fig. 5.48) and shotcrete thickness

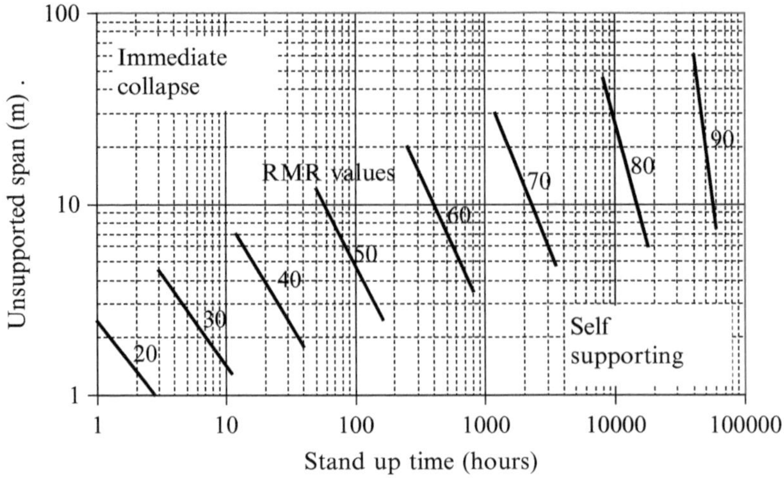

Fig. 5.47 Ranges of unsupported span and stand up time for different RMR values (Adopted from Stacey and Page 1986 after Bieniawski 1976)

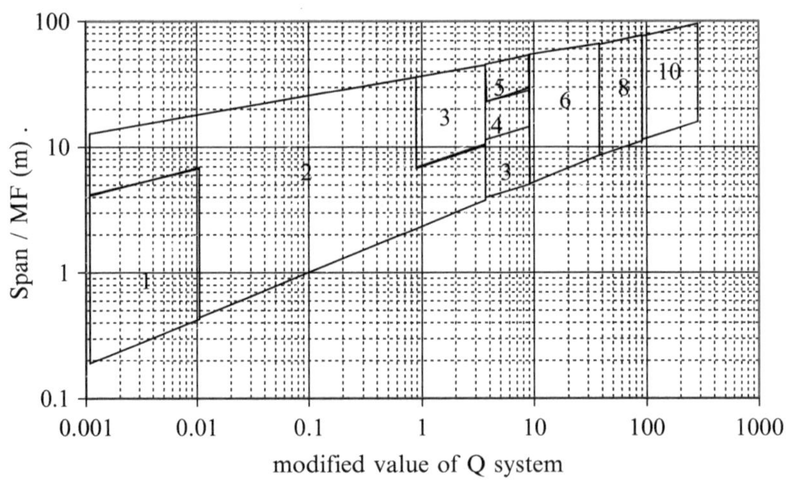

Fig. 5.48 Bolt covering area (m^2), when >6 m^2 spot bolting is implied (Adopted from Stacey and Page 1986)

(Fig. 5.49) are shown according to Stacey and Page (1986), who simplified the original data by Barton et al. (1974), for modified values of Q system (Sect. 3.2.3) and modified spans (the modification factor MF given in Table 5.10 from Barton et al. 1974). The modified values of Q after Barton et al. (1974) are for:

$Q > 10$, the modified value $= 5Q$
$0.1 < Q < 10$, the modified value $= 2.5Q$
$Q < 0.1$, the modified value $= Q$

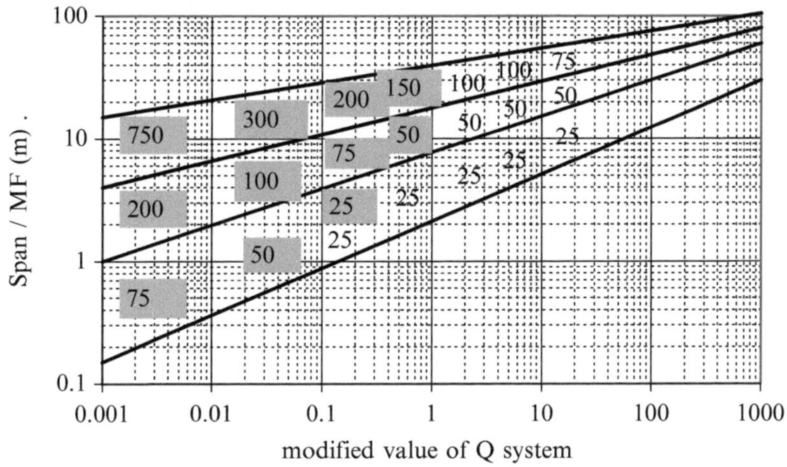

Fig. 5.49 Shotcrete thickness (mm), mesh reinforced when shaded areas (Adopted from Stacey and Page 1986)

Table 5.10 Span modification factor MF (Barton et al. 1974)

Type of excavation	MF
Temporary mine openings	3–5
Permanent mine openings, low pressure water tunnels, pilot tunnels, drifts and headings for large excavations	1.6
Storage chambers, water treatment plants, minor road and railway tunnels, surge chambers, access tunnels	1.3
Power houses, major road and railway tunnels, civil defence chambers, portals, intersections	1
Underground nuclear power stations, sport and public facilities, factories	0.8

The length L in metres of rock bolts or cables can be calculated from the formulas suggested by to Stacey and Page (1986):

$$\begin{aligned} Roof - Bolts &: L = 2 + 0.15 \cdot {}^{B}\big/_{MF} \\ - Cables &: L = 0.4 \cdot {}^{B}\big/_{MF} \\ Walls - Bolts &: L = 2 + 0.15 \cdot {}^{H}\big/_{MF} \\ - Cabls &: L = 0.35 \cdot {}^{H}\big/_{MF} \end{aligned} \tag{5.95}$$

B is span
H is wall height
MF is the modification factor in Table 5.10

Recommended maximum unsupported excavation span for different rock mass qualities and span modification factors are according to Barton (1976)

$$Max.\quad span = 2 \cdot MF \cdot Q^{0.4} \tag{5.96}$$

MF is span modification factor (Table 5.10)
Q is rock mass quality value (Sect. 3.2.3)

5.7 Pipes

5.7.1 Types

Low pressure rigid and flexible pipes in static and seismic conditions are considered.

5.7.2 Stability/Capacity

5.7.2.1 Static Condition

- For the case when the density of the fill material is smaller than that of the original soil, the soil load per unit length W_c on a **rigid pipe in a narrow trench** is according to Marston (1930), Schlick (1932) and Spangler (1947)

$$W_c = C_d \cdot \gamma \cdot B_d^{\,2} \tag{5.97}$$

  C_d is load coefficient for ditch conduits $= 1 - \exp[-2K_a\mu'H/B_d]/(2K_a\mu')$, Spangler 1947)
  K_a is the active pressure coefficient from Eq. (5.62)
  μ' is the coefficient of friction between backfill and side of ditch
  H is the thickness of fill above the top of pipe
  γ is the unit weight of fill material
  B_d is the width of ditch at the top of conduit

- In trenches wider than about two to three times the outside diameter of the pipe, the soil load per unit length W_c on a **rigid pipe in wide trench** is

$$W_c = C_c \cdot \gamma \cdot B_c^{\,2} \tag{5.98}$$

  C_c is load coefficient for wide excavations $= 2\,(H/B_c - 0.5)$ for $H/B_c > 1.5$ or $=4/3\ H/B_c$ for $H/B_c < 1.5$
  γ is the unit weight of fill material
  B_c is outside diameter of the conduit
  H is the thickness of fill above the top of pipe

The load on a conduit is the smaller value calculated from Eqs. (5.97) and (5.98).

- Load on a **flexible pipe** according to Spangler and Handy (1981) for the case when the backfill around a conduit has the same stiffness as the conduit is

$$W_c = C_d \cdot \gamma \cdot B_d \cdot B_c \tag{5.99}$$

C_d is load coefficient for ditch conduits $= 1 - \exp[-2K_a\mu' H/B_d]/(2K_a\mu')$, Spangler 1947)
K_a is the active pressure coefficient from Eq. (5.62)
μ' is the coefficient of friction between backfill and side of ditch
H is the thickness of fill above the top of pipe
γ is the unit weight of fill material
B_d is the width of ditch at the top of conduit
B_c is outside diameter of the conduit

- For **concentrated load** acting on ground surface the vertical stress acting on a tube is calculated according to Boussinesq (1885) formula given in Eq. (5.49)
- The **critical buckling pressure** at the top of a pipe is $P_{cb} = 1.15\ (E\ P_{cr})^{0.5}$, where E is the axial soil modulus, P_{cr} is the critical buckling force at top of pipe according to Meyerhof and Baike (1963)

$$P_{cr} = 2 \cdot E \cdot \frac{t^3}{(1 - \nu^2) \cdot D} \tag{5.100}$$

E is stress and time dependent modulus of elasticity of pipe
ν is Poisson's ratio for pipe (~0.45 for HDPE)
D is pipe diameter
t is pipe thickness

5.7.2.2 Seismic Conditions

- Figure 5.50 shows the linear regression that was developed between water supply cast iron, steel, ductile iron and asbestos cement pipeline repair rates and peak ground velocity (PGV) based on data from the Northridge and other US earthquakes as reported by O'Rourke and Bonneau (2007)
- Figure 5.51 shows the peak values of the largest principal ground strain (PGS) as a function of the largest absolute value of peak ground velocity (PGV) by Paolucci and Pitilakis (2007)

$$\log_{10} PGS = 0.955 \cdot \log_{10} PGV - 3.07 \tag{5.101}$$

PGV is in m/s. If multiplier of $\log_{10}$PGV is forced to be unity, the best fit line turns to be PGS = PGV/Ψ where Ψ = 963 m/s for the median value, 671 m/s and 1,382 m/s correspond to the 16 and 84 percentile, respectively.

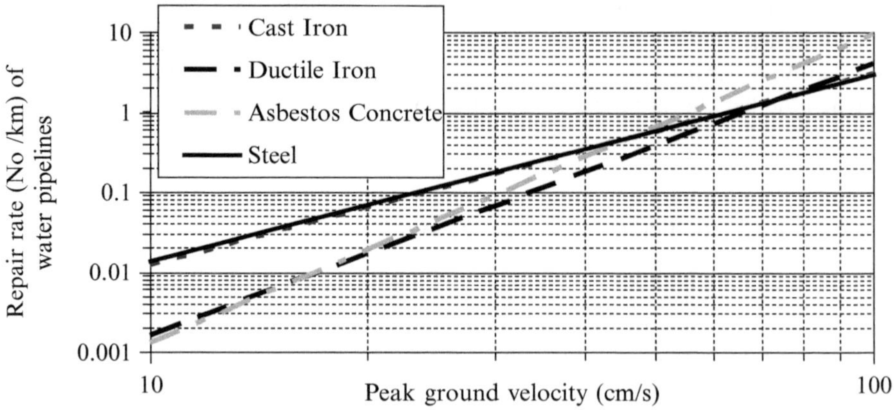

Fig. 5.50 Repair rate of water pipelines (Adapted from O'Rourke and Bonneau 2007)

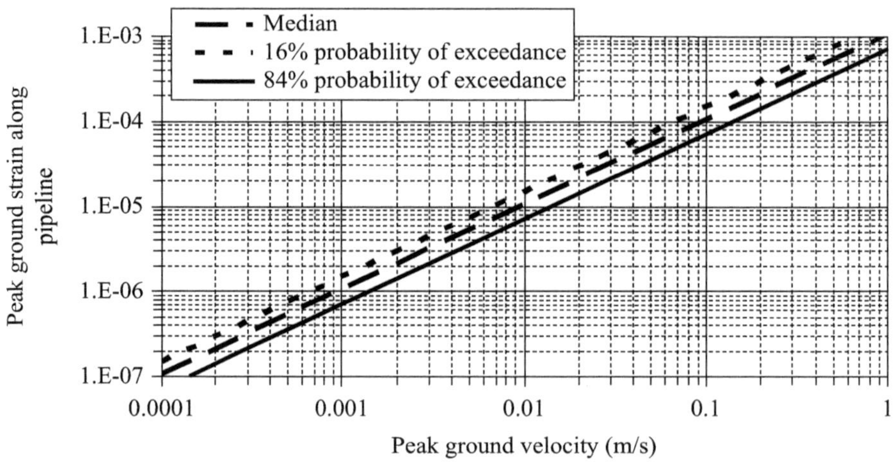

Fig. 5.51 Peak ground strain along pipelines versus peak ground velocities (Adapted from Paolucci and Pitilakis 2007)

- The interpretation of value Ψ as the propagation velocity of the prevailing wave velocity (either apparent velocity of body waves or phase velocity of surface waves) may be misleading. Abrahamson (2003) has recently proposed a model for transient ground strain evaluation, where the relative contribution of wave passage (WP), spatial incoherence (SI) and site effects (SE) are made explicit and summarised in an empirical relationship between the largest principal ground strain (PGS) and peak ground displacement (PGD).

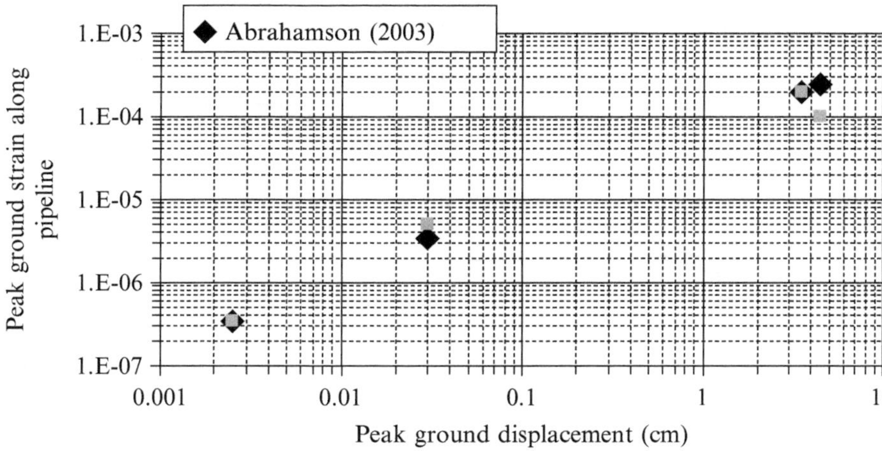

Fig. 5.52 Peak ground strain along pipelines versus peak ground displacements (Adapted from Paolucci and Pitilakis 2007)

$$
\begin{aligned}
{}^{PGS}\!\big/_{PGD} &= WP + SI + SE \\
WP &= \frac{\exp(5.8 - 0.69 \cdot M)}{C} \\
SI &= SE = 3 \cdot 10^{-5}
\end{aligned}
\qquad (5.102)
$$

PGD is measured in cm,
M is earthquake magnitude,
C is a constant with dimension of distance over time

In Fig. 5.52, the PGS-PGD pairs are shown for the four data sets considered by Paolucci and Pitilakis (2007).

- For both body and surface waves, for a fixed value of PGV, ground strain will generally be greater in soft soils (i.e. low velocity value) than stiffer soils. This has been confirmed by Nakajima et al. (2000) in a series of field measurements using strain gauges and accelerographs. As shown in Fig. 5.53, for the same value of PGV, maximum ground strain observed in soft ground (Shimonaga) is on average 3–4 times that observed in hard ground (Kansen). In this case, the predominant period of the soft ground was 1.3 s whilst the predominant period of the hard ground was around 0.4 s.
- Pipe joints may dictate behaviour of a pipeline. A gas-welded joint renders a steel pipe as vulnerable to damage as a cast iron (CI) or asbestos cement (AC) pipe, even though the tensile strength of a steel barrel is much greater than that of CI or AC. For a specific joint type (bell & spigot/ rubber gasket/ restrained/ unrestrained) steel and ductile iron pipes are less vulnerable than more brittle pipe types (polyvinyl chloride-PVC, AC, CI). A summary of data collected by Shirozu et al. (1996) is shown in Fig. 5.54.

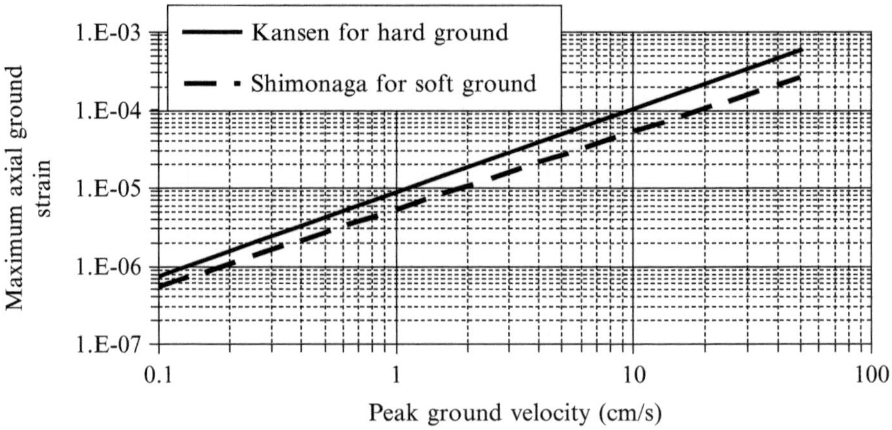

Fig. 5.53 Dependence of maximum axial ground stain on peak ground velocity (Adapted from Nakajima et al. 2000)

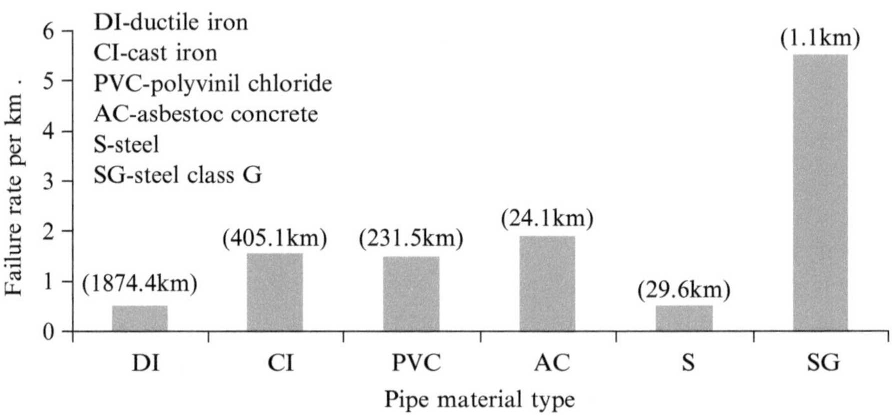

Fig. 5.54 Failure rate per km dependence on pipe material type and their lengths in km (Adapted from Shirozu et al. 1996)

5.7.3 *Movements*

The horizontal deflection of a flexible pipe is according to modified Iowa formula, Spangler and Handy (1981)

$$\Delta x = \frac{D_1 \cdot K \cdot W_c}{\frac{E \cdot I}{r^3} + 0.061 \cdot E'} \tag{5.103}$$

Δx is the horizontal deflection =0.91 of the vertical deflection
D_1 is deflection lag factor (usually 1.5)

K is bedding angle factor (0.096 for 90° angle, varying from 0.108 for 30° angle to 0.083 for 180° angle, Moser 1990)
W_c is load per unit length
r is the mean radius of tubing
E is the modulus of elasticity of tubing
I is the second moment of cross sectional area of tubing = $(D^4 - d^4)\pi/64$, D and d are the outside and inside diameters of a tubing
E' is the modulus of axial stiffness of soil

5.7.4 Execution

Manufacturers of pipes provide their recommendations for their pipes installation.

5.8 Landfills

5.8.1 Types

The following types are considered:

- Spoil hips from mining and construction industry (very heterogeneous and usually uncompacted)
- Municipal solid waste deposits capable of generating landfill gas (mainly methane)
- Hazardous (ignitable, corrosive, reactive, toxic) waste deposits

Both compacted clay layers and geomembranes are used as composite liners because some very concentrated organic wastes (e.g. water soluble liquid hydrocarbons at concentrations above 70 %) may increase clay hydraulic conductivity while leakage and diffusion through geomembranes may increase transport of contamination into surrounding environment (particularly flowing ground water). Leachate collection and monitoring drain pipes are often included and compulsory for hazardous wastes Sharma and Lewis (1994). Rowe (2005) provided a comprehensive review of long-term performance of contaminant barrier systems. Rowe et al. (1997) provide detailed analyses of clayey barrier systems for waste disposal facilities.

Sect. 4.9 contains more information on types, extent, identification and remediation of contamination within natural ground, which is applicable to landfills.

5.8.2 Stability/Capacity

- **Slope stability** of the cover layers of landfills is calculated as described in Sect. 5.1.2.4 taking into account soil-membrane friction angles for planar slides.
- Analytical solution for advection-dispersion caused **contaminant concentration** for a single homogeneous layer (barrier) of infinite depth according to several authors summarised by Sharma and Lewis (1994)

$$C(z,t) = \frac{C_o}{2} \cdot \left[\begin{array}{l} erfc\left(\dfrac{R \cdot z - v_s \cdot t}{2 \cdot \sqrt{D \cdot R \cdot t}} \right) + \exp\left(\dfrac{v_s \cdot z}{D} \right) \cdot \\ erfc\left(\dfrac{R \cdot z + v_s \cdot t}{2 \cdot \sqrt{D \cdot R \cdot t}} \right) \end{array} \right] \tag{5.104}$$

C is contaminant concentration at depth z in time t
C_o is a constant surface concentration
D is the effective diffusion coefficient obtained by steady-state, time-lag or transient method described by Sharma and Lewis (1994)
$R = 1 + \rho_d / n\ K_d$
ρ_d is dry density
n is porosity of the transport medium
K_d is distribution coefficient provided by Acar and Haider (1990) for different contaminants
$v_s = k/n$
k is the coefficient of a liquid permeability in the vertical direction

5.8.3 Movement

Charles (1993) provided index and engineering properties for a number of non-engineered fills given in Table 5.11. Charles (1993) also quotes collapse strain in various non-engineered fills as:

- Mudstone / sandstone 2 %
- Clay / shale fragments 5 %
- Stiff clay 3–6 %
- Colliery spoil 7 %

Municipal soil waste fill with around 50 % of organic matter can exhibit very large settlements because of aerobic and anaerobic decay of the organic matter. Sharma and Anirban De (2007) present settlement mechanisms and the methods for estimating settlements of municipal solid waste landfills, including bioreactor landfills.

Table 5.11 Typical non-engineered fill properties (From Charles 1993)

Fill	Unit density (Mg/m^3)	Water content (%)	Specific density (Mg/m^3)	Porosity (%)	Immediate settlement modulus (MPa)	Creep settlement (mm/m thickness) multiplier of $\log_{10}(t/t_i)$
Clay	1.53	20	2.65	42	31	0.7
Sand	1.55	6	2.65	42	–	–
Sandstone	1.60	7	2.65	40	–	–
Colliery spoil	1.56	11	2.55	39	3	0.4
Lagoon pfa	1.17	40	2.20	47	–	–
Building rubble	–	–	–	–	9	0.31
Old urban	–	–	–	–	4	0.75

5.8.4 Execution

- Geomembrane tests used are: material density (ASTM D792 or D1505), thickness (ASTM D751, D1593, D3767, D5199), tensile strength and stiffness (ASTM D412, D638, D882, D4885), tear and puncture resistance (ASTM D1004, D5884), environmental stress crack (ASTM D1693), ultraviolet light resistance (ASTM G154), carbon black content and dispersion (ASTM D1603), chemical resistance (USEPA test method 9090A). Construction tests used are: peel and shear tests (ASTM D4437) of seams formed by fillet or flat extrusion, hot air or wedge ultrasonic (for polyethylene), electric welding, solvent (for polyvinyl chloride) Sharma and Lewis (1994).

  Sharma and Lewis (1994) provide following recommendations for compacted clay liners:
- Minimum 0.6 m thickness is necessary so that any local imperfection in liner construction does not affect layer performance
- The layers are constructed in lifts about 20 cm thick before compaction and 15 cm after compaction. To provide effective bonding between two consecutive lifts, the surface of a previously compacted lift must be rough so that new and old lift blend together.
- Sheepsfoot rollers with fully penetrating feet about 23 cm long are suitable for the lift thickness of about 20 cm
- The lifts are typically placed in horizontal layers. On slopes, the minimum width of each lift equals to the width of compaction roller (about 3.6 m). The lifts can be placed parallel to the slopes not inclined more than 2.5 horizontal to 1 vertical.

Steps in the construction process of compacted clay layers are:

1. Locate clay borrow areas and investigate them using boreholes, tests pits and laboratory tests (grain size distribution, natural moisture content, liquid and plastic limits, chemical tests listed in Sect. 2.2.9, optimum moisture content for maximum compaction (ASTM D698, D1557, BS 1337-4:1990, CEN ISO/TS 17892-2:2004) with some results shown in Fig. 1.5, water or waste liquid permeability in compacted state (CEN ISO/TS 17892-11:2004) and field trial compaction tests varying thickness of lifts, moisture content, roller types, number of passes.
2. Excavate soil
3. Preliminary moisture adjustment by amendments, pulverization
4. Stockpile, hydrate, other additives such as bentonite
5. Prepare compaction surface
6. Transport from borrow areas
7. Spread lifts with breaking of clods
8. Final moisture adjustment, mixing, hydration
9. Compaction above optimum moisture content to achieve dispersed soil structure of lower permeability instead of flocculated soil structure of higher permeability with final smoothing of surface of compacted layer
10. Construction quality assurance testing (moisture content, density)
11. Protection from drying and cracking when necessary by placing temporary or permanent cover

5.9 Fill and Tailing Dams

5.9.1 Types

The following types are considered:

- Earth and rock fill
- Tailings deposits from ore extraction
- Fly ash transported hydraulically to stock piles

Typical fill dam consists of:

1. low permeability zone (central or inclined upstream core if clay is available locally or asphalt, steel sheet piles for small dams or concrete deck for rock fill only),
2. upstream and downstream sand and gravel filters around core about 1.5 m wide each for construction convenience to protection the core from erosion and improve filtration downstream of leaked water,
3. compacted fill shoulders for stability,
4. cut-off trench trough alluvium in the river bed to decrease water flow

5. grout curtain through rock formation below and besides dam in the abutments or sometimes diaphragm walls if permeation grouting is not efficient.

Typical tailings and water saturated fly ash stock pile dams can be constructed from selected deposited material wholly or in part using so called the upstream, central or downstream construction method and their combinations (e.g. Vick 1983).

5.9.2 Stability/Capacity

Foster et al. (2000a, b) compiled a data base of fill dam failures and accidents and showed that:

- piping through embankment contributed to 31 % of all failures, piping through foundations 15 % and from embankment to foundation 2 % or in total 48 %,
- overtopping 46 %,
- slope instability 4 %
- earthquakes 2 %

The water caused failures are also very problematical due to their speed of occurrence which takes place within a few hours leaving little time for intervention. Not only piping but also increased water pressure in dam abutment with filling of water reservoir in combination with the presence of unfavourably inclined tectonic fault can lead to dam failure as it has been the case for Malpasset concrete arch dam in 1959 (FMSD 2009).

- More information on types, extent, identification and remediation of erosion and piping caused by water percolation are given in Sect. 4.2.1.
- More information on stability and movement of ground slopes is provided in Sects. 5.1.2 and 5.1.3 respectively. Usually acceptable minimal factors of safety of slope stability are (Wilson and Marsal 1979)

    1. End of construction: 1.3 (1.4 for dams higher than 15 m on relatively weak foundation)
    2. Sudden drawdown from maximum pool level: 1.0 (1.5 when drawdown rate and pore water pressure developed from flow nets are used)
    3. Sudden drawdown from spillway level: 1.2 (1.5 when drawdown rate and pore water pressure developed from flow nets are used)
    4. Partial pool with steady seepage: 1.5
    5. Earthquake: 1.0

- More information on water reservoir wave surcharge and dam freeboard can be found for example in FARS (1996)
- Liquefaction of tailings during earthquakes was frequent cause of failure of the retaining dams (ICOLD 1995). When using simplified seismic stability analysis it is possible to consider either soil shear strength in cyclic condition with implicitly accounted effect of excess water pressure build-up (e.g. equivalent

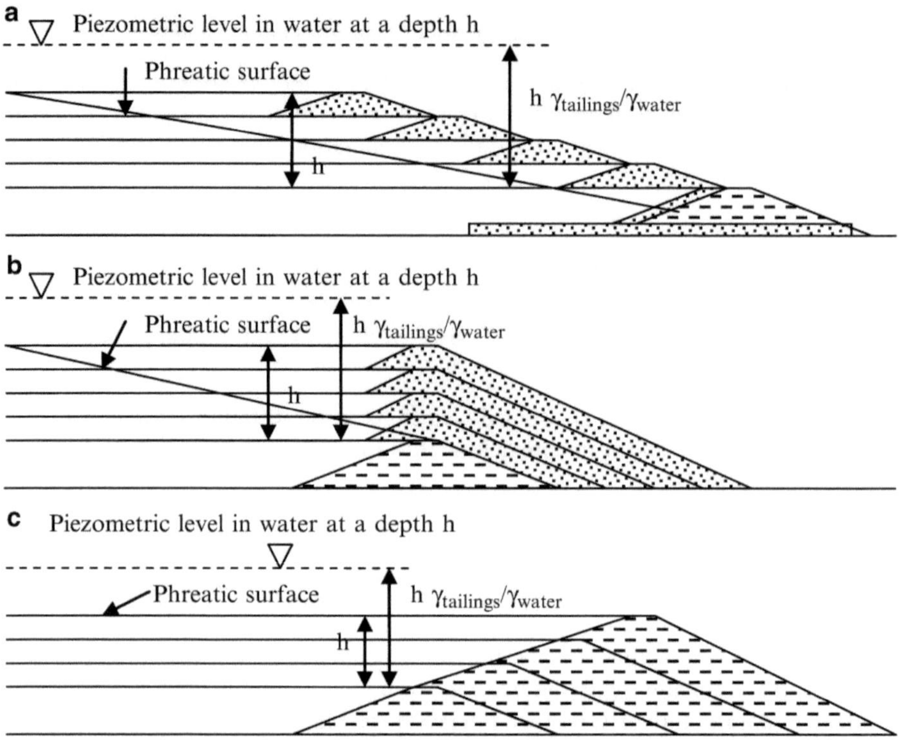

Fig. 5.55 Water piezometric level in liquefied tailings at a depth h (**a**) the upstream method of construction for a small content of hydrocycloned sand size particle from tailings, (**b**) the centreline method of construction for a large content of hydrocycloned sand size particles from tailings, (**c**) the downstream method of construction for no hydrocycloned sand size particles from tailings

friction angles in Figs. 2.2, 2.3, and 2.4) or soil shear strength in static condition with excess water pressure piezometric levels sketched in Fig. 5.55.

- The maximum horizontal acceleration at the crest of a fill dam (and a natural ridge) can be determined from acceleration response spectrum for the fundamental period of the first mode of vibration. For very long embankments (ridges), according to Gazetas and Dakoulas (1992)

$$T_1 = 2.61 \cdot {}^{H}/_{V_t} \tag{5.105}$$

T_1 is the period of the first mode of vibration
H is the maximum dam (ridge) height
V_t is averaged transversal wave velocity propagation through a dam (ridge) body

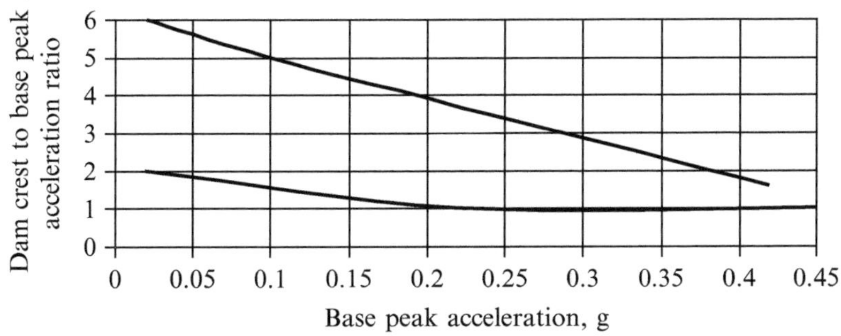

Fig. 5.56 Observed or estimated range of dam crest to base peak horizontal acceleration ratio

For dam crest length *L* to dam height *H* ratio <2.5 and rectangular shaped canyons, the fundamental period is approximately equal to $0.5(L/H)^{0.75}$ T_1. For dam crest length *L* to dam height *H* ratio <5.5 and triangularly shaped canyons, the fundamental period is approximately equal to $0.35(L/H)^{0.6}$ T_1.

Baldovin and Paoliani (1994) compiled data from 25 case histories concerning earth and rock fill dams, with heights varying from 8 m to over 200 m, affected by earthquakes, with the magnitudes range from 4.9 to 8.5. In many cases, the motions at the base and at the crest of the dams were recorded or sometimes estimated and computed. They observed that the crests to bases acceleration ratios were not significantly influenced by the geometries, heights and even upstream slope inclinations but mostly by the peak base acceleration. The upper and lower boundaries of the crest to base peak acceleration ratios based on their data are shown in Fig. 5.56.

5.9.3 Movement

- Bureau et al. (1985) provided data on observed and estimated settlement of rock fill dams during earthquakes, Fig. 5.57 and Table 5.12. In most cases, the horizontal displacements were similar to the settlements.
- Seed et al. (1978) reviewed performance of earth dams during earthquakes and concluded that while a number of hydraulically filled dams failed others performed well when they were built with reasonable slopes on good foundations for moderately strong shaking with peak acceleration up to about 0.2 g. Any well-built dam can withstand moderate earthquake shaking with peak acceleration of about 0.2 g and more, without detrimental effects. Dams constructed of clay on clay or rock foundation have withstood extremely strong shaking up to 0.8 g from magnitude 8.25 earthquake without apparent damage.

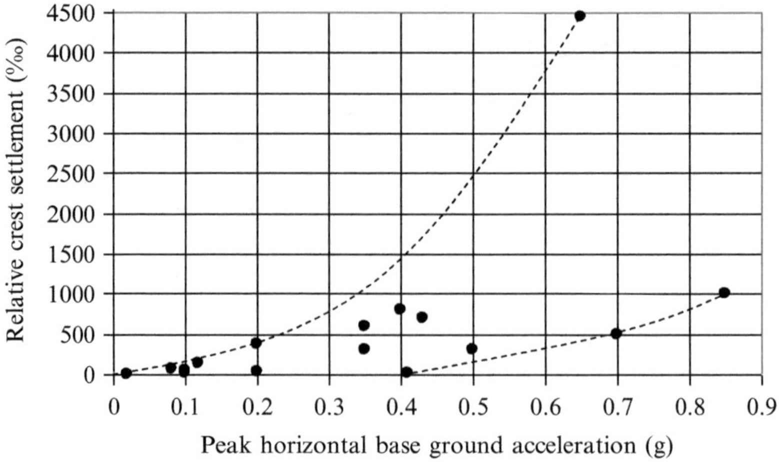

Fig. 5.57 Relative dam crest settlement (settlement/dam height) versus peak horizontal base acceleration (Data from Bureau et al. 1985)

Table 5.12 Basic data on settlement of rock fill dams (From Bureau et al. 1985)

Maximum height (m)	Relative settlement (%o)	Peak base horizontal acceleration (g)	Earthquake magnitude
84	381	0.2	8.3
131	30	0.2	7
67	7	0.02	6.9
67	61	0.08	7.5
235	9	0.1	5.7
148	130	0.12	7.6
60	45	0.1	7.6
72	15	0.41	6.2
50	600	0.35	8.5
50	300	0.5	6.5
125	300	0.35	6.5
133.5	4,450	0.65	8.25
240	1,000	0.85	7
104	800	0.4	7.5
213	700	0.43	6.5
100	487	0.7	7.1

5.9.4 *Execution*

- A guide to the identification and repair of defects in embankment dams has been provided by Charles et al. (1996). The remedial works involved:
    1. Excavation of core and installation of sealing material (clay, diaphragm wall, slurry in open trench)
    2. Grouting (permeation and jet) in central core and cut-off

3. Repair of asphaltic concrete membrane at the upstream slope
4. Installation of pressure relief wells in foundation
5. Drainage improvement within slope
6. Addition of berm (and drainage) to slope
7. New outlet

- Marcuson et al. (1996) described methods for seismic rehabilitation of earth dams. The methods are:

1. provision of berms and buttresses,
2. excavation and replacement of problematic material,
3. in-situ densification (vibrotechniques, dynamic compaction, compaction grouting),
4. in-situ strengthening (soil nailing, stone columns and deep soil mixing), increase in freeboard,
5. drainage (strip drains, stone columns, gravel trenches)
6. combinations of 1–5

More information on geo-works is provided in Chap. 6

5.10 Road and Railway Subgrade

5.10.1 Types

Road and railway natural subgrades are considered.

5.10.2 Stability/Capacity

For unpaved unreinforced roads, the effect of subgrade properties (c_u i.e. *CBR*) is given by Eq. (5.84).

For railway track, the allowed compressive stress on the subgrade according to an empirical formula reproduced by Esveld (2001) is:

$$\sigma_z = \frac{0.006 \cdot E_{v2}}{1 + 0.7 \cdot \log_{10} n} \qquad (5.106)$$

σ_z is the permissible compressive stress on subgrade (N/mm^2 i.e. MPa)
E_{v2} is modulus of elasticity taken from the second load step in a plate loading tests (poor = 10 MPa, moderate = 50 MPa, good > 80 MPa)
n is number of load cycles (=2 million)

5.10.3 Movement

For actual thickness H of the base layer of a road or the ballast layer of a railway track, the equivalent thickness H_e is according to formula reproduced by Esveld (2001):

$$H_e = 0.9 \cdot H \cdot \sqrt[3]{\frac{E_{base\ or\ ballast}}{E_{subgrade}}} \tag{5.107}$$

H is the actual thickness of the base layer of a road or the ballast layer of a railway track
$E_{base\ or\ ballast}$ is the modulus of elasticity of the base layer of a road or the ballast layer of a railway track
$E_{subgrade}$ is the modulus of elasticity of subgrade

For homogeneous elastic half-space, the settlement s under a static vertical force is calculated according to Boussinesq (1885) and under a dynamic vertical force in the near field according to Wolf (1994):

$$s = \frac{(1-\nu) \cdot P}{2 \cdot \pi \cdot G \cdot r} \tag{5.108}$$

ν is the Poisson's ratio
P is the acting force magnitude
G is the shear modulus $= E/[2(1 + \nu)]$
r is the distance between the locations of acting force and the location where settlement s is calculated

5.10.4 Execution

Loose subgrade in coarse grained material is compacted to very dense condition (D_r in Sect. 3.1.4 >85 %) and soft subgrade in fine grained material is stabilized by lime or cement mixing. More information on geo-works is provided in Chap. 6.

5.11 Offshore Foundations

5.11.1 Types

The following types are considered:

- Shallow depth – mudmats under jackets for temporary condition during construction
- Medium depth – spudcans of jack-ups (mobile platforms) and suction installed caissons
- Deep – driven steel tubular piles

5.11.2 Stability/Capacity

5.11.2.1 Shallow Depth – Mudmats

Equations (5.26) and (5.29) are applied with the coefficients according to Hansen (1970) in Tables 5.1 and 5.2 except $d_c = 0.4 \arctan(D/B')$ and $c' = 0$.

5.11.2.2 Medium Depth

- For spud-cans of jack-up mobile platform, SNAME (2008) recommendations are used
- For suction installed caissons, lateral and pull-out capacity can be considered as for piles according to API RP 2GEO (2011) although for the diameters of up to about 3 m.

5.11.2.3 Deep – Piles

The capacity is calculated according to API RP 2GEO (2011), which expressions are similar to Eqs. (5.33), (5.34), (5.35), (5.36), (5.37), and (5.38) with some differences:

- In Eq. (5.33) $K_s \tan\delta_\phi$ is replaced by shaft friction factor β, which values are provided in Table 1 of API RP 2GEO (2011)
- In Eq. (5.34) N_q values are provided in Table 1 of API RP 2GEO (2011)
- In Eq. (5.36) α_p is defined in API RP 2GEO (2011) as

$$\begin{aligned} \alpha &= \frac{0.5}{\sqrt{\psi}} \quad for \quad \psi \leq 1.0 \\ \alpha &= \frac{0.5}{\sqrt[4]{\psi}} \quad for \quad \psi > 1.0 \\ \psi &= \frac{c_u}{\sigma_v{}'} \end{aligned} \tag{5.109}$$

c_u is undrained shear strength
$\sigma_v{}'$ is effective vertical stress at a depth

API RP 2GEO (2011) Table 1 is applicable to silicious soil (less than 20 % of carbonate content). For calcareous sand having weak grains and bonds with more than 80 % carbonate content, Kolk (2000) provides design recommendations for the limited unit shaft friction of non-grouted piles of 15 kPa and ultimate unit end bearing of 3 MPa. For a carbonate content in the range from 20 % to 80 %, engineering judgment needs to be applied, e.g.

$$Q_{rec} = Q_{si} - \frac{Q_{si} - Q_{ca}}{\log_{10} 4} \cdot \log_{10} \frac{CaCO_3}{20} \tag{5.110}$$

Q_{rec} is recommended (friction or end bearing) capacity
Q_{si} is recommended (friction or end bearing) capacity for silica sand
Q_{ca} is recommended (friction or end bearing) capacity for carbonate sand with carbonate content $CaCO_3 > 80\ \%$
$CaCO_3$ is carbonate content (%)

5.11.3 Movement

5.11.3.1 Shallow Depth – Mudmats

- **In short-term condition (undrained)**, API RP 2GEO (2011) provides the following expressions for isotropic and homogeneous fine grained soil, circular and rigid base on the soil surface

$$\begin{aligned} u_v &= \left(\frac{1-\nu}{4\cdot G\cdot R}\right)\cdot V \\ u_h &= \left(\frac{7-8\cdot\nu}{32\cdot(1-\nu)\cdot G\cdot R}\right)\cdot H \\ \theta_r &= \left(\frac{3\cdot(1-\nu)}{8\cdot G\cdot R^3}\right)\cdot M \\ \theta_t &= \left(\frac{3}{16\cdot G\cdot R^3}\right)\cdot T \end{aligned} \tag{5.111}$$

u_v is the vertical displacement
u_h is the horizontal displacement
θ_r is the rotation around the horizontal axis
θt is the rotation around the vertical axis
V is the vertical load
H is the horizontal load
M is the moment around the horizontal axis
T is the moment around the vertical axis
G is shear modulus of soil
ν is Poisson's ratio
R is the (equivalent) radius of the base

- **In long-term condition (primary consolidation)**, Eq. (5.52) is used
- **In long-term condition (secondary consolidation)**, Eq. (5.54) is used

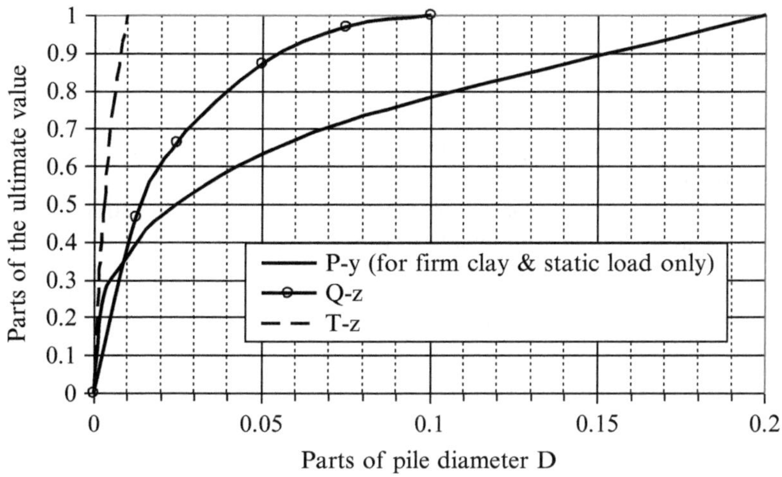

Fig. 5.58 Example plots based on API RP 2GEO (2011)

5.11.3.2 Medium Depth

SNAME (2008) provides recommendations for equivalent spring stiffness.

5.11.3.3 Deep – Piles

API RP 2GEO (2011) provides expressions for equivalent soil springs: P–y in the horizontal direction, T–z along pile shaft and Q–z at a pile tip. Figure 5.58 shows typical shapes of the equivalent soil springs for factors of safety greater than 1. For firm clay, the plot is made for $\varepsilon_{50} = 0.01$, while actual values of ε_{50} should be obtained as sketched in Fig. 2.15a). Reese (1997) provides P–y data for weak rock. Wesselink et al. (1988) provide P–y data for calcareous sand.

For mono piles used for wind turbines offshore, besides the small inclination at the pile top of less than 0.25° in operation, the period of oscillation is of importance. An equivalent period of structural vibration T_e can be calculated as (e.g. Wolf 1994):

$$T_e = \sqrt{T_s^2 + T_h^2 + T_r^2} \tag{5.112}$$

$T_s = 2\pi\ (m_s / k_s)^{1/2}$,
m_s is the mass of structure (pylon and wind turbine)
k_s is structural stiffness in horizontal direction (=ratio between applied a horizontal force at the top of a structure and achieved horizontal structural displacement for rigidly fixed structure at the end)
$T_h = 2\pi\ (m_s/K_h)^{1/2}$,

m_s is the mass of structure (pylon and wind turbine)
K_h is the horizontal stiffness coefficient from Table 5.6 (when the structure is rigid, $k_s = \infty$, and the foundation unable to rotate, $K_r = \infty$)
$T_r = 2\pi\ (m_s\ h_s^2/K_r)^{1/2}$,
m_s is the mass of structure (pylon and wind turbine)
h_s is the structural (wind turbine) height above ground level
K_r is the rotational stiffness coefficient from Table 5.6 (when the structure is rigid, $k_s = \infty$, and the foundation unable to translate, $K_h = \infty$)

Only static stiffness is considered without taking into account damping coefficients from Table 5.7 because of long vibration periods considered, about 3.5–5 s.

5.11.4 Execution

- Mudmats are fixed to jackets and lowered on the sea bed by cranes operating from barges
- Spud cans are fixed to legs of jack-ups, which use the weight of water within their hauls for pushing in spudcans to necessary depth
- Steel tubular piles are driven using different impact (or vibro) hammers. Pile drivability is checked using method described for hard clay, very dense sand and rock by Stevens et al. (1982). The main steps are:

  1. Calculation of axial pile capacity according to API RP 2A 1984 and use of multiplication coefficients suggested by Stevens et al. (1982) to obtain soil resistivity to driving (SRD).
  2. Use of GRLWEAP software (or equivalent) to obtain correlation between SRD and blow counts per unit depth (say 0.25 m).
  3. Combination of SRD versus depth and SRD versus blow count per unit depth for plotting of blow counts versus depth (upper and lower bound values for plugged and unplugged pile condition).
  4. Use of clause 12.5.6 of API RP2A-WSD (2007) for checking if pile driving refusal will occur. If the answer is yes then driving hammer properties or pile properties need to be changed. If the answer is no and the number of blow counts per unit depth is not much greater then 250 blows/0.25 m then report the maximum compressive and tensile stresses during driving calculated by GRLWEAP software (or equivalent). According to clause 6.10.5 of API RP2A-WSD (2007), the dynamic stresses should not exceed 80–90 % of yield stress.
  5. Use clause 5.2.4 of API RP2A-WSD (2007) for analysis of cumulative fatigue damage caused by pile driving, when required, and report the results (pile driving damage ~1 to 6 %). More details on pile driving are provided by Dean (2009).

5.12 Summary

In addition to the following issues that need to be considered for geo-structures:

- Stability/capacity
- Movement
- Execution (constructability)

The following issues important for sustainability need to be considered for geo-structures:

- Economy
- Social acceptance
- Environmental protection

References

Abrahamson N (2003) Model for strain from transient ground motion. In: Proceedings workshop on the effects of earthquake induced transient ground surface deformation on at-grade improvements. CUREE publication No. EDA-04, Oakland, California, USA

Acar YB, Haider L (1990) Transport of low-concentrated contaminants in saturated earthen barriers. J Geotech Eng ASCE 116(7):1031–1052

Ambraseys N, Menu J (1988) Earthquake induced ground displacements. J Earthq Eng Struct Dyn 16:985–1006

Ambraseys NN, Srbulov M (1995) Earthquake induced displacements of slopes. Soil Dyn Earthq Eng 14:59–71

API RP 2GEO (2011) Geotechnical and foundation design considerations. American Petroleum Institute, Washington, DC, USA

API RP2A-WSD (2007) Recommended practice for planning, designing and constructing fixed offshore platforms – working stress design – 21st edition. American Petroleum Institute

ASTM D1004 – 09 Standard test method for tear resistance (graves tear) of plastic film and sheeting. American Society for Testing and Materials, Philadelphia, USA

ASTM D1143/D1143M – 07e1 Standard test methods for deep foundations under static axial compressive load. American Society for Testing and Materials, Philadelphia, USA

ASTM D1505 – 10 Standard test method for density of plastics by the density-gradient technique. American Society for Testing and Materials, Philadelphia, USA

ASTM D1557 – 12 Standard test methods for laboratory compaction characteristics of soil using modified effort (56,000 ft-lbf/ft3 (2,700 kN-m/m3)). American Society for Testing and Materials, Philadelphia, USA

ASTM D1593 – 09 Standard specification for nonrigid vinyl chloride plastic film and sheeting. American Society for Testing and Materials, Philadelphia, USA

ASTM D1603 – 12 Standard test method for carbon black content in olefin plastics. . American Society for Testing and Materials, Philadelphia, USA

ASTM D1693 – 12 Standard test method for environmental stress-cracking of ethylene plastics. American Society for Testing and Materials, Philadelphia, USA

ASTM D3689 – 07 Standard test methods for deep foundations under static axial tensile load. American Society for Testing and Materials, Philadelphia, USA

ASTM D3767 – 03(2008) Standard practice for rubber—measurement of dimensions. American Society for Testing and Materials, Philadelphia, USA

ASTM D3966 – 07 Standard test methods for deep foundations under lateral load. American Society for Testing and Materials, Philadelphia, USA

ASTM D412 – 06ae2 Standard test methods for vulcanized rubber and thermoplastic elastomers—tension. American Society for Testing and Materials, Philadelphia, USA

ASTM D4435 – 08 Standard test method for rock bolt anchor pull test. American Society for Testing and Materials, Philadelphia, USA

ASTM D4436 – 08 Standard test method for rock bolt long-term load retention test. American Society for Testing and Materials, Philadelphia, USA

ASTM D4437 – 08 Standard practice for non-destructive testing (ndt) for determining the integrity of seams used in joining flexible polymeric sheet geomembranes. American Society for Testing and Materials, Philadelphia, USA

ASTM D4885–01 (2011) Standard test method for determining, performance strength of geomembranes by the wide strip tensile method. American Society for Testing and Materials, Philadelphia, USA

ASTM D4945 – 12 Standard test method for high-strain dynamic testing of deep foundations. American Society for Testing and Materials, Philadelphia, USA

ASTM D5199 – 12 Standard test method for measuring the nominal thickness of geosynthetics. American Society for Testing and Materials, Philadelphia, USA

ASTM D5882 – 07 Standard test method for low strain impact integrity testing of deep foundations. American Society for Testing and Materials, Philadelphia, USA

ASTM D5884-64a (2010) Standard test method for determining tearing strength of internally reinforced geomembranes. American Society for Testing and Materials, Philadelphia, USA

ASTM D638 – 10 Standard test method for tensile properties of plastics. American Society for Testing and Materials, Philadelphia, USA

ASTM D6760 – 08 Standard test method for integrity testing of concrete deep foundations by ultrasonic crosshole testing. American Society for Testing and Materials, Philadelphia, USA

ASTM D698 – 12 Standard test methods for laboratory compaction characteristics of soil using standard effort (12 400 ft-lbf/ft3 (600 kN-m/m3)). American Society for Testing and Materials, Philadelphia, USA

ASTM D7383 – 10 Standard test methods for axial compressive force pulse (rapid) testing of deep foundations. American Society for Testing and Materials, Philadelphia, USA

ASTM D751 – 06(2011) Standard test methods for coated fabrics. American Society for Testing and Materials, Philadelphia, USA

ASTM D792 – 08 Standard test methods for density and specific gravity (relative density) of plastics by displacement. American Society for Testing and Materials, Philadelphia, USA

ASTM D882 – 12 Standard test method for tensile properties of thin plastic sheeting. American Society for Testing and Materials, Philadelphia, USA

ASTM G154 – 12a Standard practice for operating fluorescent ultraviolet (uv) lamp apparatus for exposure of non-metallic materials. American Society for Testing and Materials, Philadelphia, USA

Baldovin E, Paoliani P (1994) Dynamic analysis of embarknest dams. In: Bull IW (ed) Soil-structure interaction numerical analysis and modeling. E & FN Spoon, London/New York, pp 121–125

Barton N (1976) Recent experiences with the Q system of tunnel support design. In: Proceedings of the symposium on exploration for rock engineering, Johannesburg. A A Balkema 1:107–117

Barton N, Lien R, Lunde J (1974) Engineering classification of rock masses for the design of tunnel support. Rock Mech 6(4):189–236

Berezantsev VG, Khristoforov VS, Golubkov VN (1961) Load bearing capacity and deformation of piled foundations. In: Proceedings of the 5th international conference SMFE, Paris 2:11–15

Bieniawski ZT (1976) Rock mass classification in rock engineering. In: Proceedings of the symposium on exploration for rock engineering, Johannesburg. AA Balkema 1:97–106

Bishop AW (1955) The use of slip circle for stability analysis. Geotechnique 5(1):7–17

Boussinesq J (1885) Application des potentials a l'etude de l'equilibre et du mouvements des solides elastique. Gauthier-Villard, Paris

Broms BB (1964a) Lateral resistance of piles in cohesive soil. J Soil Mech Found Div ASCE 90(2):27–63

Broms BB (1964b) Lateral resistance of piles in cohesionless soil. J Soil Mech Found Div ASCE 90(3):123–156

BS 1377-4:1990 Methods of test for soils for civil engineering purposes. Compaction-related tests. The British Standards Institution, UK

BS 6031 (2009) Code of practice for earthworks. The British Standards Institution, UK

BS 8002 (1994) Code of practice for earth retaining structures (no longer current). The British Standards Institution, UK

BS 8006-1 (2010) Code of practice for strengthened/reinforced soils and other fills. The British Standards Institution, UK

Bureau G, Volpe RL, Roth WH, Udaka T (1985) Seismic analyses of concrete face rockfill dams. In: Cooke JB, Sherard JLK (eds) Concrete face rockfill dams – design, construction and performance. Proceedings of Symposium of ASCE convention in Detroit, Michigan, pp 479–508

Burland JB, Burbridge MC (1985) Settlements of foundations on sand and gravel. Proc. Institute of Civil Engineering, UK 1:78

Burland JB, Potts DM, Walsh NM (1981) The overall stability of free and propped embedded cantilever retaining walls. Ground Eng Inst Civil Eng UK 7:23–38

Caquot A, Kerisel J (1948) Tables for calculation of passive pressure, active pressure and bearing capacity of foundations. Gauthier-Vilars, Paris

CEN ISO/TS 17892-11:2004, Geotechnical investigation and testing – laboratory testing of soil – Part 11: Determination of permeability by constant and falling head test. European Committee for Standardization, Brussels (ISO/TS 17892-11:2004, International Organization for Standardization, Geneva, Switzerland)

CEN ISO/TS 17892-2:2004 Geotechnical investigation and testing – laboratory testing of soil – Part 2: Determination of density of fine grained soil. European Committee for Standardization, Brussels (ISO/TS 17892-2:2004, International Organization for Standardization, Geneva, Switzerland)

Cerrutti V (1882) Richerce intorno all'equilibrio dei corpi elastici isotropi. Reale Academia dei Lincei, Roma 13

Charles JA (1993) Building on fill: geotechnical aspects. Building research establishment report S

Charles JA, Tedd P, Hughes AK, Lovenbury HT (1996) Investigating embankment dams. A guide to the identification and repair of defects. Building Research Establishment UK

Cloueterre (1991) French National Research Project – Recommendations Clouterre. English translation by Federal Highway Administration, FHWA-SA-93-026, Washington, USA

Clough GW, Smith EM, Sweeney BP (1989) Movement control of excavation support systems by iterative design procedure. ASCE Found Eng Curr Princ Pract 1:869–884

Coulomb CA (1776) Essai sur une application des regles des maximis et minimis a quelques problems de statique relatifs a l'architecture. Memoires de l'Academie Royale pres Divers Savants

Culmann C (1875) Die graphische statik. Meyer and Zeller, Zurich

Das BM (1985) Advanced soil mechanics. McGraw-Hill, Singapore

Davies TR, McSaveney MJ (2002) Dynamic simulation of the motion of fragmented rock avalanches. Can Geotech J 39:789–798

de Beer EE, Martens A (1957) Method of computation of an upper limit for the influence of heterogeneity of sand layers on the settlement of bridges. Proc 4t ICSMFE London 1:275–281

Dean ETR (2009) Offshore geotechnical engineering – principles and practice. Thomas Telford, London

EN 12063 (1999) Execution of special geotechnical work – sheet pile walls. European Committee for Standardisation, Brussels

EN 12699 (2001) Execution of special geotechnical work – displacement piles. European Committee for Standardisation, Brussels
EN 14199 (2005) Execution of special geotechnical works – micropiles. European Committee for Standardisation, Brussels
EN 14475 (2006) Execution of special geotechnical works – reinforced fill. European Committee for Standardization, Brussels
EN 14490 (2010) Execution of special geotechnical works – soil nailing. European Committee for Standardisation, Brussels
EN 1536 (2010) Execution of special geotechnical works – bored piles. European Committee for Standardisation, Brussels
EN 1537 (2000) Execution of special geotechnical work – ground anchors. European Committee for Standardisation, Brussels
EN 1538 (2010) Execution of special geotechnical works – diaphragm walls. European Committee for Standardisation, Brussels
EN 1993–5 (2007) Eurocode 3 – design of steel structures – piling. European Committee for Standardisation, Brussels
EN 1997–1 (2004) Eurocode 7: geotechnical design, part 1: general rules. European Committee for Standardisation, Brussels
EN 1997–2 (2007) Eurocode 7 – geotechnical design, part 2: ground investigation and testing. European Committee for Standardization, Brussels
EN 1998–5 (2004) Eurocode 8: design of structures for earthquake resistance – part 5: foundations, retaining structures and geotechnical aspects. European Committee for Standardization, Brussels
EN ISO/DIS 22477-1 (2005) Geotechnical investigation and testing – testing of geotechnical structures Part 1: Pile load test by static axially loaded compression. European Committee for Standardisation, Brussels. International Organization for Standardization, Geneva, Switzerland
Esveld C (2001) Modern railway track, 2nd edn. Delft University of Technology. Koninklijke van de Garde BV, Zaltbommel
FARS (1996) Wave surcharge and dam freeboard. In: Floods and reservoir safety, 3rd edn. The Institution of Civil Engineers UK. Thomas Telford
FMSD (2009) Burst of a dam 2 December 1959, Malpasset, France. French Ministry of Sustainable Development – DGPR/SRT/BARPI No 29490 1–7
Foster M, Fell R, Spannagle M (2000a) The statistics of embankment dam failures and accidents. Can Geotech J 37:1000–1024
Foster M, Fell R, Spannagle M (2000b) A method for assessing the relative likelihood of failure of embankment dams by piping. Can Geotech J 37:1025–1061
Gaba AR, Simpson B, Powrie W, Beadman DR (2003) Embedded retaining walls – guidance to economic design. Construction industry research and information Association report CIRIA C580
Gazetas G (1991) Foundation vibrations. In: Fang HY (ed) Foundation engineering handbook, 2nd edn. Chapman & Hall, New York/London, pp 553–593
Gazetas G, Dakoulas P (1992) Seismic analysis and design of rockfill dams: state-of-the-art. Soil Dyn Earthq Eng 11:27–61
Giroud JP, Noiray L (1981) Geotextile-reinforced unpaved road design. Proc ASCE J Geotech Eng 197(GT9):1233–1254
Goldberg DT, Joworski WE, Gordon MD (1976) Lateral support systems and underpinning. Report No. FHWA-RD-75-129, vol 2, Federal Highway Administration, U.S. Department of Transportation, Washington, DC
Hamada M, Yasuda S, Isoyama R, Emoto K (1986) Study on liquefaction induced permanent ground displacements. In: Report of the Association for the Development of Earthquake Prediction in Japan, Tokyo, Japan
Hammitt GM (1970) Bodenmechanik. Ferdinand Enke Verlag, Stuttgart
Hansen JB (1970) A revised and extended formula for bearing capacity. Bulletin No. 28, Geoteknisk Institut, Akademiet for de Tekniske Videnskaber, Copenhagen, Denmark

Healy PR, Weltman AJ (1980) Survey of problems associated with the installation of displacement piles. Construction Industry Research and Information Association UK Report PG8

Heezen BC, Ewing M (1952) Turbidity currents and submarine slumps, and the 1929 Grand Banks earthquake. Am J Sci 250:849–873

Hiley AE (1925) A rational pile driving formula and its application in practice explained. Engineering 119:657

Hoeg K (1968) Stresses against underground structural cylinders. J Soil Mech Found Div ASCE 94 (SM4):833–858

Hoek E, Bray JW (1981) Rock slope engineering, revised 3rd edn. The Institution of Mining and Metallurgy, London, UK

ICOLD (1995) Tailings dams and seismicity, review and recommendations. Committee on Mine and Industrial Tailings Dams. International Commission on large Dams Bulletin 98

Ingold TS (1979) The effects of compaction on retaining walls. Geotechnique 29(3):265–283

Janbu N (1953) Une analyse energetique du battage des pieux a l'aide de parametres sans dimension. Institut Technique du Bâtiment et des Travaux Publics. Annales 63–64

Janbu N, Bjerrum L, Kjaernsli B (1956) Veiledning ved losning av fundamenteringsoppgaver. Norwegian Gt. Inst. Publication No. 16

Jardine R, Chow FC, Overy RF, Standing JR (2005) ICP design methods for driven piles in sands and clays. Thomas Telford

Jenkins WM (1989) Theory of structures, Chapter 3. In: Blake LS (ed) Civil engineer's reference book, 4th edn. Butterworth, Oxford, pp 3–16

Jewell RA (1992) Strength and deformation in reinforced soil design. In: Hoedt D (ed) Proceedings of the 4th international conference on geotextiles, geomembranes and related products, Hague, Netherlands, pp 913–946

Jewell RA (1996) Soil reinforcement with geotextiles. Construction Industry Research and Information Association, CIRIA special publication 123, UK

JRASHB (2002) Specifications for highway bridges, part V seismic design. Japan Road Association, English version translated by PWRI, Japan 2003

Koerner RM (1998) Designing with geosynthetics, 4th edn. Prentice Hall, Englewood Cliffs

Kolk HJ (2000) Deep foundations in calcareous sediments. In: Al-Shafei (ed) Proceedings of the conference on eng for calcareous sediments, Balkema, Rotterdam 2:313–344

Kontoe S (2009) Case study: geotechnical structures. The Society for Earthquake and Civil Engineering Dynamics, UK, Newsletter 21(3):12–14

Kuenen PH (1952) Estimated size of the Grand Banks turbidity current. Am J Sci 250:874–884

Kulhaway F, Carter JP (1994) Settlement and bearing capacity of foundations on rock masses. In: Bell FG (ed) Engineering in rock masses. Butterworth Heinemann, pp 231–245

Liyanapathirana DS, Poulos HG (2005) Pseudostatic approach for seismic analysis of piles in liquefying soil. J Geotech Geoenviron Eng ASCE 131(12):1480–1487

Lord JA, Clayton CRI, Mortimore RN (2002) Engineering in chalk. Construction Industry, Research and Information Association, UK, report CIRIA C574

Makdisi F, Seed HB (1978) Simplified procedure for estimating dam and embankment earthquake induced deformations. J Geotech Eng 104(7):849–867

Maksimovic M (2008) Mehanika tla. AGM knjiga, Belgrade

Marcuson WF, Hadala PF, Ledbetter RH (1996) Seismic rehabilitation of earth dams. J Geotech Eng ASCE 122(1):7–20

Marston A (1930) The theory of external loads on closed conduits in the light of the latest experiments. Iowa Eng Exp Sta Bull 96

McDougall S, Hungr O (2005) Dynamic modelling of entrainment in rapid landslides. Can Geotech J 42:1437–1448

Meyerhof GG (1963) Some recent research on the bearing capacity of foundations. Can Geotech J 1(1):16–26

Meyerhof GG (1965) Shallow foundations. Proc J Soil Mech Found Eng ASCE 91 (SM2):21–31

Meyerhof GG, Baike CD (1963) Strength of steel culverts sheets bearing against compacted sand backfill. Highway Research Board proceedings 30, Highway Research Board, National Academy of Sciences, Washington, DC
Midlin RD (1936) Forces at a point in the interior of a semi-infinite solid. Physics 7:195–202
Mononobe N, Matsuo H (1929) On the determination of earth pressures during earthquakes. In: The World Engineering Congress 9:177–185
Moser AP (1990) Buried pipe design. McGraw-Hill, New York
Nakajima Y, Abeki N, Watanabe D (2000) Study on the stability of H/V spectral ratio of micro tremor in short period range for the estimation of dynamic characteristics of surface geology. In: Proceedings of the 12th World conference on earthquake engineering, Auckland, New Zealand. CD-ROM, Paper No. 2904
Newmark NM (1965) Effect of earthquakes on dams and embankments. Geotechnique 15:139–160
Nikolaou S, Mylonakis G, Gazetas G, Tazoh T (2001) Kinematic pile bending during earthquakes: analysis and field measurements. Geotechnique 51(5):425–440
Okabe S (1926) General theory of earth pressures. J Jpn Soc Civil Eng 12:1
O'Rourke TD, Bonneau AL (2007) Lifeline performance under extreme loading during earthquakes. In: Pitilakis KD (ed) Proceedings of 4th international conference on earthquake geotechnical engineering. Springer, Thessaloniki, pp 407–432
Paolucci R, Pitilakis K (2007) Seismic risk assessment of underground structures under transient ground deformations. In: Pitilakis KD (ed) Proceedings of 4th international conference on earthquake geotechnical engineering. Springer, Thessaloniki, pp 433–459
Peck RB (1969) Deep excavations in soft ground. Proceedings of the international conference on soil mechanism and foundations engineering. State-of-the Art volume, Mexico City, Mexico, pp 225–250
Penzien J (2000) Seismically induced racking of tunnel linings. Earthq Eng Struct Dyn 29:683–691
Poulos HG (2005) pile behaviour – consequences of geological and construction imperfections. J Geotech Geoenviron Eng ASCE 131(5):538–563
Poulos HG, Davis EH (1980) Pile foundation analysis and design. Wiley, New York
Reese LC (1997) Analysis of laterally loaded piles in weak rock. J Geotech Geoenviron Eng ASCE 123(11):1010–1017
Richards R, Huang C, Fishman KL (1999) Seismic earth pressure on retaining structures. J Geotech Geoenviron Eng ASCE 25:771–778
Rowe RK (2005) Long-term performance of contaminant barrier systems. Geotechnique 55 (9):631–678
Rowe RK, Armitage KK (1984) The design of piles socketed into weak rock. Report GEOT-11-84, University of Western Ontario, London, Canada
Rowe RK, Quigley RM, Booker JR (1997) Clayey barrier systems for waste disposal facilities. E & FN Spon
Schlick WJ (1932) Loads on pipe in wide ditches. Iowa Eng. Exp Sta Bull 108
Schlosser F, Unterreiner P (1991) Soil nailing in France: research and practice. In: Proceedings of the transportation research board, annual meeting, Washington, DC
Schmertmann JH (1970) Static cone to compute settlement over sand. J Soil Mech Found Div ASCE 96(SM3):1011–1043
Schmertmann JH, Hartman JP, Brown PR (1978) Improved strain influence factor diagrams. J Geotech Eng Div ASCE 104(GT8):1131–1135
Seed HB, Whitman RV (1970) Design of earth retaining structures for dynamic loads. In: Proceedings of ASCE specialty conference, lateral stresses in the ground and design of earth retaining structures. Cornell University, Ithaca, New York, pp 103–147
Seed HB, Makdisi FI, De Alba P (1978) Performance of earth dams during earthquakes. J Geotech Eng Div ASCE 104(7):967–994
Sharma HD, Anirban D (2007) Municipal solid waste landfill settlement: postclosure perspectives. J Geotech Geoenviron Eng ASCE 133(6):619–629
Sharma HD, Lewis SP (1994) Waste containment systems, waste stabilisation, and landfills: design and evaluation. Wiley, New York

Shirozu T, Yune S, Isoyama R, Iwamoto T (1996) Report on damage to water distribution pipes caused by the 1995 Hyogoken-Nanbu (Kobe) earthquake. In: Proceedings from the 6th Japan-US workshop on earthquake resistant design of lifeline facilities and countermeasures against soil liquefaction. Technical Report NCEER-96-0012, pp 93–110

Skempton AW (1964) Long-term stability of clay slopes. Geotechnique 14:75–102

Skempton AW, Hutchinson JN (1969) Stability of natural slopes and embankment foundations. State-of-the-Art Report. In: 7th international conference of the soil mechanisms and foundations engineering, Mexico, pp 291–335

SNAME (2008) Guidelines for site specific assessment of mobile jack-up units. The Society of Naval Architects and Marine Engineers, Technical and Research Bulletin 5–5A:68–69

Spangler MG (1947) Underground conduits – an appraisal of modern research. Proc ASCE 73:855–884

Spangler MG, Handy RL (1981) Soil engineering, 4th edn. Harper & Row, New York

Srbulov M (1987) Limit equilibrium method with local factors of safety for slope stability. Can Geotech J 24(4):652–656

Srbulov M (1991) Bearing capacity of a strip footing on brittle rock. Rock Mech Rock Eng 24(1):53–59

Srbulov M (1995) A simple method for the analysis of stability of slopes in brittle soil. Soils Found 35(4):123–127

Srbulov M (1997) On the influence of soil strength and brittleness and nonlinearity on slope stability. Comput Geotech 20(1):95–104

Srbulov M (1999) Force-displacement compatibility for reinforced embankments over soft clay. Geotextiles Geomembranes 17:147–156

Srbulov M (2001a) Design of soft clay reinforcement beneath unpaved roads and embankments. In: Proceedings of the 3rd international conference on soft soil engineering, Hong Kong, pp 493–498

Srbulov M (2001b) Analyses of stability of geogrid reinforced step slopes and retaining walls. Comput Geotech 28:255–269

Srbulov M (2003) On the fast damage propagation through marine soil offshore Newfoundland following Grand Banks earthquake of 1929. Eur Earthq Eng XVII(3):3–9

Srbulov M (2008) Geotechnical earthquake engineering – simplified analyses with case studies and examples. Springer. The examples are at: http://extras.springer.com/2008/978-1-4020-8684-7

Srbulov M (2009) An elementary model of soil and rock avalanches, debris run-out and fast spreads triggered by earthquakes. Ingegneria Sismica XXVI(4):5–22

Srbulov M (2010a) Simple considerations of liquefaction due to earthquakes and flow failures of slopes in sandy soil. Ingegneria Sismica XXVII(3):7–16

Srbulov M (2010b) Simple considerations of kinematic and inertial effects on piles during earthquakes. Ingegneria Sismica XXVII(2):7–16

Srbulov M (2010c) Ground vibration engineering – simplified analyses with case studies and examples. Springer. The examples are at: http://extras.springer.com/2010/978-90-481-9081-2

Srbulov M (2011) Practical soil dynamics – case studies in earthquake and geotechnical engineering. Springer, Dordrecht

Stacey TR, Page CH (1986) Practical handbook for underground rock mechanics. Trans Tech Publications, Clansthal-Zellerfeld

Stevens RS, Wiltsie WA, Turtom TH (1982) Evaluating pile drivability for hard clay, very dense sand and rock. In: Proceedings of the 14th annual OTC, Huston, Texas

Terzaghi K (1956) Varieties of submarine slope failures. In: Proceedings of the 8th Texas conference on soil mechanics and foundation engineering, pp 41–55

Terzaghi K, Peck RB (1948) Soil mechanics in engineering practice. Wiley, New York

Thornburn S, Thorburn JQ (1977) Review of problems associated with the construction of cast-in-place concrete piles. Construction Industry Research and Information Association UK Report PG2

Tomlinson MJ (2001) Foundation design and construction, 6th edn. Chapman & Hall, London

Towhata I, Islam S (1987) Prediction of lateral movement of anchored bulkheads induced by seismic liquefaction. Soils Found 27(4):138–147

USEPA test method 9090A. Compatibility test for wastes and membrane liners. United States Environmental Protection Agency

Vick S (1983) Planning, design, and analysis of tailings dams. Wiley, New York

Wesselink BD, Murff JD, Randolph MF, Nunez IL, Hayden AM (1988) Analysis of centrifuge model test data from laterally loaded piles in calcareous sand. Eng Calcareous Sediment Perth 1:261–270

Westergaard H (1931) Water pressure on dams during earthquake. Trans ASCE 1835:418–433

Whitaker T (1957) Experiments with model piles in groups. Geotechnique 7:147–167

Williams BP, Waite D (1993) The design and construction of sheet piled cofferdams. Construction Industry Research Information Association UK special publication 95

Wilson SD, Marsal RJ (1979) Current trends in design and construction of embankment dams. American Society of Civil Engineers

Wolf JP (1994) Foundation vibration analysis using simple physical models. PTR Prentice Hall

Wong RCK, Kaiser PK (1988) Design and performance evaluation of vertical shafts: rational shaft design method and verification of design. Can Geotech J 25:320–337

Youd TL, Hansen CM, Bartlett SF (2002) Revised multilinear regression equations for prediction of lateral spread displacement. J Geotech Geoenviron Eng ASCE 128(12):1007–1017

Young WC, Budynas RG (2002) Roark's formulas for stress and strain, 7th edn. McGraw-Hill, New York

Chapter 6
Geo-works

Abstract The following geo-works are considered:

- excavation and compaction
- drainage
- grouting
- mixing
- separation
- freezing
- blasting
- underpinning
- soil washing and waste solidification/stabilisation
- field measurements and observational method
- remote sensing
- asset management
- forensic investigation

6.1 Excavation and Compaction

6.1.1 Description

6.1.1.1 Excavation

Excavation is usually performed to enable access to a place (such as shafts/galleries for inspection/testing ground at depths, ore deposits in mines) or to construct an underground structure. The following points should be noted:

- Excavation should be avoided or minimised and the excavated material reused on site whenever possible to eliminate/minimise possibility of triggering terrain instability and environmental impacts (such as noise and vibration, dust generation,

M. Srbulov, *Practical Guide to Geo-Engineering: With Equations, Tables, Graphs and Check Lists*, Geotechnical, Geological and Earthquake Engineering 29,
DOI 10.1007/978-94-017-8638-6_6,

Table 6.1 Typical bulked-up unit volumes on excavation (Adopted from Horner 1981)

Material	Bulked-up unit volume (%)
Soil	
Sand and gravel	10–15
Clay	20–40
Loam	25–35
Peat and top soil	25–45
Rock	
Granite, Basalt/Dolerite, Gabbro	50–80
Gneiss, Schist and Slate, Shale	30–65
Quartzite, Sandstone, Chert and Flint	40–70
Limestone, Marble	45–75
Marl	25–40
Chalk	30–40
Coal	35

wildlife, local community) according to local legislations. In some countries like the U.K., landfill tax and aggregate levy exist to discourage movement of construction materials along public roads and bridges into landfill sites.

- Table 6.1 contains typical bulked-up unit volumes on excavation from Horner (1981).
- Geo-hazards (listed in Chap. 4) need to be considered together with access routes, stockpiling areas, adjacent land use and limitations to excavation, effects on protective species, bird nesting period, weather conditions
- Sufficient time needs to be allowed for archaeological finds, removal of unforeseen obstacles or remedial measures of previously unknown/undetected underground features such as natural cavities, tectonic fault zones, perched water levels, abandoned mine shafts etc.

6.1.1.2 Compaction

All fills should be compacted whenever possible to increase their stiffness and strength and minimise possibility of their erosion or instability.

6.1.2 Execution

6.1.2.1 Excavation

Excavations are performed by mechanic and hydraulic means (water jet cutting, air lifts under water) or by blasting, onshore and offshore (dredging) in soil and rock. The following points should be noted:

- During terrain clean-up operations, various invasive plant species need to be carefully managed (e.g. Booy et al. 2008)
- Top soil (humus layer) need always be stripped when exists and stock piled for reuse and re-vegetation on completion of construction works.
- Excavated surfaces will need protection from drying/wetting/freezing during their exposures by their covering with geo-membranes, sprayed concrete or vegetation.
- Control of ground/rainfall water needs to be arranged in order to prevent flooding/contamination/instability/accidents.
- Sequencing of excavation may need to be prescribed
- Contamination testing need to be arranged prior to transportation to landfills or reuse of excavated material if comprehensive testing of contamination has not been performed during site investigation stage. More on land contamination is given in Sect. 4.9.
- Table 6.2 contains a list of earthmoving plants with common and basic operational properties compiled by Horner (1981). Whenever possible, pedestrian and moving plant routes should be separated for safety reasons.
- Figure 6.1 indicates preferable treatment methods for rock excavation adopted from Franklin et al. (1971).
- More details for execution of excavated slopes are given in Sect. 5.1.4.
- Specification and method statement to be prepared by designer and contractor respectively

6.1.2.2 Compaction

Compactions are performed by mechanic means, preloading (using fill weight or vacuuming) and blasting onshore and offshore in soil and rock fill.

Table 6.3 contains a list of compacting plants with common and basic operational properties compiled by Horner (1981) and Kramer (1996). Specification and method statement to be prepared by designer and contractor respectively

6.1.3 Control

6.1.3.1 Excavation

Technical specification defines:

- construction method or
- end-product or
- performance

Table 6.2 Earthmoving plants and their basic common operational properties (Adopted from Horner 1981)

Type	Notes on use
Excavators	
Rippers	Fitted to dozers
Drill and blasters	Hand operated drills or track-mounted rotary-percussion rigs. Explosives are usually medium strength nitro-glycerine gelatines (Opencast Gelignite, ANFO) or slurry (Supergel) initiated by electric detonators or detonating fuse, short delay detonators or detonating relays
Impact hammers	Compressed air or diesel powered attached to the boom of crawler-mounted excavators
Hydraulic breakers	Compressed gases (Cardox) or hydraulic splitters are alternatives to explosives.
Graders	Commonly used to maintain haul roads
Skimmers	–
Excavators and loaders	
Draglines	For excavation of soft or loose materials at a level beneath or slightly above the level of its tracks because it operates by pulling of a bucket suspended on a cable towards the machine by a second cable
Face shovels	Extensively used in quarries, pits and construction sites up to a height of about 10 m
Forward loaders	Uses a wide bucket at and above wheel level and can be used to push or haul material over short distances
Grabbers	Used for the excavation of pits or trenches and loading to and from stockpiles by cable or hydraulically controlled bottom opening bucket suspended from the boom of a crane
Back-hoes	Bucket on a boom set on a tractor can excavate to 6 m depth
Bucket wheel excavators	Used for the excavation of linear features such as canals and trenches by a series of buckets set on a circular wheel or in a closed loop on a boom that can move laterally and vertically
Haulers and depositors	
Dumpers	From 1 to 77 tonnes except in large mines and quarries
Dump trucks	As above
Lorries	Up to 32 tonnes capacity
Conveyors	Loaded generally via a hopper commonly ends in a stacker. Generally used in quarries, areas with poor or problematic access or very steep terrain. Involve large cost to set up but the operational costs are generally low
Excavators, loaders, haulers and depositors	
Dozers	Crawler units ranging between 60 and 700 horse powers. Blades may be attached to other plant
Scrapers	Towed or self-propelled. Towed size range from 5.3 to 16.8 m^3 to a distance of 400 m. Single engine motorized range in size from 10.7 to 24.5 m^3 struck capacity (15.3 to 33.6 m^3 heaped). Double engine motorized scrapers are similar, economically operated on hauls of up to 2.6 km each way, optimum approximately 800 m. Elevating scrapers range in size from 7.2 to 26 m^3 heaped capacity self-propelled or towed units
Dredgers	Used for excavation below water level usually purpose made floating vessels such as cutter-suction, bucket wheel; grab and dipper (face shovel) dredgers. Excavated material can be pumped away or transported by barge or the dredger to deposition or off loading place

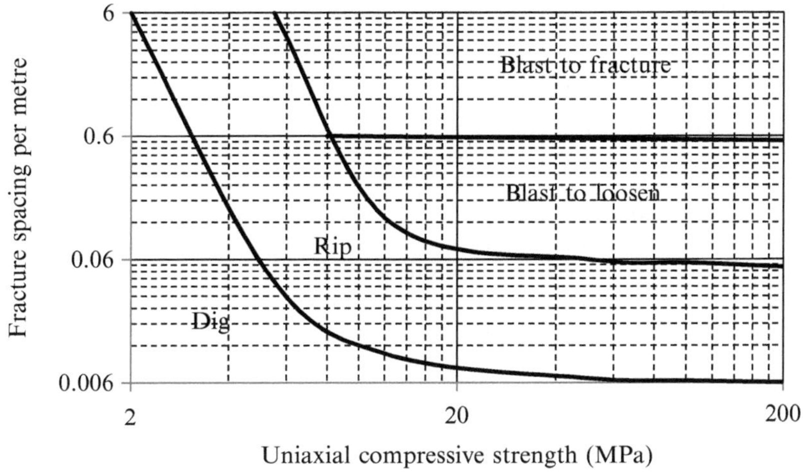

Fig. 6.1 Preferable excavation method for rock (Adopted from Franklin et al. 1971)

of earthworks together with bill of quantities, acceptable tolerances, test types and frequencies, monitoring, procedure for non-compliance cases with remedial measures, time scales, measurements, approval, payment, reporting, as built documentation, disposal of material, finishes, etc.

Basis for writing technical specifications are codes of standards such as BS 6031:2009.

6.1.3.2 Compaction

Technical specifications define:

- construction method or
- end-product or
- performance

of compacted fill together with bill of quantities, acceptable tolerances, test types and frequencies, monitoring, procedure for non-compliance cases with remedial measures, time scales, measurements, approval, payment, reporting, as built documentation, disposal of material, finishes, etc.

Basis for writing technical specifications are codes of standards such as BS 6031:2009; EN 14731:2005, books like by Monahan (1994), brochures from manufacturers of lightweight fill, polystyrene blocks etc. The following points should be noted:

- Table 6.4 contains a list of common compaction data compiled by Horner (1981) as a guide. Filed tests for varying layer thickness, number of passes, moisture content etc. need to be performed with available compaction equipment and materials before construction start.

Table 6.3 Compaction plants and their basic common operational properties (Adopted from Horner 1981)

Type	Notes on use
Rollers	
Smooth-wheeled	Mass range from 1.7 to 17 tonnes deadweight without ballast. Speed 2.5 to 5 km/h
Pneumatic-tyred	Towed or self-propelled, one or two axles. Speed 1.6 to 24 km/h
Grid	Have a cylindrical steel mesh roller and may be ballasted with concrete blocks. Speed between 5 and 24 km/h, mass between 5.5 tonnes net and 15 tonnes ballasted
Tamping	Sheepsfoot and pad rollers. Speed between 4 and 10 km/h
Construction traffic	Similar to pneumatic-tired roller. Can lead to over compaction, rutting and degradation of the fill
Vibrating compactors	
Rollers	Mass varies between 0.5 and 17 tonnes (static). Speed 1.5 to 2.5 km/h if not manual then 0.5 to 1 km/h. Frequency varies from 20 to 3 Hz for larger and 45 to 75 Hz for smaller units although frequency can be variable
Plates	Mass varies from 100 kg to 2 tonnes, plate area from 0.16 to 1.6 m^2. Usual speed 0.7 km/h
Vibrotampers	Mass from 50 to 100 kg
Vibrofloats	Torpedo like probes 0.3 to 0.5 m in diameter, 3 to 5 m long. Incrementally withdrawn in 0.5 to 1 m intervals at about 30 cm/min while vibrating. Water or air may be jetted in the upper parts to loosen soil. Alternatively, bottom feed by granular material to form a densified column. When gravel or crushed stone is used, stone columns are formed with additional benefits of reinforcement and drainage. Effective when fines content less than 20 % and clay content below 3 %. Grid pattern spacing of 2 to 3 m to depths of 35 m
Vibro rods	Uses vibratory pile driving hammer and a long probe. Several types exist, grid spacing is smaller than for vibrofloats because of vertical vibrations, effectiveness variable with depth
Blasting	Explosive charges 3 to 6 m apart in boreholes spaced at 5 to 15 m and backfilled prior to detonation. Tine delays used to increase efficiency. Two or three round of blasting are used in soil that contains less than 20 % silt and less than 5 % clay. In partially saturated soil not effective because of capillary tension and gas bubbles. Quite economical but its effectiveness difficult to predict in advance
Stone columns	Installed by vibroflotation, by Franki pile method etc. Stone columns installation increases soil density, columns add their strength, stiffness and permeability
Compaction piles	Displacement piles made of pre-stressed concrete or timber driven and left in place at a distance from 7 to 12 pile diameters to depths of about 18 m
Impact	
Power rammers	Mass about 100 kg uses internal combustion engine
Weight dropping rammers	Mass 180 kg or more dropped from variable heights of about 3 m
Dynamic consolidation	Mass 6 to 500 tonnes dropping from 10 to 40 m. Several passes used decreasing drop weight/height; each pass involves 3 to 8 weight drops. Effective to depth of 12 m

Table 6.4 Typical compacted layer thicknesses and number of passes (Adopted from Horner 1981)

Equipment	Cohesive soil maximum compacted layer thickness (mm)	Cohesive soil minimum number of passes	Coarse graded and dry cohesive soil maximum compacted layer thickness (mm)	Coarse graded and dry cohesive soil minimum number of passes
Smooth wheel roller	150	4	150	8
Grid roller	150	4	150	12
Tamping roller	225	4	150	12
Pneumatic-tired roller	450	4	175	6
Vibrating roller	275	4	275	4
Vibrating plate compactor	200	6	200	5
Vibro tamper	200	3	150	3
Power rammer	275	8	275	12
Dropping weight compactor	600	2	600	4

- Typical tests include particle size distribution, liquid and plastic (Atterberg) limits, moisture content and unit density for clayey soil (CBR, plate test, etc. occasionally), particle size distribution and relative density for coarse grained soil (other tests occasionally). More on soil testing can be found in Chap. 2 and on soil properties in Sect. 3.1.
- Before placement of compaction fill, the surface on which a fill is placed needs to be clean without debris and water, compacted and roughened to achieve a good bond between previous and subsequent layers.
- Clayey soil should be compacted to the moisture content above optimum to minimise its potential for collapsibility on wetting and to decrease water permeability by dispersed rather than flocculated grain structure. Coarse grained soil should be compacted to a relative density greater than 80 % (to achieve very dense condition). Examples of density-water content relationships for clayey soil are shown in Fig. 1.5.

6.2 Drainage

6.2.1 Description

Ground drainage is usually performed for supplies of ground water, decrease of ground water levels/pressures or flow gradients and collection of leachate from waste disposal sites. Intake from or discharge to ground water reservoirs is frequently subjected to permits by local environmental agencies. Drainage needs to fulfil two basic requirements: extraction of water and prevention of soil internal erosion by water flow towards drains. Several factors influence clogging of drains in time:

- Siltation (movement by water flow of fines towards drains)
- Cementation (formation of calcium, iron, magnesium and manganese combined with carbonate and sulphur)
- Grow of algae and bacteria colonies
- Freezing of drainage outlets exposed to cold air
- Ground movements (land sliding, large subsidence, heave, etc.)

Several types of drainage exist, such as:

- Vertical drains
- Water pumping wells
- Trenches (filled with well permeable soil, geotextiles or both)
- Near horizontal perforated pipes
- Near vertical water permeable zones and horizontal layers made of well permeable soil, geotextiles or both

6.2.1.1 Vertical Drains

Sand drains (with typical diameter range from 0.15 to 0.6 m, spacing 1 to 5 m and the maximum length of 35 m) and prefabricated drains (with typical diameter range from 0.05 to 0.15 m, spacing from 1.2 to 4 m and the maximum length of 60 m, e.g. Jamiolkowski et al. 1983) are mainly used for ground water discharge into lower more permeable and less pressurised soil layers, for speeding up of consolidation during ground preloading by fill or vacuuming and for decrease of liquefaction potential during earthquakes (e.g. EN 15237:2007(E)).

- The **discharge capacity of a drain** q_w (m^3/year) at 20 °C is calculated as according to EN 15237:2007(E) and tested for pre-fabricated drains according to EN ISO 12958. Typical discharge capacity for prefabricated drains are in the range from 500 to 100 m^3/year according to Holtz et al. (1991).

$$q_w = \frac{q_p \cdot b \cdot R_T}{i \cdot f_{cr}} = \frac{\theta \cdot b \cdot R_T}{f_{cr}} \tag{6.1}$$

q_p is in-plane flow capacity (m^2/year), which is the volumetric flow rate of water and/or liquids per unit width of the drain at defined gradients in the plain of the drain
b is drain width (m)
i is hydraulic gradient

$R_T = 1.763/(1 + 0.03771\,T + 0.00022\,T^2)$
T is temperature (°C)
θ is transmissivity (m^2/year), which is in-plane laminar water flow capacity of a drain at a hydraulic gradient equal to 1
f_{cr} is creep factor (between 10 for 2-days testing period and 1 for 30-days testing period)

- **Degree of consolidation** U_h due to drainage in the horizontal direction towards drains is according to EN 15237:2007(E)

$$U_h = 1 - \exp\left(\frac{8 \cdot c_h \cdot t}{\mu \cdot D^2}\right)$$
$$\mu = \frac{D^2}{D_2 - {d_w}^2} \cdot \left[\ln_e\left(\frac{D}{d_s}\right) + \frac{k_h}{k_s} \cdot \ln_e\left(\frac{d_s}{d_w}\right) - \frac{3}{4}\right] + \frac{k_h}{q_w} \cdot \pi \cdot z \cdot (2 \cdot l - z) \qquad (6.2)$$

c_h is the coefficient of consolidation in the horizontal direction from CPT or PPP dissipation test; $c_h \sim k_h/k_v\ c_v$, where the coefficients k_h, k_v, c_v are described in Sects. 3.1.8 and 3.1.9
t is time
$D = 2(A/\pi)^{1/2}$, A is the horizontal area between adjacent drains
d_s is the diameter of disturbed (smeared) zone by drain installation
d_w is the diameter of drain
k_h is the coefficient of water permeability in the disturbed (smeared) zone
k_h is the coefficient of water permeability of drain
z is depth below ground surface
l is half-length of a drain

Holtz et al. (1991) provide more information on design and performance of prefabricated vertical drains.

- **Equivalent shear modulus G_{eq}** of soil-vertical drain system. Stone columns may have appreciated influence because of their diameter and stone stiffness. From compatibility of shear deformation and the static equilibrium of vertical forces acting on soil and columns follows:

$$G_{eq} = G_s \cdot (1 - n) + n \cdot G_{sc} \qquad (6.3)$$

G_s is shear modulus of soil
n is the proportion of total volume occupied by stone column
G_{sc} is the shear modulus of stone column material

The shear stress distribution is the same as for the shear modulus distribution because of a direct proportion between shear stresses and shear modulus. For example, when cyclic shear stress during earthquake is calculated according to Eq. (4.11) and soil shear strength determined from the boundary line in Fig. 4.8 is not sufficient to resist it then necessary proportion of the total volume occupied by stone columns can be determined from the following equation:

$$\tau = \tau_s \cdot (1 - n) + n \cdot \tau_{sc} \tag{6.4}$$

τ is cyclic shear stress from Eq. (4.11), for example
$\tau_s = \sigma'_v \, tan\phi_a$ is the maximum shear stress that can be resisted by soil
σ'_v is the effective overburden stress at a depth
ϕ_a is apparent soil friction angle during earthquakes from Figs. 2.2, 2.3, and 2.4
n is the proportion of total volume occupied by stone column
$\tau_{sc} = \sigma'_v \, tan\phi_{sc}$ is the maximum shear stress that can be resisted by stone column
ϕ_{sc} is the friction angle of stone column (min. 35°)

6.2.1.2 Water Pumping Wells

- The flow rate Q (m^3/s) according to BS 5930:1999+A2:2010 for **confined permeable layers** between two less permeable (impermeable) layers is

$$Q = \frac{k \cdot 2 \cdot \pi \cdot h_o \cdot (s_1 - s_2)}{2.3 \cdot \log_{10} \frac{r_2}{r_1}} \tag{6.5}$$

k is the coefficient of water permeability (m/s)
h_o is the thickness (m) of permeable layer
$s_{1,2}$ are depressions (m) of ground water levels at the distances $r_{1,2}$ from well, $r_2 \geq r_1$

- For **unconfined conditions** (thick permeable layer over an impermeable layer) the flow rate Q (m^3/s) according to BS 5930:1999+A2:2010 is

$$Q = \frac{k \cdot \pi \cdot \left(h_2{}^2 - h_1{}^2\right)}{2.3 \cdot \log_{10} \frac{r_2}{r_1}} \tag{6.6}$$

k is the coefficient of water permeability (m/s)
$h_{1,2}$ are water piezometric heights (m) above an impermeable layer at the distances $r_{1,2}$ from well, $r_2 \geq r_1$

6.2.1.3 Trenches and Near Vertical Water Permeable Zones

The flow rate Q per unit length of a trench, which bottom rest in an impermeable layer, is according to Cedergren (1989)

$$Q = k \cdot \frac{2 \cdot h_{\max}{}^2}{L} \tag{6.7}$$

k is the coefficient of water permeability
h_{max} is the piezometric height above an impermeable bottom layer at the distance $L/2$
L is the centre to centre spacing between trenches

6.2.1.4 Near Horizontal Perforated Pipes

- For **steady state flow**, according to Schwab et al. (1993) citing Colding's solution in 1872.

$$Q = \frac{4 \cdot k \cdot \left(b^2 - d^2\right)}{S} \tag{6.8}$$

Q is flow rate into a drain from two sides per unit length of drain
k is the coefficient of water permeability
b is the piezometric height above a deeper impermeable ground layer at S/2
d is the height of drain above a deeper impermeable ground layer
S is the horizontal spacing between drain pipes

- For **unsteady flow**, Schwab et al. (1993) provide equation and graph.

More information on pipes is provided in Sect. 5.7.

6.2.1.5 Near Horizontal Water Permeable Layers (Blanket Drains)

- For a **horizontal upstream slope** on a permeable base like in the case of tailings dam, Nelson-Skornyakov (1949) defined the minimum drain length L_{dr} from downstream slope toe and the hydraulic exit gradient I_E as

$$\begin{aligned} L_{dr} &= \frac{h}{\pi} \cdot \sinh^{-1}\frac{2 \cdot h}{\pi \cdot L} + \frac{L}{2} \cdot \left[1 - \cosh\left(\sinh^{-1}\frac{2 \cdot h}{\pi \cdot L}\right)\right] \\ I_E &= \left(1 - \frac{\pi \cdot L}{2 \cdot h} \cdot \sinh\frac{\pi \cdot \psi}{k \cdot h}\right)^{-1} \end{aligned} \tag{6.9}$$

h is the hydraulic head decrease between the upstream slope and downstream slope toe layer
L is the minimal horizontal distance between the upper hydraulic head and the zero head within permeable layer
k is the coefficient of water permeability of the slope
$\psi = 0.1\ k\ h, 0.2\ k\ h, \ldots$

The quantity of water inflow into the drain layer per its unit length is proportional to the product of $k\ I_E$ and the length of water entry into water permeable layer (about three layer thickness)

- For **upstream slope approximately vertical** on a permeable base like in the case of a river dykes on alluvium, Nelson-Skornyakov (1949) defined the minimum drain length L_{dr} from downstream slope toe and the hydraulic exit gradient I_E as

$$L_{dr} = 2 \cdot \left[\frac{L}{2} + \frac{h}{\pi} \cdot \sinh^{-1}\frac{2 \cdot h}{\pi \cdot L} - \frac{L}{2} \cdot \cosh\left(\sinh^{-1}\frac{2 \cdot h}{\pi \cdot L}\right)\right]$$
$$I_E = \left(1 - \frac{\pi \cdot L}{2 \cdot h} \cdot \sinh\frac{\pi \cdot \psi}{2 \cdot k \cdot h}\right)^{-1} \tag{6.10}$$

h is the hydraulic head decrease between the upstream slope and downstream slope toe layer
L is the minimal horizontal distance between the upper hydraulic head and the zero head within permeable layer
k is the coefficient of water permeability of the slope
$\psi = 0.1\ k\ h, 0.2\ k\ h, \ldots$

The quantity of water inflow into the drain layer per its unit length is proportional to the product of $k\ I_E$ and the length of water entry into water permeable layer (about three layer thickness)

- The solutions for the cases of impermeable base layers are provided by Harr (1990).

6.2.2 *Execution*

Designer prepares specifications for the works. EN 15237:2007(E) lists information needed for the execution of the work such as:

- Site conditions:
  - Geometrical data (boundary conditions, topography, access, slopes, headroom restrictions, trees, fills, etc.)
  - Ground and ground water properties (soil description with types, classification, existence and extent of sand, silt and hard layers, penetration test resistances, presence of cobbles or boulders or cemented layers, ground water level and flow direction, ground water category for protection)

  - Climatic and environmental information (weather, currents, tidal movements, wave heights, water and soil contamination, hazardous gas, unexploded ordnance, restrictions concerning noise, vibration and pollution)
  - Existing underground structures, services, archaeological facts
  - Planned or on-going construction activities such as dewatering, tunnelling and deep excavations
  - Previous experience from drain installation adjacent to site
  - Conditions of structures, roads, services at and adjacent to drains

- Setting out locations and lengths
- Legal/statutory restrictions
- Method statement containing:

  - Equipment and installation method
  - Control procedure
  - Testing methods
  - Health and safety measures
  - Field trials if contracted
  - Reporting procedures for unforeseen circumstances/conditions which are different from those in the contract documents, reporting procedure if an observational method is adopted

- Physical and hydraulic properties of drains
- Specifications for drains and materials with the schedule for testing and acceptance procedure
- Description of quality management system with supervision and monitoring

Filter material used for drains should fulfil the criteria specified in Sect. 4.2.4. Granular material needs to be compacted to minimise its erodibility by flowing water.

Jamiolkowski et al. (1983) list the following common installation methods for

sand drains:

- Driven or vibratory closed-end mandrel
- Hollow steam continuous flight auger
- Jetting

prefabricated sand drains ('sandwicks'):

- Driven or vibratory closed end mandrel
- Flight auger
- Rotary wash boring

prefabricated band shaped drains:

- Driven
- Vibratory closed end mandrel

Holtz et al. (1991) list various installation methods of sand drains in order of decreasing efficiency, as follows:

- Closed end cross shaped mandrel (pushing in rather than vibrating)
- Jetting
- Augering
- Closed end circular mandrel
- Closed end star shaped mandrel
- Open end circular mandrel

For prefabricated drains, the most common installation procedure is by closed end mandrel. The amount of soil disturbance caused by installation is dependent on the size and shape of the mandrel, dimensions and shape of the detachable shoe or anchor at the mandrel tip. Because the volume of prefabricated drains is smaller than for sand drains, prefabricated drains are appropriate for sensitive cay and soil having high degree of anisotropy of the permeability. Other advantages of prefabricated over sand drains is the simplicity and high speed of the installation, which results in relatively low unit installation costs according to Holtz et al. (1991).

6.2.3 Control

EN 15237:2007(E) lists requirements for testing, of band drains, supervision and monitoring of vertical drains. The following properties of prefabricated drains are required to be specified and tested:

- Tensile strength according to EN ISO 10319
- Elongation at maximum tensile force according to EN ISO 10319
- Tensile strength of filter according to EN ISO 10319
- Tensile strength of seams and joints according to EN ISO 10321
- Velocity index of filter according to EN ISO 11058
- Characteristic opening size of filter (O_{90}) according to EN ISO 12956
- Discharge capacity of drain according to EN ISO 12958
- Durability in years (EN 13252:2000)

Suitably qualified and experienced personnel should be in charge of supervising, verification, control (methods and frequency) and acceptance of work according to procedures established before work commencement.

Identification of prefabricated drains on site shall be performed according to EN ISO 10320.

More information on control of compacted granular layer is given in Sect. 6.1.3.2.

6.3 Grouting

6.3.1 Description

The geotechnical grouting can be classified into:

- **Permeation/contact/bulk filling** (without displacement of the grouted medium) for decreasing of fluid permeability and some increase in strength and stiffness, provision of intimate contact between structure and ground and infill of cavities such as abandoned mine shafts
- **Compaction/hydraulic fracturing** (with displacement of the grouted medium) for increase in density by compaction and structural uplift after ground subsidence because of tunnelling or creating of fractures for increase of fluid flow ('fracking' in gas industry)
- **Jet** (with cutting of the grouted medium by high pressure jet of grout 'single', water 'double' or water within air sleeve 'triple') for formation of columns and zones of mixed soil and binder (cement) for ground improvement (increase in strength and stiffness and decrease of permeability).

Grouting is mainly experience based and involves several disciplines from soil/rock/fluid mechanics to mechanical equipment and electronic data acquisition systems. Several useful publications are by ASCE Press (1997), Widman (1996), Byle and Borden (1995), Karol (1990). Houlsby (1990), Baker (1982).

6.3.1.1 Permeation/Contact/Bulk Filling

EN 12715:2000 provides information concerning site investigation, materials and products and design among others. The following **materials** can be used for the grouting:

- Portland cement (for the coefficient of water permeability of grouted medium $>5 \times 10^{-3}$ m/s – coarse gravel), sulphate resistant cement (in sea water), micro fine (ultra-fine) cement (for the coefficient of water permeability of grouted medium $<5 \times 10-^{3}$ m/s – coarse gravel and $>5 \times 10^{-5}$ m/s –coarse silt)
- Clay composed of activated or modified bentonite can also be added to cement based grouts to reduce 'bleeding' and filtration under pressure, to vary viscosity and cohesion of the grout, or to improve pumpability of the grout. 'Bleeding' is referred to the sedimentation of particles from water suspension once grout flow stops. Bentonite grout can penetrate coarse sand.
- Sand, gravel, fillers as bulking agents or for varying grout consistency, its resistance to wash-out, or its mechanical strength and deformability
- Water for grout transport and activation of hydrophobic components such as cement

- Silicates and their reagents, lignin based materials, acrylic (in granular soil, finely fissured rock), phenolic (in fine sand and sandy gravel) or epoxy (in fissured rock) resins, polyurethanes (in large voids) or others (for the coefficient of water permeability of grouted medium $>5 \times 10^{-6}$ m/s – medium silt and $<10^{-4}$ m/s – coarse sand)
- Additives for modification of grout properties and to control grout parameters such as viscosity, setting time, stability, strength, resistance, cohesion and permeability after placement (e.g. super plasticizers, water retaining agents, air entrainers)
- Calcareous or siliceous fillers, pulverized fuel ash, pozzolans and fly ash from thermal power plants provided that they are chemically compatible with each other and satisfy immediate and long term environmental requirements

The following **consistencies** can be used for the grouting:

- Suspensions (particulate or colloidal) characterized by:
  - Grain size distribution with/without flocculation
  - Water/solid ratio
  - Rate of sedimentation and 'bleeding'
  - Water retention capacity under pressure filtration
  - Rheological properties (viscosity, creeping) and behaviour with time
- Solutions (true or colloidal) characterized by:
  - Stability with time if silicate based
  - Resistance to proliferation of bacteria if organic silica gels
  - Syneresis effect (for silicate based grouts). Syneresis is the extraction or expulsion of a liquid from a gel.
  - Temperature differences effect
  - Toxicity of individual resin grout components
  - Dilution of the grout mixture
- Mortars flowing under their own weight must be stable and their rheological behaviour (similar to suspensions) characterized by flow cones

Table 1 of EN 12715:2000 contains a list of parameters characterizing grout properties before and after setting. For solutions, suspensions and mortars, the following are common **parameters before setting**:

- setting time
- density
- pH
- viscosity

For solutions, suspensions and mortars, the following are common **parameters after setting**:

- hardening time
- final strength

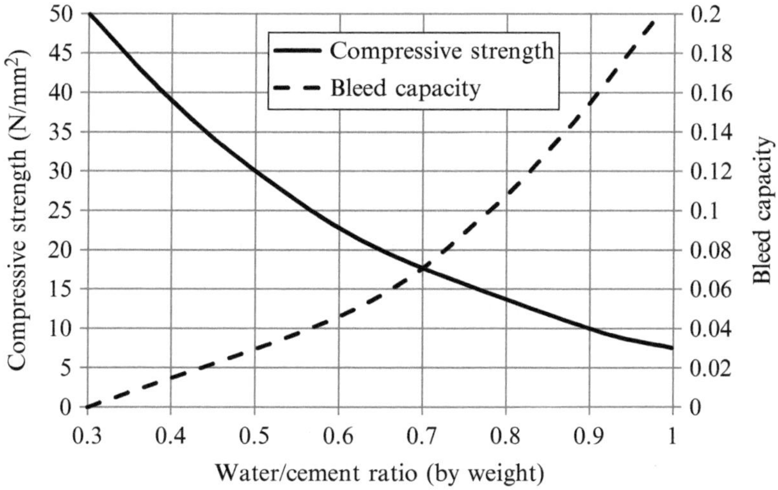

Fig. 6.2 Effect of water content on cement grout properties (Adopted from Littlejohn 1982)

- deformability
- durability
- shrinkage
- expansion

The following **intrinsic properties** are considered when selecting a grout material:

- Rheological (viscosity, cohesion), setting time, stability
- Particle size (when applicable)
- Strength and durability. Effects of water content on cement grout properties are shown in Fig. 6.2 from Littlejohn (1982)
- Toxicity

6.3.1.2 Compaction Grouting

- Kovacevic et al. (2000) showed that if the compacted soil is not sufficiently permeable for consolidation to occur as it is treated, excess pore pressures may be generated, which will dissipate after treatment. It is found that the efficiency of treatment may be reduced substantially.
- Mortars used for should contain a minimum of 15 % of fines passing 0.1 mm according to EN 12715:2000.
- The approach used for this type of grouting is essentially based on experience and observation method when applied to uplift of structures.

6.3.1.3 Jet Grouting

According to EN 12716:2001 jet grouting process consists of erosion of soil or weak rock and its mixing with, and partial replacement by, a cementing agent. The erosion is achieved by means of grout/water jet at pressures up to 70 MPa and flow up to 650 l/min. Mixed soil columns/zones/slabs can be formed in any direction.

- Water: cement ratio by weight should range between 0.5 and 1.5
- Water reducing, stabilizing, plasticising, waterproofing or anti-washing additives can be used
- Bentonite, filler, fly-ash can be added. When water-bentonite suspension is used it should be prepared before adding cement
- Reinforcement can be installed in fresh routed material or in a borehole drilled after hardening.

6.3.2 Execution

Designer prepares specifications for the works. EN 12715:2000 lists information needed for execution of **permeation and compaction grouting** as:

- Site conditions and limitations (e.g. size, gradients, access)
  - Geological (including total and solid rock core recovery, rock core and drill fluid loss, drill parameters and rate of drill advance) and geotechnical (including: water permeability, temperature, chemistry, organic and bacteriological content of ground water and ground, ground water levels and gradients with time, presence of obstructions, loose, soft, unstable, soluble, collapsible or swelling ground on drilling and grouting)
  - Existence, location and condition of any adjacent structure and its foundation, roads, utilities, services
  - Concurrent or subsequent activities which could affect the works (e.g. ground water extraction or recharge, tunnelling, deep excavation)
  - Previous experience with grouting or underground work on o adjacent to the site or comparable work under similar conditions
- Restrictions (environmental, legal or statutory)
- Underground contamination or hazard (e.g. cavities, UXO)
- Personnel qualification and experience
- Field grouting trials tests (for borehole spacing, grouting pressure, grout take and type)
- Method statement including:
  - Drilling method (rotational, percussion, with/without casing and muds), pattern (borehole number, positions, spacing, depths, diameter, inclination and orientation) and sequences
  - Flushing of boreholes in rock after drilling unless it is detrimental

  - Storage of materials (protection especially against temperature and humidity)
  - Batching and mixing of grout
  - Pumping and delivery of grout
  - Grout placement (volume, pressure, rate, rheology), with/without packers or sleeve
  - Grout sequences (single/multiple stages, top-down/bottom-up, split spacing (primary, secondary, tertiary, quaternary) or onwards/outwards)

EN 12716:2000 lists information needed for execution of **jet grouting** as:

- Ground profile and geotechnical properties within range of jet grouting particularly:
  - Grain size distribution (detrimental presence of cobbles/boulders), Atterberg limits, consistency (detrimental presence of firm/stiff)
  - Density and moisture content
  - Shear strength (detrimental presence of cemented layers/lenses)
  - High organic content
  - Swelling or highly sensitive (quick) clay which presence id detrimental
  - Aggressive soil or water
  - Large voids
- Hydrogeological conditions particularly:
  - Artesian or confined aquifers
  - High hydraulic gradients or water permeability
- Boundary conditions (adjacent/buried structures and services, overhead power lines, access, other restrictions)
- Disposal of spoil return
- Acceptable deformation of adjacent structures
- Method statement including:
  - Objective and scope
  - Geometry
  - Grouting system (single, double, triple)
  - Procedures (for drilling, jetting, sequencing)
  - Grouting parameters
  - Materials
  - Precautionary measures to avoid excessive settlement/heave especially in silty and clayey soil
  - Plant and equipment (for drilling, mixing and pumping, grouting, monitoring pressure, flow rate and volume, rate of rotation and withdrawal, depth)
  - Spoil management
  - Quality control procedures including possible interruptions, modifications
  - Testing methods and reports
  - Field trial results
  - Qualification and experience of the personnel

6.3.3 Control

EN 12715:2000 lists information needed for monitoring and control of **permeation and compaction grouting** as:

- Grout properties during preparation and placement such as
  - Density
  - Marsh viscosity (suspensions), grain size (micro fine suspensions), workability (mortars)
  - Setting time (suspensions and solutions)
  - Bleeding (suspensions)
- Tolerances in location and direction of boreholes
- Criteria for ending injection after each pass in soil
  - Limiting pressure and/or volume
  - Ground movement due to grouting
  - Escape of grout to surface and elsewhere
  - Bypassing of packers when used

  In rock:

  - Limit pressure (refusal) and/or volume
  - Ground movement
  - Escape of grout
  - Unacceptable loss of grout into adjoining areas
- Results achieved after each pass and the end of work
- Ground movement and deformations
- Chemistry of water
- Water levels in existing wells or observation boreholes

EN 12715:2000 lists information needed for monitoring and control of **jet grouting** as:

- Geometry
- Strength, deformability, permeability where appropriate on four samples for each 1,000 m^3 volume
- Density (twice a shift), bleeding (daily), Marsh viscosity (daily), setting time
- Coring when required
- In situ loading/pumping tests when required after setting time

For all grouting types, a continuous control of the amount of used cement or other expensive ingredient is of vital importance.

6.4 Mixing

6.4.1 Description

Mechanical dry or wet mixing of clay, peat or silt with binder (lime or cement) to increase overall mass strength and stiffness and decrease of water permeability for:

- Settlement reduction
- Increase of bearing capacity
- Increase of slope stability
- Decrease of liquefaction potential during earthquakes
- Solidification of waste deposits and containment of polluted soil
- Reduction of vibration effects
- Reduction of active loads on retaining walls and of the horizontal displacements

When only a part of soil mass is mixed and overall mass properties are considered in design then the differences in stress-strain relationships for mixed and non-mixed soil must be taken into account because cemented soil can behave like brittle and non-mixed soil like ductile materials. Figure 5.10 illustrates for slope stability analyses different combinations of shear strengths (peak and residual) of brittle and ductile materials at different displacements. Equation (6.3) can be used for calculation of equivalent shear modulus and Eq. (6.4) for calculation of equivalent shear stress acting on combined mixed and non-mixed soil.

Typical data for wet mixing technique in Europe and Japan from EN 14679:2005 (E) are:

- Number of mixing rods 1–4 (8 offshore)
- Diameter of mixing tool 0.4–1.6 m
- Maximum depth of treatment 25 m (Europe), 48 m (Japan), 70 m below sea level
- Injection pressure 300–1,000 kPa
- Penetration speed of mixing shaft 0.5–1.5 m/min
- Retrieval speed of mixing shaft 0.7–7 m/min
- Rotation speed of mixing blades 20–60 rpm
- Amount of binder injected 70–450 kg/m^3

EuroSoilStab (1996) soil mixing advantages, disadvantages and comparisons with other ground improvement/treatment/reinforcement methods are given in Table 6.5.

6.4.2 Execution

Designer prepares specifications for the works. According to EN 14679:2005(E), necessary information includes:

- Legal/statutory restrictions

Table 6.5 Soil mixing advantages, disadvantages and comparisons with other ground improvement/treatment/reinforcement methods (Adopted from EuroSoilStab 1997)

Advantages	Disadvantages	Vertical drains	Piling	Excavation and replacement
Economical Flexible Saves materials and energy Exploits soil properties at site Zero spoil production No soil transport	Embankment high limited Incompletely stabilises soil Curing time required Depth limited to 5 m for mass stabilisation, 40 m for columns	Less expensive More time consuming More mass consuming (for preload fill) More stability problems Larger settlements	More expensive Bigger differential settlement (piled/no piled) Faster Possible deeper	Cost case dependent Significantly more mass consuming Higher risk of failure Larger impact on environment

- Conditions of structures, roads, services etc. adjacent to the work
- Quality management system (supervision, monitoring, testing and acceptance procedure)
- Site conditions
  - Boundary conditions, topography, access, slopes, headroom restrictions etc.
  - Existing underground structures, services, known contamination, archaeological constrains, other hazards (such as UXO)
  - Environmental restrictions (including noise, vibration, pollution, contamination)
  - Future or on-going construction activities (such as dewatering, tunnelling, deep excavations)
- Previous experience of deep mixing or special geotechnical works adjacent to the site
- Procedure for unforeseen circumstances or conditions
- Reporting procedure if an observational method is adopted
- Restrictions such as construction phasing
- Geotechnical conditions
  - Subsurface layering and ground properties (consistency limits, grains size distribution, density, natural moisture content, chemistry contents, shear strength, stiffness, water permeability), tree roots, fill, etc.
  - Presence of cobbles/boulders, cemented layers, underlying rock
  - Presence of swelling soil, cavities, voids or fissures
  - Ground water levels, variations, artesian pressure

- Method statement including:
  - Method
  - Tools
  - Procedure (penetration and retrieval, mixing and sequence of execution)
  - Tolerances
  - Binder properties (type and composition, content, water ratio, admixture, filler, protection from temperature and moisture)
  - Precautions against heave/settlement
  - Working areas
  - Plant and equipment
  - Spoil management
  - Quality control
  - Interruption procedure
  - Possible modification
  - Verification testing method, extent and frequency
  - Control and acceptance criteria
  - Reports and drawings list
  - Safety and environmental risk assessment
  - List of qualification and experience of the personnel involved
- Results of field trials
- Reinforcement if needs to be installed into fresh mix

6.4.3 Control

EN 14679:2005(E) specifies what should be checked and reported:

- Penetration and retrieval speed of mixing tool
- Rotation speed of mixing tool
- Depth of penetration
- Air pressure (for dry mixing)
- Feed rate of binder/slurry
- Quantity of binder for each column
- Recording equipment calibration
- Spoil collection and disposal
- Strength, stiffness and permeability of mixed soil when required to be tested must be distributed uniformly in space and time
- pH value, carbonate, chlorite, sulphate and sulphide content when relevant and required
- Load plate/pressure meter/cone penetration test results when required
- Depth and extent of mixing (overlap, column inclination, diameter, etc.)

6.5 Separation

6.5.1 Description

Several types exist such as:

- Horizontal and inclined **filters** (granular or geotextile) for allowing ground water flow through when preventing migration of fine soil particles (internal erosion) in one layer into voids between coarser soil particles of an adjacent layer. More information on internal erosion is provided in Sect. 4.2.
- Horizontal or inclined **liners** (compacted clayey or geosynthetic such as geomembranes and geotextile/geomembrane clay liners) for prevention of water/leachate flow across interfaces between soil layers or from water reservoirs/waste deposits/landfills into ground. More information on low permeability liners is provided in Sect. 5.8 for both types of liners and Sect. 6.1 for compacted clay.
- Vertical **slurry (diaphragm) walls** and **mixed soil barriers** for containment of hazardous contaminated land and waste deposits and prevention/minimization of contamination leak laterally. More information on diaphragm wall is provided in Sect. 5.3 and on mixed soil in Sect. 6.4.
- Horizontal **jet grouted blankets** under hazardous contaminated land and waste deposits for containment and prevention/minimization of contamination leak vertically. More information on jet grouting is provided in Sect. 6.3.1.3
- **Combined systems** for encapsulation and leakage drainage/detection including pumping for very hazardous contaminated land and waste deposits

Typical ranges of some **geomembrane** properties from GFR (1992) are:

- Tensile strength in machine direction (ASTM D638 for polyethylene – PE, D882 for polyvinyl chloride – PVC):
  - 39.4 kN/m for 60-mil high density PE – HDPE to 56.9 kN/m for 80-mil HDPE,
  - 21.9 kN/m for 40-mil very low density PE – VLDPE to 36.8 kN/m for 60-mil VLDPE,
  - 7.9 kN/m for 20-mil PVC to 16.6 kN/m for 40-mil PVC
- Puncture resistance (FTMS 101C, Method 2065)
  - 0.31 kN for 60-mil HDPE to 0.47 kN for 80-mil HDPE,
  - 0.22 kN for 40-mil VLDPE to 0.38 kN for 60-mil VLDPE
- Tear resistance (ASTM D1004)
  - 0.18 kN for 60-mil HDPE to 0.27 kN for 80-mil HDPE,
  - 0.07 kN for 40-mil VLDPE to 0.13 kN for 60-mil VLDPE,
  - 0.03 kN for 20-mil PVC to 0.05 kN for 40-mil PVC

Typical properties for **geotextiles** used as separators from Rowe et al. (1995) at 50 % elongation at failure and for high survivability are:

- Grab strength (ASTM D4632) from 0.8 to 1.2 kN
- Puncture resistance (FTMS 101C, Method 2065) and tear strength (ASTM D638 for polyethylene – PE, D882 for polyvinyl chloride – PVC) from 0.3 to 0.45 kN

Typical ranges of some **nonwoven geotextile** properties from GFR (1992) for a range of geotextile weight are:

- Apparent opening size (ASTM D4751; EN ISO 12956:1999) from 50 to 200
- Puncture resistance (FTMS 101C, Method 2065) from 0.11 to 1.33 kN
- Mullen burst strength (ASTM D3786) from 966 to 6,210 kPa
- Trapezoid tear strength (ASTM D4533) 0.16 to 1 kN
- Grab strength (ASTM D4632) 0.4 to 2.22 kN
- Wide width tensile strength (ASTM D4595; EN ISO 10319:2008) from 8.8 to 36 kN/m
- Permittivity (ASTM D4491; EN ISO 11058:1999) from 0.05/s to 2/s

Maximum leakage through an open hole in geomembrane is calculated as (e.g. Giroud and Bonaparte 1989):

$$Q = 0.6 \cdot a \cdot \sqrt{2 \cdot g \cdot h_w} \tag{6.11}$$

a is area of a hole in geomembrane
g is the gravitational acceleration
h_w is liquid depth above a hole

Radius of wetted area R is (e.g. Giroud and Bonaparte 1989):

$$R = 0.39 \cdot d \cdot \frac{\sqrt[1/4]{2 \cdot g \cdot h_w}}{\sqrt{k_s}} \tag{6.12}$$

d is diameter of a hole in geomembrane
g is the gravitational acceleration
h_w is liquid depth above a hole
k_s coefficient of hydraulic conductivity of soil underlying a geomembrane

6.5.2 Execution

Designer prepares specifications for the works. More details are as follows:

- Granular and geotextile filter criteria are provided in Sect. 4.2.4.1.
- Construction notes for compacted clay liners and geo-membranes are provided in Sect. 5.8.4 for both types of liners and Sect. 6.1.2.2 for compacted clay.

- Construction notes for jet grouting are provided in Sect. 6.3.2 and for soil mixing in Sect. 6.4.2.
- Construction notes for diaphragm walls in general are given in Sect. 5.3.4

6.5.2.1 Geotextiles

EN 14475:2006(E) lists necessary information as:

- Procedure to deal with unforeseen circumstances or conditions worse than assumed in design
- Reporting procedure if an observational method is used or monitoring is required
- Notice of restrictions such as construction phasing, site access, environmental or statutory requirements
- Positions, levels and co-ordinates of fixed reference points
- Schedule of testing and acceptance procedure for materials
- Protection against high/low temperature, ultraviolet radiation, mechanical damage etc.
- Locations of services (electricity, telephone, water, gas, drains, sewers)
- Geotechnical data with respect to:
  - Electro-chemical, chemical, mechanical and biologic aggressivity of the ground or fill
  - Interface friction angle and reinforcing effect on the stiffness of composite soil-geotextile material
- Long term properties of geotextile and reduction coefficients (e.g. for reinforced soil provided in Sect. 5.5.2.1)
- Qualification and experience of personnel involved
- Method statement

6.5.2.2 Slurry Diaphragm Walls

EN 1538:2010 lists necessary information as:

- Legal or statutory restrictions
- Conditions of structures, roads, services etc. adjacent to the work, including any necessary surveys
- Management system including supervision, monitoring and testing
- Site conditions
  - Boundary conditions, topography, access, slopes, headroom restrictions, previous use, etc.
  - Existing underground structures, services, known contamination, archaeological constraints
  - Environmental restrictions including noise, vibration, pollution
  - Future or on-going activities such as dewatering, tunnelling, deep excavations, underpinning, pre-treatment of soil, dewatering

- Specific requirements concerning tolerances, quality of materials, permeability and type of joints, strength and stiffness of slurry
- Previous experience with diaphragm walls or underground works on or adjacent to the site
- Qualifications and experience of personnel involved
- Geotechnical conditions
  - Ground profile and classification properties
  - Piezometric levels and permeability of ground
  - Presence of coarse, highly permeable ground or cavities/mine openings
  - Presence, strength and stiffness of soft soil or peat
  - Presence, size and frequency of boulders or obstructions
  - Presence, position, strength of hard rock
  - Chemistry of ground water, soil/rock, waste/landfill and temperatures when required
  - Slope stability
- Bentonite sources and properties of calcium, natural sodium or activated from natural calcium bentonite such as chemical and mineralogical composition. Table 6.6 contains list of properties with their limited values for bentonite suspensions
- Constituents of hardened slurry such as silt, clay or bentonite, cement or another binder, water, other additives and admixtures
- Method statement detailing all stages of construction, including equipment, guide walls, testing/control/monitoring and reporting including emergency situations such as sudden loss of supporting fluid and excavation instability

6.5.3 Control

- Control notes for fills (granular filters, clayey liners) are provided in Sect. 6.1.3.2
- Control notes for jet grouting are provided in Sect. 6.3.2 and for soil mixing in Sect. 6.4.2.

6.5.3.1 Geotextiles

EN 14475:2006(E) lists the following:

- Site preparation: topography, set-up, geometry
- Geotextiles conformity with design, reception, handling, storage, placing, damage during installation

Table 6.6 Characteristics for bentonite suspensions (From EN 1538:2010)

Property	Fresh	Re-used
Density (g/cm^3)	<1.1	<1.25
Marsh value (s)	32–50	32–60
Fluid loss (cm^3)	<30	<50
pH	7–11	7–12
Filter cake (mm)	<3	<6

6.5.3.2 Slurry Diaphragm Walls

EN 1538:2010 lists the following items for supervision, testing and monitoring:

- Excavation method, dimensions and alignment
- Cleaning the excavation
- Joints formation
- Material properties according to Table 6.6
- Integrity/permeability

6.6 Freezing

6.6.1 Description

Frozen ground is created by installing freeze pipes in which a cooling medium circulates down an inner pipe and returns within the space between an inner and outer pipe resulting in heat extraction from surrounding ground and transformation of free water in cooled ground into ice causing ground expansion or additional pressure if the expansion is prevented. With continued heat extraction, the initial frozen columns formed around the cooling pipes increase in diameter until they merge and form a frozen zone, providing that ground water flow velocity does not exceed 3 m/day, in which case ground permeability must be reduced by grouting for example. Ground freezing is a versatile technique used in soil and rock to increase their strength and stiffness and decrease water permeability depending more on time and temperature than ground properties. A significant content of dissolved salts in ground water reduces strength of frozen ground.

Section 4.13 contains notes on description, extent, identification and remediation of frozen ground.

Andersland and Ladany (1994) provide a summary of construction ground freezing for typical applications such as: shafts, deep excavations, tunnels, ground water control, structural underpinning, and containment of hazardous waste in the case that ground water is present naturally or artificially.

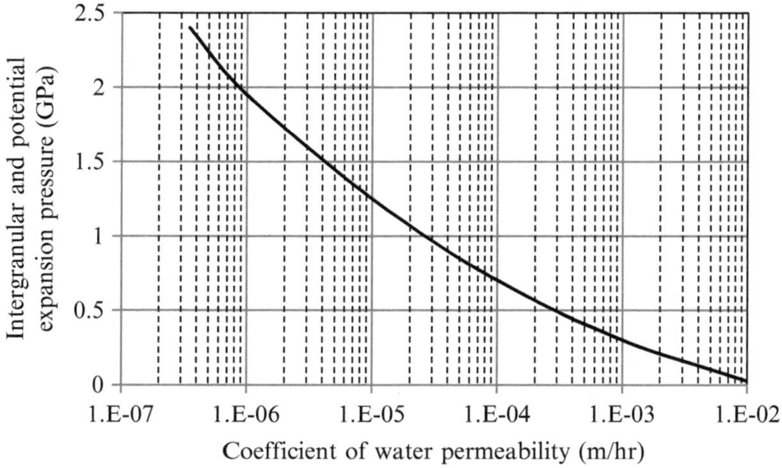

Fig. 6.3 Typical effect of coefficient of water permeability on frost related expansion pressure (Adopted from Shuster 1972)

6.6.2 Execution

Values of additional pressure due to ground freezing when ground expansion is prevented are shown in Fig. 6.3, which is adopted from Shuster (1972).

Change in soil column height *ΔH* due to conversion of ground water to ice during freezing (with about 9 % increase in volume) for a constant water content and under the assumption that one half of the volume expansion is in the vertical direction is according to Sanger and Sayles (1979):

$$\Delta H = \frac{0.5 \cdot H \cdot 0.917 \cdot (w - w_u)}{1/G_s + w/S_r} \tag{6.13}$$

H is ground column height
w and w_u are the total and unfrozen water contents, respectively; $w_u = 0$ for temperatures <-10 °C
0.917 is the specific gravity of ice
G_s is the specific gravity of ground (~2.65)
S_r is degree of saturation

Necessary information for execution includes (Andersland and Ladany 1994):

- Geometry of excavation or frozen barrier
- Soil and ground water conditions (mainly thermal conductivity which is smaller in silt and clay and higher for rock, content of dissolved salts in ground water and ground water flow velocity and temperature). Freezing soil columns will not

merge for a critical ground water velocity u_c as described by Andersland and Ladany (1994):

$$u_c = \frac{k_f \cdot V_s}{4 \cdot S \cdot V_o \ln_e\left({}^{S}/_{4 \cdot r_o}\right)} \tag{6.14}$$

r_o is the radius of freeze pipe (m)
k_f is the thermal conductivity of frozen soil (W/m °C)
S is the spacing of freeze pipes (m)
V_s is the difference between freeze pipe surface temperature and the freezing temperature of ground water (°C)
V_o is the difference between ambient ground temperature and the freezing temperature of ground water (°C)

- Proximity of adjacent streets, utilities and structures
- Properties of freezing method

Installation of a cooling system involves the following steps:

- Surface grading to ensure that surface water is collected and drained away from frozen ground
- Insulating utility pipes usually with sprayed polyurethane foam
- Installation of freeze pipes welded when necessary and pressure tested to ensure that no coolant leaks into ground
- Spacing of refrigeration pipes should not exceed about 13 times their diameter based on projects with freeze pipe diameters varying in the range from 50 to 150 mm
- Alignment of all freeze pipes needs to be verified after their installation particularly for lengths greater than 20 m
- Installation of freeze pipes in drilled holes of larger diameter requires backfill of the space between the pipes and the hole walls
- Refrigeration plant and coolant distribution manifold are selected and installed
- Each freeze pipe and their group in series should be provided with positive air-bleed valves to allow relief of trapped air in the system during operation. A temperature difference between inlet and outlet pies of 4–5 °C often indicates the presence of air pockets. Valves are also useful to turn off a portion of a system on its damage by excavation machinery.

6.6.3 Control

Andersland and Ladany (1994) list the following monitoring requirements for ground freezing:

- Freeze hole deviation (by an inclinometer)
- Temperatures within the coolant distribution system and frozen ground (in observation boreholes)

- Frost boundary location and zone thickness (by geophysical methods e.g. ASTM D4428)
- Insulation of exposed frozen excavated surface particularly in unsaturated soil such as coarse clean sand and gravel (by a single layer of reinforced reflective plastic, fiberglass or foam blankets with reflective plastic on both sides)
- Removals of ground/rainfall water particularly discharge hoses passing over frozen ground. A small hole in a discharge line can emit a jet of water sufficient to thaw, erode and ultimately destroy a frozen zone.
- Blasting may result in fracturing of refrigeration pipes and loss of coolant into ground while water jetting can inadvertently create a hole through a frozen zone in minutes.

6.7 Blasting

6.7.1 Description

Explosives are most frequently used in mines, collieries, quarries, and rock excavations. Detonation weakens rock strength by brisance, helped by high gas pressure created in the charge hole. A compressive strain pulse created by detonation travels in all directions from the charge hole and attenuates to zero unless and until it reaches a free face, in which case it is reflected as a tensile strain pulse that breaks rock in tension, aided and displaced by high gas pressure.

Gregory (1984) describes special techniques such as:

- "Bulling" or chambering of holes
- Tunnel or coyote blasting (centre, wedge, draw, burn, large-hole, Cormant cuts)
- Secondary blasting
- Seismic blasting for geophysical testing
- Over break control (line drilling, cushion blasting, smooth wall blasting, presplitting)
- Controlled trajectory blasting
- Blasting in plastic rock

6.7.2 Execution

Gregory (1984) describes operations involved in charging a usual short hole with conventional cartridge explosives as:

1. After drilling, the hole is cleaned out with compressed air through a blow pipe
2. Explosive cartridges are inserted into a hole one by one and firmly squeezed into position by wooden tamping pole. Metal tamping rod must not be used.

3. The primer cartridge (if of conventional safety fuse/plain detonator type) is placed last. The primer cartridge should not be tamped or squeezed with the tamping pole.
4. The hole is sealed by stemming with a cartridge of sandy clay squeezed against the primer
5. Check is made that no damage is caused to the safety fuse, which end is protruding from the hole
6. If electrical firing is used, then the electric primer is the first cartridge to be inserted (inversely) in the hole. The leg wires are then held taut to one side of the hole while the ordinary cartridges are tamped into position.
7. With either type of primer the live end of the detonator should face the bulk of the charge, i.e. with the ordinary primer – inwards, with the electric primer – outwards.

When charging soft plastic cartridges of slurries, adjusted techniques need to be adopted.

Gregory (1984) provides a detailed description of blasting theories and quarry blast design.

Field tests need to be arranged with available explosive types and for actual site conditions to check the initial blast design.

6.7.3 Control

Gregory (1984) describes the following stages with reference to U.S.:

- **Manufacturing** (or import) of a complete range of explosives and accessories must be by licensed companies as well as for mixing of blasting agent components at or near blast site by mining companies, quarry operators, demolition contractors or explosive manufacturers.
- **Storage** places at factories, distribution depots, mixing depots, mines and quarries and demolition sites must fulfil various regulations concerning location, construction, maintenance, operation and surveillance by licensed operators subjected to periodic inspection. The maximum weight of high explosive is subjected to limitations concerning distances to inhabited places and other sensitive locations. Detonators must be stored separately from explosives.
- **Transportation** regulations covering the packaging exist in separate countries. For example, U.S. Department of Transportation classifies explosive materials for transportation as:
  - Class A: Maximum hazardous such as dynamite, nitro-glycerine, picric acid, lead azide, fulminate of mercury, black powder, blasting caps, and detonating primers.
  - Class B: Possessing flammable hazard, such as propellant explosives

- Class C: Minimum hazardous such as detonating cord, boosters, safety fuse, fuse lighters, ignite cord
- Flammable solids: Oxidizers such as nitrates, ammonium nitrate prills, ammonium nitrate carbonaceous mixtures (AN/FO)

- **Use** is regulated by various state laws, health and safety acts, industry codes and company procedures. Regulations exist concerning certification and permits required for personnel authorized for blasting operations.

Gregory (1984) provides a list of **common causes of incidents** with explosives (involving lack of care, disregard of regulations or established safe practice):

1. Delaying too long in lighting fuse
2. Drilling into explosives
3. Premature firing of electric blasts
4. Returning too soon after blasting
5. Inadequate guarding
6. Unsafe practice during transport, handling and storage
7. Improper handling of misfires
8. Using too short fuse
9. Improper tamping procedure
10. Smoking during handling of explosives

Other causes include lightning strikes, electric storms, static electricity, radio frequency currents, and stray currents.

6.8 Underpinning

6.8.1 Description

Underpinning involves support of existing shallow foundation to improve its capacity and serviceability because of expected changes of loads/other conditions in the future or to remediate problems caused by loss of foundation capacity and serviceability.

Need for underpinning arises when:

- **new deeper shallow foundation** is to be constructed adjacent to an old shallow foundation and the old foundation will lose its bearing capacity and serviceability (ability to limit settlement)
- **load is added to existing shallow foundation** (for example due to addition of storeys)
- **ground properties under an existing shallow foundation have deteriorated** and the existing foundation can no longer provide necessary bearing capacity and limited settlement (for example due to collapse of loess structure because of wetting caused by leaking drainage pipes).

- **environmental changes occur** (for example ground water level increases/decreases, frost penetration depth increases, local ground subsidence is caused by adjacent tree roots extraction of ground moisture, regional subsidence is caused by landfill settlement or nearby excavation)

Different types of underpinning exist:

- **Excavation and concreting** in segments under existing shallow foundation
- **Addition of bored mini/micro piles** (micro piles diameter is less than 300 mm) through or around existing shallow foundation or segmented piles (Mega) by pushing them into soil under exiting shallow foundations
- **Grouting** under existing shallow foundations (compaction and permeation grouting although jet grouting is not excluded)

6.8.2 Execution

Hunt et al. (1991) list sources of information used:

- Geological maps and memoirs
- Soil survey maps and memoirs
- Land utilisation survey
- Agricultural land classification maps
- Planning maps
- Ordnance survey maps – historical coverage
- Mining maps
- Hydrological maps
- Air photographs
- Local people/organisation surveys
- Previous site investigation reports
- Statutory bodies (local, regional, state)

Necessary information includes:

- Topography (elevation, slopes, hillsides, watercourses, nearby structures and roads)
- Geology (coal seams, unstable slopes, ground types, aquifers)
- Features (mining activity, land use, drainage, solifluction, valley bulging)
- Local knowledge on history of problems
- Local services and structures

Site survey can reveal the following:

- Vegetation (trees)
- Marshy condition
- Topographical anomalies
- Local erosion

- Ponds, drainage ditches, flood plains, springs
- Nearby excavations
- Type and extent of damage to structures if any

Existing foundation and ground investigations information include:

- Shape, size, depth and condition of existing shallow foundation and buried foundations if any
- Ground profile with ground identification properties
- Ground strength and stiffness parameters, swelling and collapse properties
- Ground water level and its changes throughout a year
- Chemical properties of ground and ground water (organic, sulphate, chloride content, pH)

6.8.2.1 Excavation and Concreting

It is restricted to depth of about 2 m below existing shallow foundations and is performed in non-contiguous sequences. Requires hand excavation and protection of workers by propping for safety reasons. Other limitations can be applicable.

When reinforcement is used to form a continuous beam, the reinforcement needs to be bended back to the horizontal position from its temporary vertical position in the primary segment after excavation of adjacent secondary segment.

6.8.2.2 Piles

This method is used for greater loads and depths and for unstable ground conditions when open excavation is not safe. Micro piles (with their diameter less than 300 mm) are usually bored from the ground surface and inclined through existing shallow foundation. Mini piles (with their diameters greater than 300 m and smaller than 600 mm) are bored or augered in vertical direction with the pile cap doweled into existing foundation or constructed through existing walls in sequences. Segmented piles are jacked into ground under existing foundations, require manual handling and access under foundation and for this reason are less frequently used than other types of piles.

Notes for piling are provided in Sect. 5.2.4.2.

6.8.2.3 Grouting

It is used when piling is not possible, for example because of limited space or obstructions. It requires the use of drilling rigs sometimes within limited headroom and other equipment for grouting. The effect is more difficult to evaluate than for piling but may be useful when other methods are not applicable.

Notes for grouting are provided in Sect. 6.3.2.

6.8.3 Control

Basic requirements are given as follows:

- **For excavation** in Sect. 6.1.3 while for concreting is beyond the scope of this book.
- **For piling** in EN 1536:2010 and 14199:2005, among others, which involve:
  - Excavation method (tools and equipment), dimensions and depth, ground conditions and ground water levels, obstructions
  - Execution (level and characteristics of the support fluid, installation of casing, pile sockets, cleaning of the borehole)
  - Placing (depth, position) of reinforcement
  - Concreting (characteristics, placement, quantity, duration, rise and final level, recovery of the tremie pipe)
  - Post concreting phase (recovery of temporary casings, shaft/base grouting including the grout properties)
  - Integrity and load tests
- **For grouting** in Sect. 6.3.3.

6.9 Soil Washing and Waste Solidification/Stabilisation

6.9.1 Description

6.9.1.1 Soil Washing

It originated in 1970s for remediation of beach sand contaminated by oil spills. Grasso (1993) lists compounds that have been reported to be amenable to washing including: xylene, styrene, phenol, aromatic hydrocarbons, total PAH, pyrene, naphthalene, arsenic, cadmium, chromium, copper, lead, nickel, zinc. The process is based on the following assumptions:

- Contaminants are concentrated in the fine fraction of soil matrix, due to the high surface area to volume ratio
- Contaminants in the coarse and sandy fraction are physically bonded to the particles due to compaction and adhesion

Ex situ soil washing is most common. Grasso (1993) lists the advantages and limitations as shown in Table 6.7.

Table 6.7 Advantages and limitation of soil washing and chemical extraction (Adopted from Grasso 1993)

Advantages	Limitations
Clean-up level of <1 ppm achievable	Contaminant not destroyed
Favourably viewed by the public	Fines and washing fluid require secondary treatment or must be disposed of as a hazardous waste
Relatively low-cost	Sludge from wastewater treatment must be treated
Can utilise a closed-treatment system	Large organic fraction may require use of additional pre-treatment steps
Can effectively pre-treat soil for bioremediation	

6.9.1.2 Waste Solidification/Stabilisation (S/S)

It was originally developed in 1950s, it involves,

- **Stabilisation** aims at reducing the hazard potential of waste by converting contaminants into their least soluble form.
- **Solidification** aims at encapsulation of waste in a monolithic solid of high structural integrity.

Grasso (1993) lists compounds that have been reported to be amenable to washing including: metals (aluminium, barium, chromium, lead, antimony, beryllium, cobalt, mercury, arsenic, cadmium, copper, tin) inorganics (fluorides, cyanides, sulphides). Operative mechanisms for S/S are:

- Chemical reaction
  - Precipitation with pH control via precipitation of metals
  - Redox reactions to change oxidation state of metals, rendering them less soluble
  - Complexation
  - Passivation
- Sorption
  - Adsorption
  - Ion exchange
  - Diodochy
- Encapsulation (micro and macro, embedment)

Inorganic treatment by cement and pozzolans is most common as organic treatment by thermoplastics and organic polymers maybe hydrophobic and may suffer long-term stability problems. Grasso (1993) lists the advantages and limitations as shown in Table 6.8.

Table 6.8 Advantages and disadvantages of waste solidification/stabilisation (Adopted from Grasso 1993)

Advantages	Disadvantages
Can immobilize certain compounds (which allows delisting as hazardous)	Does not destroy contaminants
Can utilize process by-products (e.g. flyash)	Can deteriorate over time
Results in a stable structurally sound end product	Typically requires excavation and material handling
	Requires curing period
	Contaminants will diffuse slowly (by leaching)
	May increase volume

6.9.2 Execution

6.9.2.1 Soil Washing

Grasso (1993) lists **ex situ equipment** used:

- Size reduction (primary crushers to +100 mm and secondary crushers 1,000 mm to 5 mm)
- Screening (grizzly, trammel, screen, vibrators, shakers, rotary screens)
- Separation (classifiers, clarifiers)
- Mixing and extraction (impellers, water knife, high pressure jet pipe)
- Washing fluid treatment (metal precipitation, concentration/separation of contaminants, biodegradation, polishing)
- Air emission
- Secondary treatment of fines (secondary extraction, rinsing, dewatering)

In situ equipment used:

- Injection wells
- Extraction wells
- Wastewater treatment systems
- Pumps and instrumentation
- Containment systems

Necessary information for feasibility evaluation (from Hsieh and Raghu 1990):

- Soil properties
  - Highly variable can produce inconsistent flushing
  - High organic content soil can inhibit desorption of contaminant
  - Large organic fraction in soil may require use of additional pre-treatment steps
  - Grain size, specific gravity, density and water content determine the extent of volume reduction with soil washing
  - Low permeability (high clay or salt content) reduces percolation and leaching

- Buffer capacity and soil pH affect neutralisation and possibly precipitation reaction
- A high cation exchange capacity may decrease contaminant mobility and can attenuate treatment of metals and their salts

- Site hydrology (if in situ)
- Contaminants (solubility and adsorption affinity)
- Extracting agent. The fluids used should have:
  - Favourable separation coefficient for extraction
  - Low volatility under ambient conditions
  - Low toxicity for handling
  - Reuse capability
  - Treatability
  - High water solubility
- Mixing (time and efficiency)

6.9.2.2 Waste Solidification/Stabilisation (S/S)

A typical S/S process involves addition of reagents (dry or liquid) to waste in a mixing vessel. Subsequent to mixing, the mixed waste is transferred by either pumps or mechanical conveyors to curing areas. Pilot studies are essential in the optimisation of the process.

Necessary information for feasibility evaluation from Grasso (1993):

- Soil and site properties
  - Permeability of soil
  - Depth to groundwater
  - Locations of sensitive environmental receptors (such as flood plains and marshes)
  - Access
  - Space for processing (area)
  - Health and safety issues (existing groundwater contamination, dust generation)
- Contaminant location and properties (content of oil and grease, aromatic and halogenated hydrocarbons, alcohols, phenols, organic acids, glycols)

6.9.3 Control

6.9.3.1 Soil Washing

Grasso (1993) lists the most important factors that impact the process:

- Type of contaminant
- Contaminant solubility

- Chemical kinetics
- Solvent
- Site heterogeneities and site hydrology (if in situ)
- Soil gradation
- Contact time
- Flow path of solvent versus solute (if in situ)

6.9.3.2 Waste Solidification/Stabilisation (S/S)

Grasso (1993) lists the most important factors that impact the process:

- **Particle size and shape**. Small particles can release alkalinity quicker, reduce diffusion limitation, and increase area available for sorption.
- **Free water content** (available for setting reactions)
- **Solids content**. A low solid, low viscosity, high specific gravity waste requires either a fast setting S/S process or the addition of bulking agent.
- **Specific gravity**. Large differences between waste and reagent specific gravities may result in phase separation until setting begins.
- **Viscosity**. Rheological agents can be added to alter viscosity. These agents may subsequently interfere with setting reactions.
- **Wetting**. For some hydrophobic additives and soil it may be necessary to add surfactants, which may interfere with subsequent cementation reactions.
- **Mixing**. Some S/S processes are sensitive to energy input and shear rate. Over mixing may interfere with initial gel formation of cementations S/S process, causing delayed set, slow curing, and possibly loss of final physical properties.
- **Temperature and humidity**. Below freezing, gel structures can be broken. At high temperatures, steam release can break up solid mass and reduce available water reactions. Kiln dust and fly ash contain CaO, which can hydrate violently, releasing heat.
- **Chemical content**. Additives/constituents can retard, inhibit, and accelerate setting and curing as well as final strength, permeability and durability.

6.10 Field Measurements and Observational Method

6.10.1 Description

Filed measurements are undertaken when:

- an observational method is specified to decrease cost/increase safety
- for control of ground behaviour during construction
- for control of ground performance in operation
- when required by authorities for protection of public and properties
- for research and development purposes

Frequent measurements conducted during construction and operation and the instruments used are:

- Precise levelling by theodolites (0.1 mm accuracy)
- Ground water level measurements by piezometers (open tube)
- Ground water pressure measurements by hydraulic/pneumatic/vibrating wire cells
- Water flow measurements by open weirs/flow meters in tubes
- Ground pressure measurements by total cells
- Crack width/axial strain measurements by extensometers and fibre optic cables
- Horizontal displacements along depth by inclinometers
- Tilt with electrolytic levels
- Settlement along depth by gauges (magnetic plates in fills/spring anchors in boreholes)
- Peak particle velocity by geophones for smaller amplitudes and frequency of ground vibration in general
- Peak ground accelerations by accelerometers for greater amplitudes and frequency of ground vibration in general

Optical type instruments are preferred to mechanical ones which in turn are preferred to hydraulic, pneumatic, and electrical types concerning their simplicity and reliability (Dunnicliff 1988).

Keep it simple and straight (KISS) principle applies to measurements as well as other geo-works.

Basic properties of instruments/transducers are (Dunnicliff 1988):

- **Conformance** so not to alter the value of the parameter being measured
- **Accuracy** is the closeness of a measurement (correctness) to the true value measured
- **Precision** is closeness of a number of similar measurements (repeatability) to the mean value
- **Resolution** is the smallest division on an instrument readout scale
- **Sensitivity** is the amount of output response when an input is applied
- **Linearity** is direct proportionality between indicated measured values to the quantity being measured
- **Hysteresis** is difference in response to increasing/decreasing values of measured quantity
- **Noise** indicates random measurement variations caused by external, creating lack of precision and accuracy
- **Error** is the deviation between measured and true value
  - Gross error is caused by carelessness, fatigue and inexperience
  - Systematic error is caused by improper calibration, or its alteration in time
  - Conformance error is caused by poor selection of installation procedure and by limitations in instrument performance
  - Environmental error arises because of the influence of heat, humidity, vibration, shock waves, moisture, pressure, corrosion etc.

– Observational error arise when different observers use different observation techniques
– Sampling error occurs due to heterogeneity of ground and limited number of instruments

In general, more accurate instruments are expected to have smaller measurement range although that problem is largely eliminated in modern electronic instruments.

6.10.1.1 Observational Method

The method is a continuous, managed, integrated, process of design, construction control, monitoring and review that enables previously defined modifications to be incorporated during or after construction as appropriate. All these aspects have to be demonstrably robust. The objective is to achieve greater overall economy without compromising safety. The method cannot be used when there is insufficient time to implement fully and safely complete the planned modification or emergency plans (Nicholson et al. 1999).

EN 1997-1:2004 lists requirements to be met before start of construction:

- **Acceptable limits** of behaviour are established
- The **range of possible behaviour** is assessed and shown that there is an acceptable probability that the actual behaviour will be within acceptable limits
- A **plan of monitoring** is devised, which will reveal if the actual behaviour lies within the acceptable limits. The monitoring must be made early and within short intervals to allow contingency actions to be undertaken successfully.
- The **response time of instruments and the results analyses** must be sufficiently rapid for evaluation of system behaviour.
- A **plan of contingency action** must be devised.

6.10.2 Execution

- **Reference points** must be stable and not influenced by moisture and temperature changes (heating/cooling, freezing/thawing) or their movement accounted for in the results
- Modern **total stations** incorporate measurements of angles as well as distances using electromagnetic waves
- **Piezometer** reaction time depend on borehole preparation and washing is necessary if wash boring has not been used

- Penman (1960) defined the **response time *t* of open stand pipe piezometers** as:

$$t = 3.3 \times 10^{-6\cdot} \cdot \frac{d^2 \cdot \ln_e \left[L/D + \sqrt{1 + (L/D)^2} \right]}{k \cdot L} \tag{6.15}$$

t is the time (days) required for 90 % response
d is the inside diameter (cm) of stand pipe
L is the length (cm) of intake filter (or sand zone around the filter)
D is the diameter (cm) of intake filter (or sand zone around the filter)
k is the coefficient of ground permeability (cm/s)

- **Water pressure measurement cells** should have fully saturated high air entry filters for low air blow-through occurrences
- **Total pressure cells** measurements can be affected by soil-structure stiffness differences and arching effects
- **Fibre optic cables** are cheap but the readout units are expensive for routine measurements at present (e.g. Mair 2006)
- **Inclinometer** readings are taken in two opposite positions to eliminate misalignment error
- **Electrolytic levels** are affected by ambient vibrations and by temperature
- **Geophones** measure ground velocity, which peak value is used frequently as an indicator of severity of damage to structures, effect on humans and sensitive equipment. The ratio between particle velocity and ground wave velocity equals to strain. The product of ground unit density, particle velocity and ground wave velocity equals to pressure on a surface. More details are provided by Srbulov (2010) among others.
- **Accelerometers** measure ground acceleration, which peak value is used frequently as an indicator of effect on humans, structures and sensitive equipment. The product of structural mass and acceleration equals to inertial force acting on the mass. More details are provided by Srbulov (2010) among others.

Not only that repeated measurements are frequently used but also the same type of measurements from different types of instruments for cross checking.

6.10.2.1 Observational Method

Nicholson et al. (1999) describe that primary and secondary monitoring systems are usually installed:

- **The primary system** is simple and controlled and reviewed by site staff, used for routine monitoring, and the results are checked against the trigger criteria.
- **The secondary system** provides additional data for designer, acts as a check and as a back-up system and supplement to the primary system.

The green-amber-red traffic light system is often used:

- **Green condition** represents safe state
- **Amber condition** represents transition to undesirable state during which monitoring frequency of the primary instrumentation system is increased and the results of secondary instrumentation system are reviewed and additional reading taken. Minor modification plans may be executed.
- **Red condition** indicates closeness to undesirable limit state (concerning strength/deformation) at which stage modification must be implemented.

6.10.3 Control

- Required instruments are delivered to a construction site on time for the installation
- Instruments are safely stored and protected from damage at the construction site
- Transducer calibration certificates are not expired or recalibration is required
- Operators qualification and experience is satisfactory
- Installation procedure is according to the manufacturer recommendations and design specifications
- Initial (zero) reading(s) had been taken before changes to be measured occurred
- Readings are taken at desired intervals and are complete
- Results are stored and processed as prescribed
- Reports are delivered as required concerning completeness and timing

6.10.3.1 Observational Method

Nicholson et al. (1999) define the following management considerations:

- Culture involving:
  - Quality adherence
  - Health and safety application
  - Value management consideration
  - Risk management consideration
  - The method requirements and limitations
  - Co-operation among parties involved
  - Integrated design and construction
  - Research and development issues
  - Training and education on job
  - Business partnering
  - Enthusiasm for implementation

- Strategy involving:
  - Contract requirements
  - Risk-based control
  - Team building and work
  - Resource planning
- Competence involving:
  - Skills of personnel involved
  - Knowledge about the method and construction
  - Experience in dealing with different situations
- Systems involving:
  - Clear communication
  - Reliable and timely information gathering
  - Reliable and timely information processing
  - Reliable and timely information reviewing
  - Auditing (the method procedures and requirements, quality, health and safety)

6.11 Remote Sensing

6.11.1 Description

Campbell (2006) provides the following definition relevant to geo-engineering: *"Remote sensing is the practice of deriving information about the Earth's land and water surfaces using images acquired from an overhead perspective, by employing electromagnetic radiation in one or more regions of the electromagnetic spectrum, reflected or emitted from the Earth's surface"*.

The use of remote sensing is wide spread from subjective examination by ayes of stereographic projection of landform photographs taken from airplanes since the beginning of twenty century to interferometric synthetic aperture radar (InSAR) computerised analyses of phase differences in reflected waves emitted by satellites to obtain vertical displacements of ground surface due to active tectonics and human activities with millimetre accuracy, as shown in Figs. 4.16 and 4.17 for example.

Developments in remote sensing are dependent on developments in many technological disciplines. For example, light detection and ranging (Lidar) imagery is based on application of light amplification by stimulated emission of radiation (laser) that penetrates through vegetation; digital image processing is dependent on the increase of speed and capacity of computers and developments in modern mathematical fields such as classification and regression tree analysis, fuzzy clustering, artificial neural networks, contextual and object-oriented classifications, and iterative guided spectral class rejection in addition to traditional mathematical disciplines such as multiple correlation and Bayes' and maximum likelihood classifications.

Gathering and use of ground data is essential not only for the confirmation of the interpreted information gathered remotely but also for increase in accuracy of the interpretation. For example, the use of local differential global positioning system (DGPS) when a GPS receiver can be stationed at a fixed position of known location on the ground, can increase the accuracy of measured location to within 1–5 m from the actual location in comparison with the use of GPS from satellites alone that provides accuracy of less than about 10 m.

6.11.2 Execution

Campbell (2006) provides the following examples of the use of remote sensing for earth sciences:

- **Photo geology** for providing information concerning:
  - Lithology (physical and chemical properties of sedimentary, igneous and metamorphic rocks)
  - Structure (deformations experienced by rocks including folding, fracturing and faulting)
  - Vegetation patterns (relationships between plant cover and the lithology of underlying rock)
- **Drainage patterns** (Fig. 6.4)
  - Dendritic – for uniformly resistant surface materials, gentle regional slopes, and the absence of major faults or structural systems.
  - Parallel – for uniformly resistant surface materials but for landscapes with significant regional slopes.
  - Entrenched or incised – for landscapes uplifted by tectonic forces or if for lowered based level. Sharp, deep, well defined edges suggest the presence of strong, cohesive surface material. More gently sloping terrain near the channel suggests the presence of less cohesive, weaker, strata.
  - Trellis – for linear structures or lithological features when tributaries often follow the strike of the structure while the main stream cut across the principal structure so that the tributaries and the main stream are oriented perpendicular to each other.
  - Radial – for a large central peak like in the case of volcano dome.
  - Braided – for arid region or in glacial melt water and high sediment transport. A wide, sparsely vegetated strip of open sand or gravel, with a network of ephemeral, anastomosing channels and elongated bars and islands.
- **Lineaments** such as tectonic fault break of the ground surface, axis of an anticline fold, wind-blown dunes

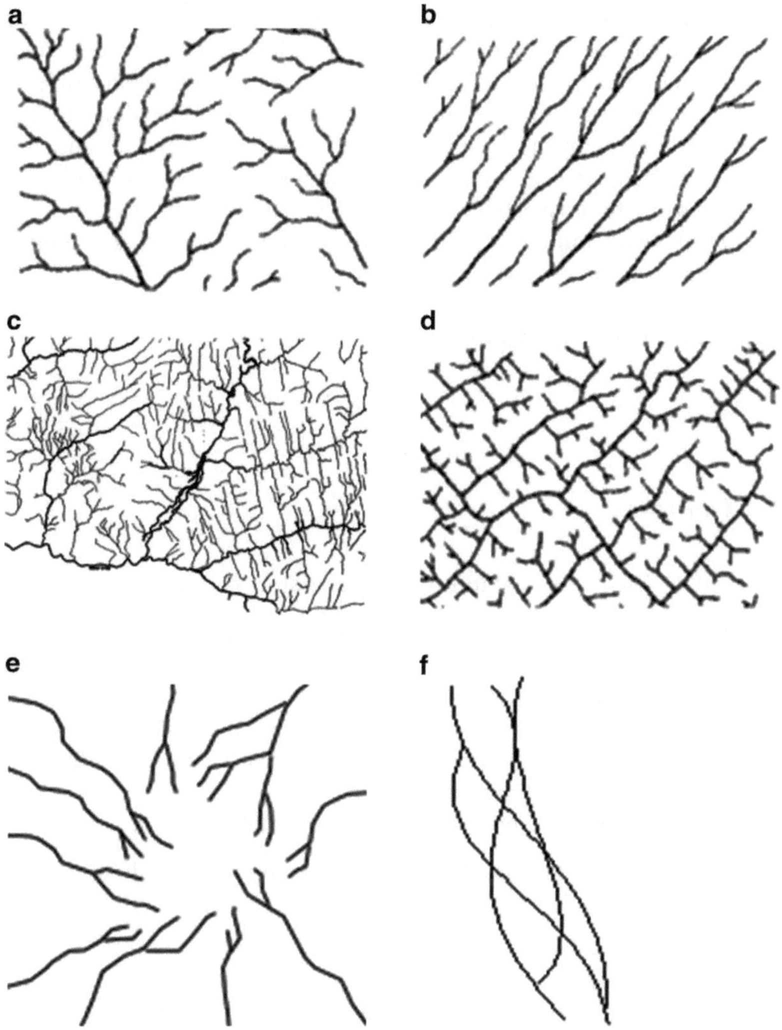

Fig. 6.4 Drainage patterns (**a**) dendritic, (**b**) parallel, (**c**) entrenched/incised, (**d**) trellis, (**e**) radial, (**f**) braided

- **Geo-botany** in connection with the location of major nutrients (phosphorus, potassium and nitrogen) and micronutrients (barium, magnesium, sulphur, calcium) or toxicants (nickel, copper, chromium, lead) and pollutants (hydrocarbons)
- **Direct multispectral observation of rock and minerals**
- **Photo-clinometry** for relationship between image brightness and the orientation of the surface that generated the brightness
- **Band ratios** for conveying spectral information by moving effects of shadowing.

- **Soil and landscape mapping** involving the following steps: examination by a stereoscope of aerial photographs, characterization by sample collection from each prospective mapping unit, classification by measurements of physical, chemical and mineralogical properties, correlation for matching the mapping units within the mapped region to those in adjacent regions and those in ecologically analogous areas, interpretation for evaluation of each mapping unit with respect to prospective agricultural and engineering uses.
- **Integrated terrain units** mapping of soil, vegetation, hydrology and physiography based on comprehensive examination
- **Wetland inventory**
- **Radar imagery for exploration**

6.11.3 Control

Campbell (2006) lists the key abilities of the operators such as:

- Sufficient knowledge of several interrelated disciplines such as electromagnetic radiation, photographic sensors, digital data, image interpretation, geology, hydrology, geomorphology, soil mechanics, forestry
- A good detailed knowledge in an area of expertise
- Understanding of scope of remote sensing and broad overview of the interrelationships of earth oriented disciplines
- Experience in operating if not programming computer software

As with other measurements, determination, equipment availability and sufficient time are necessary for insurance of:

- Accuracy and precision of measurements
- Appropriate processing of gathered information with sufficient resolution
- Correctness and uniqueness of interpretation of processed data

6.12 Asset Management

6.12.1 Description

Geo-structures may or may not need maintenance. Foundations are designed to be maintenance free during their useful life but others like ground slopes and fill dams may need at least observation if not maintenance.

Hooper et al. (2009) define asset management as "*the systematic and co-ordinated activities and practices through which an organisation optimally and sustainably manages its assets and asset systems, their associated performance, risk and expenditures over their life cycles for the purpose of achieving its organisational strategic plan. The combination of management, financial, economic, engineering and other practices applied to physical assets with the objective of providing the required level of service in the most cost-effective manner."*

In general:

- **Hazard** is considered as something with the potential to do harm.
- **Risk** is a function of the probability of harm occurring and severity of its consequence.

6.12.2 Execution

6.12.2.1 Infrastructure Cuttings

Perry et al. (2003) list typical performance requirements as:

- Safety and reliability
- Operational efficiency
- Satisfaction of statutory and regulatory obligations
- Value for money and business improvement
- Minimisation environmental impact and maximisation of environmental value
- Satisfaction of customer and employee expectations and perceptions.

Benefits of an effective asset management system include:

- Conversion of owner policy and objectives into appropriate actions
- Assistance with the privatisation of expenditure at regional and national level
- Provision of comparative analyses between regional and national assets
- Support of submissions to financial sponsors for the funding of maintenance works
- Allowance of progress against strategic and financial targets to be monitored and reported
- Provision of information regarding the serviceability and rate of improvement or deterioration of the assets
- Assistance with the qualification and mitigation of risks
- Identification of immediate and future investment requirements.

A simple risk matrix is given in Table 6.9.

6.12.2.2 Infrastructure Embankments

Perry et al. (2001) list typical performance requirements and benefits of an effective asset management system as in Sect. 6.12.2.2. A simple risk matrix is given in Table 6.9.

Table 6.9 A simple risk matrix example (Adopted from Perry et al. 2001, 2003)

	Hazard probability		
Hazard consequence	Low	Medium	High
LOW (no risk to people or property)	NEGLIGIBLE risk (routine inspection)	LOW risk (routine inspection)	MEDIUM risk (increased inspection frequency)
MEDIUM (minimal risk to people, but high cost of disruption)	MEDIUM risk (routine inspection)	MEDIUM risk (increased inspection frequency)	HIGH risk (assessment required)
HIGH (high risk to people, property and high costs of repair or disruption)	MEDIUM risk (routine inspection)	HIGH risk (assessment required)	UNACCEPTABLE risk (assessment and mitigation required)

6.12.2.3 Tunnels

McKibbins et al. (2009) list examples for need of tunnel management due to:

- Modification caused by change in use
- Change in load conditions such as urban development
- Ground movement and ground water variations
- Weathering and decay
- Inaccessible structural element for inspection

Several features and properties of ageing tunnels require special considerations in their management:

- Passed their originally intended useful life
- Very individual in their character, behaviour and maintenance needs
- Lack information about their design, construction, hidden structure and ground conditions
- Structural response and performance is complex
- Access is restricted and conditions within tunnel may be poor
- Visual inspection is limited
- The effectiveness of repairs and alternations and their likely influence on the long-term performance and maintenance of the structure are not well known
- Hidden features, such as tunnel shafts, may be difficult to access and inspect, or their presence may not be known, which can be a hazard to safety

The maintenance of a tunnel includes:

- **Condition appraisal** (inspections, testing and monitoring, structural assessment)
- **Routine maintenance** (like-for-like replacement)
- **Interventions** (vital repairs to and modification of the structure in response to deterioration and loss of performance, or adaptations to meet new requirements. e.g. higher loads, health and safety or control equipment)
- **Emergency action** (response to unforeseen incidents, like fire)

6.12.3 Control

Hooper et al. (2009) mention that asset management organisations are required to monitor their asset management process in addition to monitoring of the condition and performance of their assets. Asset management organisations are required to have processes and defined authority and responsibilities for:

- Handling and investigating asset related failures (including failures to meet required functions, performance and condition, incidents and emergencies and non-conformance)
- Taking mitigating actions.
- Initiating corrective or preventive action and confirming the effectiveness of the corrective or preventive actions.

It is essential that even experienced staff use the decisions making process to:

- Confirm the validity of the decision making process.
- Provide an audit trail to demonstrate that the asset management procedures have been successfully carried out
- Enable engineers to rationalise and explain decisions clearly to others
- Ensure that opportunities for low risk innovation are identified and exploited.

6.13 Forensic Investigation

6.13.1 Description

Day (1999) defines forensic engineering as the investigation of a damaged or deteriorated structure. Shuirman and Slosson (1992) have developed a routine check list that should be completed before accepting an assignment involving civil litigation:

1. There are no conflicts of interest with any of the parties connected with the lawsuit, directly or indirectly.
2. Acquire as much background information on the disaster or failure as possible, mindful that such information may be biased or lacks pertinent data.
3. Attempt to separate facts from opinions so as to be able to form an objective early picture of the main issues.
4. Inquire about the status of the case and its tentative schedule to determine if there is sufficient time for a thorough investigation.
5. Make sure that the subject matter is within the appropriate areas of expertise for you or your company.
6. Discuss fee schedules and determine when and by whom payments are to be made.
7. Check out the reputation of the attorney or law firm if it is not already known.

Attorneys state that in giving testimony to a jury the "keep it simple" principle should be followed using photographs and charts rather than calculations and Tables.

6.13.2 Execution

Greenspan et al. (1989) list the following activities when planning the investigations:

- Budget and schedule constraints consideration
- Selection of an interdisciplinary team with appropriate expertise
- Site observations and testing requirements
- Document collection
- Analysis and synthesis of data
- Development of failure hypothesis

It should be borne in mind that courts require proof without reasonable doubt so that evidence based conclusions are crucial in the process.

A detailed knowledge and understanding of geo-hazards (Chap. 4) and geo-structures (Chap. 5) as well as basic mechanisms, factors and assumptions in the analyses performed is important in the evaluation process to reach a correct conclusion.

6.13.3 Control

Team work and peer reviews are essential to build a robust defence and explanation of investigated failure/deterioration.

6.14 Summary

- Large diversity of local conditions and practices influence the need that filed trial tests are performed for geo-works.
- Observational method has potential to decrease cost and increase safety for most challenging geo-works.
- Temporary works are not subjected to regulations by many codes but they must follow health and safety procedures in force. Most frequent causes of death and serious injuries in the construction industry are caused by fall from height in over 50 % cases, by falling objects/ground collapse, by machinery/transport, working in confined spaces, due to increased pressure and heat, dangerous substances (such as lead, asbestos).
- Health and safety risk assessment should be performed for all construction works at the beginning and updated regularly during the works. An example of health and safety risk assessment for ground investigations is given in Table 1.4.

References

Andersland OB, Ladany B (1994) An introduction to frozen ground engineering. Chapman & Hall, New York/London

ASCE Press (1997) Chemical grouting. Adopted from Technical Engineering and Design Guides US Army Corps of Engineers, No. 24. American Society of Civil Engineers

ASTM D1004-13 Standard test method for tear resistance (graves tear) of plastic film and sheeting. American Society for Testing and Materials, Philadelphia, PA

ASTM D3786/D3786M-13 Standard test method for bursting strength of textile fabrics – diaphragm bursting strength tester method. American Society for Testing and Materials, Philadelphia, PA

ASTM D4428/D4428M-07 Standard test methods for crosshole seismic testing. American Society for Testing and Materials, Philadelphia, PA

ASTM D4491-99a(2009) Standard test methods for water permeability of geotextiles by permittivity. American Society for Testing and Materials, Philadelphia, PA

ASTM D4533-11 Standard test method for trapezoid tearing strength of geotextiles. American Society for Testing and Materials, Philadelphia, PA

ASTM D4595-11 Standard test method for tensile properties of geotextiles by the wide-width strip method. American Society for Testing and Materials, Philadelphia, PA

ASTM D4632/D4632M-08(2013) Standard test method for grab breaking load and elongation of geotextiles. American Society for Testing and Materials, Philadelphia, PA

ASTM D4751-12 Standard test method for determining apparent opening size of a geotextile. American Society for Testing and Materials, Philadelphia, PA

ASTM D638-10 Standard test method for tensile properties of plastics. American Society for Testing and Materials, Philadelphia, PA

ASTM D882-12 Standard test method for tensile properties of thin plastic sheeting. American Society for Testing and Materials, Philadelphia, PA

Baker WH (ed) (1982) Grouting in geotechnical engineering. In: Proceedings of the conference on grouting in geotechnical engineering. New Orleans, LA

Booy D, Wade M, White V (2008) Invasive species management for infrastructure managers and the construction industry. Construction Industry Research and Information Association UK report C679

BS 5930:1999+A2:2010 Code of practice for site investigations. British Standards Institution, London

BS 6031:2009 Code of practice for earthworks. British Standards Institution, London

Byle M, Borden R (eds) (1995) Verification of geotechnical grouting. ASCE Gt Spec Publ 57:1–66

Campbell JB (2006) Introduction to remote sensing, 4th edn. Taylor & Francis, London/New York

Cedergren HR (1989) Seepage, drainage and flow nets. Wiley, New York

Day RW (1999) Forensic geotechnical and foundation engineering. McGraw-Hill, New York

Dunnicliff J (1988) Geotechnical instrumentation for monitoring field performance. Wiley, New York

EN 1536:2010 Execution of special geotechnical works – bored piles. European Committee for Standardization, Brussels

EN 1538:2010 Execution of special geotechnical works – diaphragm walls. European Committee for Standardization, Brussels

EN 1997-1:2004 Eurocode 7: geotechnical design – Part 1: General rules. European Committee for Standardization, Brussels

EN 12715:2000 Execution of special geotechnical work – grouting. European Committee for Standardization, Brussels

EN 12716:2000 Execution of special geotechnical work – Jet grouting. European Committee for Standardization, Brussels

EN 12716:2001 Execution of special geotechnical works. Let grouting European Committee for Standardization, Brussels

EN 13252:2000 Geotextiles and geotextile-related products – Characteristics required for use in drainage systems. European Committee for Standardization, Brussels
EN 14199:2005 Execution of special geotechnical works – micropiles. European Committee for Standardization, Brussels
EN 14475:2006(E) Execution of special geotechnical works – reinforced fill. European Committee for Standardization, Brussels
EN 14679:2005(E) Execution of special geotechnical works – deep mixing. European Committee for Standardization, Brussels
EN 14731:2005 Execution of special geotechnical works – ground treatment by deep vibration. European Committee for Standardization, Brussels
EN 15237:2007(E) Execution of special geotechnical works – vertical drainage. European Committee for Standardization, Brussels
EN ISO 10319:2008 – Geosynthetics – wide-width tensile test. European Committee for Standardization, Brussels/International Organization for Standardization, Geneva, Switzerland
EN ISO 11058:1999 – Geotextiles and geotextile-related products – determination of water permeability characteristics normal to the plane, without load. European Committee for Standardization, Brussels/International Organization for Standardization, Geneva, Switzerland
EN ISO 12956:1999 – Geotextiles and geotextile-related products – determination of the characteristic opening size. European Committee for Standardization, Brussels/International Organization for Standardization, Geneva, Switzerland
EN ISO 12958:1999 – Geotextiles and geotextile-related products – determination of water flow capacity in their plane. European Committee for Standardization, Brussels/International Organization for Standardization, Geneva, Switzerland
EuroSoilStab (1996) Design guide for soft soil stabilisation CT97-0351. European project for Development of Design and Construction Methods to Stabilise Soft Organic Soils for the Construction of Rail, Road and other Infrastructure BE 96-3177. European Commission 4th Framework Program
EuroSoilStab (1997) Design guide for soft soil stabilization. CT97–0351, Project No: BE 96–3177. European Industrial & Materials Technologies Programme (Brite – Eu Ram III)
Franklin JA, Broch E, Walton G (1971) Logging the mechanical character of rock. Trans Inst Min Metall 80(A):A1–A9
FTMS 101C, Method 2065 Test method for puncture resistance and elongation (1/8 in. radius probe)
GFR (1992) Geotechnical fabrics report 1993 Specifier's guide. In: Sharma HD, Lewis SP (1994) Waste containment systems, waste stabilization, and landfills. Wiley
Giroud JP, Bonaparte R (1989) Leakage through liners constructed with geomembranes. Geotext Geomembr 8(1):27–67
Grasso D (1993) Hazardous waste site remediation – source control. Lewis Publishers, Boca Raton
Greenspan HF, O'Kon JA, Beasley KJ, Ward JS (1989) Guidelines for failure investigation. ASCE, New York
Gregory CE (1984) Explosives for North American engineers, 3rd edn. Trans Tech Publications, Clausthal
Harr ME (1990) Groundwater and seepage, 2nd edn. Dover Publications, Inc., New York
Holtz RD, Jamiolkowski MB, Lancellotta R, Pedroni R (1991) Prefabricated vertical drains: design and performance. Construction Industry Research and Information Association UK report, Butterworth-Heinemann
Hooper R, Armitage R, Gallagher A, Osorio T (2009) Whole-life infrastructure asset management: good practice guide for civil infrastructure. Construction Industry Research and Information Association report C677, UK
Horner PC (1981) Earthworks, ICE works construction guides. Thomas Telford, London
Houlsby AC (1990) Construction and design of cement grouting. Wiley, New York
Hsieh H, Raghu D (1990) Site cleanup with soil washing technology. In: U.S. EPA Review, U.S. Environmental Protection Agency, Cincinnati, OH

Hunt R, Dyer RH, Driscoll R (1991) Foundation movement and remedial underpinning in low-rise buildings. Building Research Establishment Report BR 184, UK

ISO 10320:1999 Geotextiles and geotextile-related products – identification on site. European Committee for Standardization, Brussels/International Organization for Standardization, Geneva, Switzerland

ISO 10321:2008 Geosynthetics – tensile test for joints/seams by wide-width strip method. European Committee for Standardization, Brussels/International Organization for Standardization, Geneva, Switzerland

Jamiolkowski M, Lancellotta R, Wolski W (1983) Precompression and speeding up consolidation. In: Proceedings of the 8th European conference on soil mechanics and foundation engineering, Helsinki, vol 3, pp 1201–1226

Karol RH (1990) Chemical grouting, 2nd edn. Marcel Dekker, Inc., New York/Basel

Kovacevic N, Potts DM, Vaughan PR (2000) The effect of the development of undrained pore pressure on the efficiency of compaction grouting. Geotechnique 50(6):683–688

Kramer SL (1996) Geotechnical earthquake engineering. Prentice Hall, Upper Saddle River

Littlejohn GS (1982) Design of cement based grouts. In: Baker WH (ed) Proceedings of the conference on grouting in geotechnical engineering, New Orleans, pp 35–48

Mair RJ (2006) Tunnelling and geotechnics – new horizons. Geotechnique 58(9):695–736

McKibbins L, Elmer R, Roberts K (2009) Tunnels: inspection, assessment and maintenance. Construction Industry Research and Information Association report C671, UK

Monahan EJ (1994) Construction of fills, 2nd edn. Wiley, New York

Nelson-Skornyakov FB (1949) Seepage in homogeneous media. Gosudarstvenoe Izdanie Sovetskaya Nauka, Moscow

Nicholson D, Tse V-M, Penny C, O'Hana S, Dimmock R (1999) The observational method in ground engineering: principles and applications. Construction Industry Research and Information Association report 185, UK

Penman ADM (1960) A study of the response time of various types of piezometers. In: Proceedings of the conference on pore pressure and suction in soils. Butterworth, London, pp 53–58

Perry J, Pedley M, Reid M (2001) Infrastructure embankments – condition appraisal and remedial treatment, 2nd edn. Construction Industry Research and Information Association report C550, UK

Perry J, Pedley M, Brady K (2003) Infrastructure cuttings – condition appraisal and remedial treatment. Construction Industry Research and Information Association report C591, UK

Rowe RK, Quigley RM, Booker JR (1995) Clayey barrier systems for waste disposal facilities. E & FN Spon, New York

Sanger FJ, Sayles FH (1979) Thermal and rheological computations for artificially frozen ground construction. Eng Geol 13:311–337

Schwab GO, Fangmeier DD, Elliot WJ, Frevert RK (1993) Soil and water conservation engineering, 4th edn. Wiley, New York

Shuirman G, Slosson JE (1992) Forensic engineering – environmental case histories for civil engineers and geologists. Academic, New York

Shuster JA (1972) Controlled freezing for temporary ground support, Chap. 49. In: Lane KS, Garfield LA (eds) Proceedings of the first North American rapid excavation and tunnelling conference, Chicago, ASCE-AIME Port City Press, Baltimore, vol 2, pp 863–895

Srbulov M (2010) Ground vibration engineering – simplified analyses with case studies and examples. Springer, London

Widman R (ed) (1996) Commission on rock grouting. International Society for Rock Mechanics. Int J Rock Mech Min Sci Geomech Abstr 33(8):803–847

Index

M. Srbulov, *Practical Guide to Geo-Engineering: With Equations, Tables, Graphs and Check Lists*, Geotechnical, Geological and Earthquake Engineering 29, DOI 10.1007/978-94-017-8638-6, © Springer Science+Business Media Dordrecht 2014

D

E

MIX
Papier aus verantwortungsvollen Quellen
Paper from responsible sources
FSC® C105338

If you have any concerns about our products,
you can contact us on
ProductSafety@springernature.com

In case Publisher is established outside the EU,
the EU authorized representative is:
Springer Nature Customer Service Center GmbH
Europaplatz 3, 69115 Heidelberg, Germany

Printed by Libri Plureos GmbH
in Hamburg, Germany